INTRODUCTION TO CIVIL ENGINEERING CONSTRUCTION

ROY HOLMES

The College of Estate Management

First published 1975
Second edition 1983
Third edition November 1995

ISBN 1899 769 30 7

Published by The College of Estate Management
Whiteknights, Reading RG6 6AW

Contents

Preface to the third edition

Since this book was published over twenty years ago I have received many complimentary remarks, together with suggestions about the text and the way in which the subject matter has been simplified to assist students. I am grateful for the comments received and have attempted in this major revision to maintain the simple style that has been very popular in the teaching/learning realm.

The first edition included technology that was at the cutting edge of civil engineering construction and it was some years before many of the techniques became widespread. That meant that the revision for the second edition was less onerous than one might have expected. However, some twelve years have passed since the second edition and it was felt that it was time to review not only current techniques but also the broader content and its usefulness to practising engineers and others involved in civil engineering.

As a result, almost every chapter has undergone major changes. Information from some chapters has been moved and integrated with others to prevent duplication of thought and to provide ease of reading. This is particularly true of chapter 8, which deals with concrete and steel. The emphasis in chapter 8 is now on the whole life-span of concrete, from placing to demolition.

The very latest technology has been checked in terms of major works such as tunnelling and, as a result, the New Austrian Method has been included. At the other end of the scale microtunnelling has also been expanded.

Some aspects have been omitted from the new edition; these relate mainly to practical specifications that were offered in previous editions – for example, the specification for the placement of diaphragm walls.

I trust that this third edition will meet the needs of the growing number of students in this field of study and also serve as a good reference book for practitioners.

Roy Holmes
University of the West of England, Bristol.
September 1995

CHAPTER 1

General Considerations in Civil Engineering

1.1 SITE INVESTIGATION

1.1.1 Objects of investigation

Civil engineering works and building structures are alike in having some form of foundation which is supported by the ground. The interaction between a structure and the soil beneath it is complex, and therefore knowledge of site and soil conditions is an essential prerequisite to sound design. This has not always been appreciated or exemplified in practice – understandably perhaps before the turn of the century, when studies first began to be made of soil conditions and soil behaviour. The earliest indication of this, with the advent of tall buildings, was the construction of three buildings in Chicago during the period 1880 to 1894, which were founded successfully on steel grillage bases and large rudimentary piles formed by filling wells with concrete. There are many other instances where the construction of buildings and engineering structures was not so successful and where monumental failures have, in some cases, become notable tourist attractions.

The first published data giving comprehensive methods for the analysis of soil was not available until 1925, when Terzaghi's classic work *Erdbaumechanik*, was eventually written. It is now standard practice to examine any site on which a structure is intended to be erected, in order to determine:

- The suitability of the site for the proposed works.
- An adequate and economic foundation design.
- The difficulties that may arise during the construction period.
- The occurrence or cause of all changes in site conditions.

The practical outcome of such investigation depends upon the nature and magnitude of the proposed works together with the availability of reliable sources of information concerning the site or area. In some cases the area may have been investigated several times and data may be available relating to substrata details. Such data will enable the engineers to determine the extent to which further investigation may be warranted.

If the proposed work is not directly connected with the erection of new structures, there may be other factors to be considered, some of which are discussed in section 1.1.3 (Types of investigation).

1.1.2 Classification of rocks and soils

Rocks and soils are classified so that a systematic and concise record of their characteristics can be kept to provide engineers and others with basic information for design and other purposes. The classification offers a basic comparison against other similar materials.

The simplest classification of geological deposits falls generally into two major classes:

- 'Rock', which refers mainly to a hard rigid and strongly cemented deposit; and
- 'Soil', which refers to the soft, or loose and uncemented deposits.

When using such definitions it must be appreciated that many deposits fall between these extremes, which are difficult to define. For example, rock may also be classified under the term 'hard stratum' which is penetrated only by breaking, blasting, chiselling or drilling.

Rock types

The geological classification of 'rocks' is very complex, but from an engineering point of view they can be simply classified either by method of excavation, ie whether blasting is required or not, or by load-bearing qualities and other physical properties.

Solid rocks normally provide excellent foundation support; they exhibit high load-bearing capacities and negligible settlement.

Rock can be placed into one or other of three basic categories: Sedimentary, Metamorphic and Igneous.

Sedimentary includes sandstone, limestone and some shales. The quality of sedimentary rocks depends on the angle of stratification and cementation (Figure 1.1) together with their behaviour in wet conditions. Deep fissures and swallow holes (swallow holes are cavities formed at some time by the passage of water through soft rock) (Figure 1.2) are common in limestone and are difficult to detect during site investigation.

Metamorphic includes slates, schists, gneisses and some shales. These rocks include any sedimentary deposit or igneous rock which, after consolidation, has become altered by heat or pressure; eg limestone may become completely re-crystallised into marble. With few exceptions the metamorphic rocks are hard but subject to faults which allow movement. This is found in tunnelling where overbreak is high ('overbreak' refers to the amount of extra rock excavation beyond the neat lines of tunnel excavation).

Igneous includes granites, dolerites and basalt, and they are formed by solidification of molten material which has ascended towards the surface from the hot lower levels of the earth's crust. Their bearing capacity is very high, in some cases three times greater than that of hard sedimentary rocks and forty times greater than that of alluvial clays and sands.

FIGURE 1.1
Rock formation

FIGURE 1.2
Cavities in rock formation

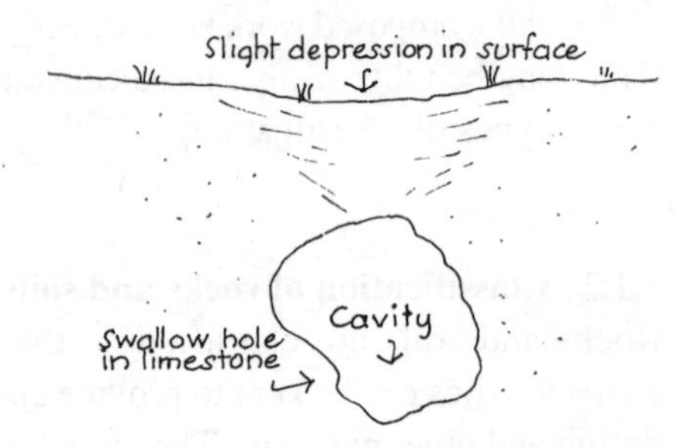

The load-bearing capacity of all rocks is greatly reduced if they are not in a sound condition. Unsound conditions are created by weathering where the rock becomes decomposed, by shattering caused by earth movements and by having steeply dipping bed joints. Some sedimentary rocks have low bearing capacity due to the presence of soft clayey material in the bed joints.

Soil types

Soils include materials of various origins, such as residual soils (topsoil), detrital sediments (sands, gravel, silts, organic deposits (peat), calcarious deposits (shell, coral) and pyroclastics (uncemented volcanic dust).

These soils are identified by two essential characteristics: firstly the size and nature of the soil particles, and secondly the density and structural properties. Basic soil types and their characteristics are shown in Table 1.1. The particle size is particularly important when considering pumping operations since silts are difficult or impossible to drain, whereas sands drain readily.

Where soils are used in the actual construction works (eg roadworks and earth dams), the particle arrangement and moisture content of the soil may alter drastically. For this to be foreseen, accurate preliminary classification is essential. Classification and testing of soils should be subject to BS 1377.

1.1.3 Types of investigation

As indicated in section 1.1.1, site investigation is carried out in most cases as a preliminary to new works. However, there are other reasons for site investigation:

- Investigation of defects or failure of existing works.
- Investigation relating to the safety of existing works.
- Investigation relating to the suitability and availability of materials for constructional purposes.

Companies may specialise in any one or more of these types of investigation.

Investigation of sites for new works

This type of investigation has broader implications and a much wider scope than the other types of investigation above, owing to the amount of information required and the economic solutions involved. Considering excavation methods, the investigation will reveal such things as:

- Ground water condition and implications such as the necessity of ground water lowering.
- Whether excavation of the soil will be difficult.
- Whether the sides of the excavation will be stable if unsupported.

On the design side, analysis of test samples and insitu tests will indicate such things as the bearing capacity of foundations, the amount of settlement likely to occur and the stability of cuttings and embankments. It is particularly important, in the case of new works, to consider the least favourable site conditions encountered when designing the proposed works – this will take into account the possible variations in strata.

Such information on sub-surface conditions is especially valuable in the case of virgin sites, particularly where there is evidence of failure or distortion in similar structures in the locality. Special precautions are necessary where it is known that underground cavities, such as mine working and swallow holes, are likely to exist.

TABLE 1.1 Field identification and description of soils

	Basic soil type	Particle size, mm	Visual identification	Particle nature and plasticity
Very coarse	BOULDERS		Only seen complete in pits or exposures.	Particle shape:
		200		
	COBBLES		Often difficult to recover from boreholes	
		60		Angular Subangular Subrounded Rounded Flat Elongate
Coarse soils (over 65% sand and gravel sizes)	GRAVELS	coarse		
		20		
		medium		
		6		
		fine		Texture:
		2		
	SANDS	coarse	Visible to naked eye; very little or no cohesion when dry; grading can be described.	Rough Smooth Polished
		0.6		
		medium	Well graded; wide range of grain sizes, well distributed. Poorly graded; not well graded. (May be uniform; size of most particles lies between narrow limits; or gap graded; an intermediate size of particle is markedly under-represented.)	
		0.2		
		fine		
		0.06		
Fine soils (over 35% silt and clay sizes)	SILTS	coarse	Only coarse silt barely visible to naked eye; exhibits little plasticity and marked dilatancy; slightly granular or silky to the touch. Disintegrates in water; lumps dry quickly; possess cohesion but can be powdered easily between fingers.	Non-plastic or low plasticity
		0.02		
		medium		
		0.006		
		fine		
		0.002		
	CLAYS		Dry lumps can be broken but not powdered between the fingers; they also disintegrate under water but more slowly than slit; smooth to the touch; exhibits plasticity but no dilatancy; sticks to the fingers and dries slowly; shrinks appreciably on drying usually showing cracks. Intermediate and high plasticity clays show these properties to a moderate and high degree respectively.	Intermediate plasticity (Lean clay) High plasticity (F at clay)
Organic soils	ORGANIC CLAY, SILT or SAND	Varies	Contains substantial amounts of organic vegetable matter.	
	PEATS	Varies	Predominantly plant remains usually dark brown or black in colour, often with distinctive smell; low bulk density.	

Composite soil types (mixtures of basic soil types)

Scale of secondary constituents with coarse soils		
Term		% of clay or silt
slightly clayey slightly silty	} GRAVEL or SAND	under 5
– clayey – silty	} GRAVEL or SAND	5 to 15
very clayey very silty	} GRAVEL or SAND	15 to 35
Sandy GRAVEL Gravelly SAND	} Sand or gravel and important second constituent of the coarse fraction	

(See 41.3.2.2)
For composite types described as:
clayey: fines are plastic, cohesive;
silty: fines non-plastic or of low plasticity

Scale of secondary constituents with fine soils		
Term		% of sand or gravel
sandy gravelly	} CLAY or SILT	35 to 65
– CLAY:SILT		under 35

Examples of composite types
(Indicating preferred order for description)
Loose, brown, subangular very sandy, fine to coarse GRAVEL with small pockets of soft grey clay
Medium dense, light brown, clayey, fine and medium SAND
Stiff, orange brown, fissured sandy CLAY
Firm, brown, thinly laminated SILT and CLAY
Plastic, brown, amorphous PEAT

Compactness/strength

Term	Field test
Loose	By inspection of voids and particle packing.
Dense	
Loose	Can be excavated with a spade; 50 mm wooden peg can be easily driven.
Dense	Requires pick for excavation; 50 mm wooden peg hard to drive.
Slightly cemented	Visual examination; pick removes soil in lumps which can be abraded.
Soft or loose	Easily moulded or crushed in the fingers.
Firm or dense	Can be moulded or crushed by strong pressure in the fingers.
Very soft	Exudes between fingers when squeezed in hand.
Soft	Moulded by light finger pressure.
Firm	Can be moulded by strong finger pressure.
Stiff	Cannot be moulded by fingers. Can be indented by thumb.
Very stiff	Can be indented by thumb nail
Firm	Fibres already compressed together.
Spongy	Very compressible and open structure.
Plastic	Can be moulded in hand, and smears fingers.

Structure

Term	Field identification
Homogeneous	Deposit consists essentially of one type.
Inter-stratified	Alternating layers of varying types or with bands or lenses of other materials. Interval scale for bedding spacing may be used.
Heterogeneous	A mixture of types.
Weathered	Particles may be weakened and may show concentric layering.
Fissured	Break into polyhedral fragments along fissures. Interval scale for spacing of discontinuities may be used.
Intact	No fissures
Homogeneous	Deposit consists essentially of one type.
Inter-stratified	Alternating layers of varying types. Interval scale for thickness of layers may be used.
Weathered	Usually has crumb or columnar structure
Fibrous	Plant remains recognizable and retain some strength.
Amorphous	Recognizable plant remains absent.

Interval scales

Scale of bedding spacing	
Term	Mean spacing, mm
Very thickly bedded	over 2000
Thickly bedded	2000 to 600
Medium bedded	600 to 200
Thinly bedded	200 to 60
Very thinly bedded	60 to 20
Thickly laminated	20 to 6
Thinly laminated	under 6

Scale of spacing of other discontinuities	
Term	Mean spacing, mm
Very widely spaced	over 2000
Widely spaced	2000 to 600
Medium spaced	600 to 200
Closely spaced	200 to 60
Very closely spaced	60 to 20
Extremely closely spaced	under 20

Colour

Red
Pink
Yellow
Brown
Olive
Green
Blue
White
Grey
Black
etc.

Supplemented as necessary with:
Light
Dark
Mottled
etc.

and

Pinkish
Reddish
Yellowish
Brownish
etc.

Extract from BS 5930:1981. Reproduced with the permission of BSI

The data obtained will require careful recording and plotting, the extent of which will depend on the nature of the site.

Sites may be classified into two broad groups:

- Compact sites such as those to contain buildings, bridges, dams, docks and airfields (see section 1.1.5).
- Extended sites covering a long narrow strip of land required for roads, railways, tunnels, sewers, pipe and transmission lines and coastal defence.

The type of site and site works determine the extent of the investigation to be made and the number of records and plans to be established.

In many cases the construction of new works affects adjacent properties or interests and therefore a desirable factor in the investigation is that of making a record of existing works. This will usually involve a photographic record in addition to the normal survey.

Investigation of defects or failure of existing works
This type of investigation is necessary to establish the cause of the failure and to provide information indicative of a remedy. Measurements and observations of the structure in question are taken to indicate whether or not the ground conditions are involved; if so then soil investigation should take place. This investigation will reveal the level of ground water and the true state of sub-strata. Faults can often be traced to a weakness in a particular stratum of soil.

Investigation relating to the safety of existing works
When proposed new works are planned, it may be necessary to investigate existing works to decide whether they will be adversely affected by changes in ground conditions brought about by the new works. Existing works may be affected by the following:

- Excavations which may reduce ground support.
- Tunnelling or mining which may cause subsidence.
- Vibrations (eg from piling operations) which may cause fractures.
- Extra load created by new works may overload stratum supporting existing works.
- Soil movement due to heat or freezing induced by proximity to plant installations.
- Ground water lowering may cause settlement.
- Disturbed drainage path may cause flooding and instability of slopes.

All these factors have to be considered if danger is to be avoided from the construction of new works.

Investigation relating to the suitability and availability of material for construction
There are two quite different problems with the mass movement of earth. The first is disposal, for example in the case of spoil from cuts, and the second is acquisition, for example for large fill projects such as reclamation. However, in both cases investigation is necessary to establish the quantity and suitability of the soil for the purpose for which it is to be used. In some cases the suitability of the material can be established by visual inspection. The main problem is the sheer volume involved and this normally requires investigation to establish the quantity available. In other cases the quantity is not in question but detailed testing of the quality is necessary.

It should now be appreciated that the type of investigation could greatly affect the cost of works at design stage.

1.1.4 General enquiries and preliminary work

The importance of site investigation has been highlighted by the comprehensive report funded by The Institution of Civil Engineers – *Site Investigation in Construction*, by Thomas Telford (1993). The report, in four parts:

- raises the awareness of the importance of ground conditions;
- presents guidelines for planning, procurement and quality management of soil investigation;
- provides a guide for specifying ground investigation;
- provides a methodology for dealing with landfills and contaminated land.

The report makes a significant contribution to site investigation work and should be treated as essential reading for surveyors involved with civil engineering.

The investigation should proceed in a logical manner and include the following:

- A desk study.
- Site reconnaissance.
- Detailed studies for design purposes.

A desk study. This should be carried out to review the available information relating to the site. Information required includes geological data, which may be available from local sources. This study may reveal records of previous or nearby soil investigations. Site boundaries, ground contours and obstructions above ground level can also be established. Information on the previous use of the site is invaluable and should be obtained; this might be facilitated by the location of old maps and records in libraries and museums. Geological data, aerial photographs, historical information and local knowledge will help to establish the existence of particular problems. In some cases the desk study will reveal sufficient information to determine the extent of the detailed studies.

If the site has been used in earlier years it is important to establish the use it has had and the position and type of foundations if any. It is important also to obtain further details of sites which have been previously used as pits, quarries, coal mines or brine-pumping. A particular problem is that where depressions have been subsequently filled in with waste, such fill areas are not only subject to excessive settlement, but can also be subject to spontaneous combustion and harmful chemical content.

Where loadings are high and substrata investigation is necessary, reference should be made to the County Geological Survey. This will show the main strata and any superficial deposits. The Geological Survey Office also keep records of boreholes, which are available for perusal.

In mining areas British Coal keeps records of active works and details of some abandoned workings are recorded and filed in the Mining Records Office.

Aerial photography is a valuable means of obtaining topographical maps, especially of extended sites. This work is undertaken by specialist firms and the photographs can be geologically interpreted so as to highlight areas for detailed investigations.

Site reconnaissance. This is normally carried out by 'walking the site'. A check should be made on the position of buildings, boundaries and other obstructions and related to the desk study. Adjacent property and the likely effects of the proposed works should be recorded. The general topography will often be indicative of the soil conditions below which will require further investigation. Typical examples of this are:

- Stepped ground, which may be caused by geological faults.
- Broken or terraced ground, which may indicate land-slip.
- Depressions in a limestone area, which may indicate the existence of swallow holes.
- The nature of the vegetation, which may give some indication of the moisture and acidity characteristics of the soil.
- Low-lying areas, which may indicate the presence of soft material deposited due to the earlier presence of water.

Detailed studies. On completion of the desk study and the site reconnaissance, a programme of detailed studies can be prepared. This may include a detailed land survey and further investigation regarding the earlier use of the site. These studies will result in a programme of investigation and soil testing.

1.1.5 Procedure for investigation

Compact sites

In the case of compact sites for buildings, bridges, dams, reservoirs, docks etc, the investigation will require deep and closely spaced borings. The actual number and position of borings depends on the type of structure and nature of site. However, the number of bore holes should be such as to give a clear picture of all significant variations in the soil over the site, and the depth of such bore holes should be such as to reach all strata likely to influence the stability of the works. Particular attention should be given to soil at or just below the proposed foundation level, because weak strata below the foundation level may fail even though only subject to reduced 'bulb pressure' (see Figures 1.3 and 1.4).

It should be noted that the pressure bulb below strip foundations is greater than that underneath pad foundations; this is due to the plan continuity of the foundation. Pressure bulbs of less than 20 percent of the original loading at foundation level can

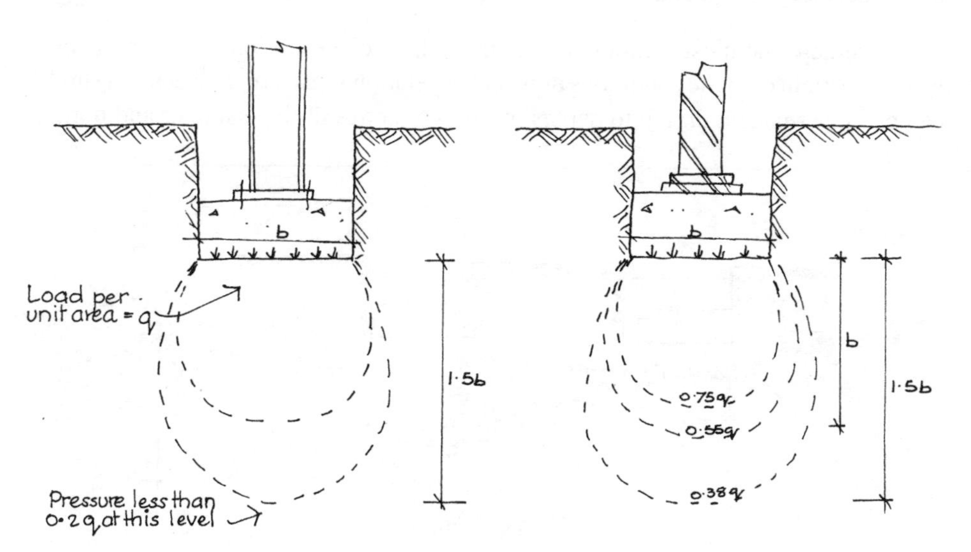

FIGURE 1.3 *(Left)* *Bulb pressure under a pad foundation showing distribution of load*

FIGURE 1.4 *(Right)* *Bulb pressure under strip foundation showing distribution of load*

be ignored when considering the effects of loading on various strata. Where two foundations are near to each other, the overlap in stresses may call for further investigation of lower strata (see Figure 1.5).

Extended sites

The detailed work of exploring extended sites, such as roadworks, will vary according to nature and complexity of the proposed works. A preliminary survey is of particular importance in such cases because the evidence thus gained will assist in planning the bore hole frequencies and patterns. In the main, shallow borings will suffice to reveal ground water conditions and the nature of upper strata. Borings are usually made by means of machine augers, often mounted on a small mobile vehicle.

The design of works on extended sites requires the following data:

- moisture content of the soil;
- compaction factor;
- liquid and plastic limits;
- resistance to deformation.

The last can be achieved by the California Bearing Ratio (CBR) test, which can be carried out in the laboratory or insitu – this test is discussed further in section 1.1.7.

The ground water should be tested for the presence of harmful substances such as sulphates, which might adversely affect the strength of concrete, steel or other materials used in the development.

Soil movements other than those due to loading

So far, the procedure for investigation has dealt with normal considerations in soil investigation, but it must be stated that not all soil movement is due to loading from new works, or to the collapse of underground cavities. Movements may be caused by any or all of the following:

- Drying and shrinking of soils.
- Frost heave.
- Artesian pressure.
- Seismic phenomena.

Perhaps the most common is the first, where clays are found. The clay loses water in certain climatic conditions or where vegetation is excessive. Trees and shrubs can cause permanent drying to depths of up to 5 metres during summer and normal

FIGURE 1.5
Showing overlapping of foundation stresses

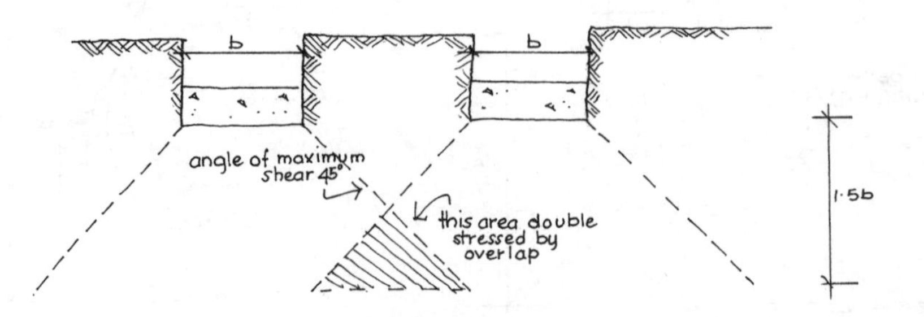

vegetation can cause drying out up to 2 metres deep. Vertical movements of the soil up to 100 mm have been recorded in the vicinity of large trees. These factors should be borne in mind when the investigation takes place.

1.1.6 Methods of site exploration

When discussing the procedure for investigation, reference was made to borings as a means of investigation. This is perhaps the most common method of site exploration, but certainly not the only one. Considering new works, from the very small to the very large contracts, a general guide to exploration would be as follows:

- Small works – trial pits up to 3 m deep.
- Medium to large scale works – borings up to 30 m deep.
- Very large scale works (eg dams, power stations) – a combination of deep boring, pits and insitu examinations from headings and shafts.

It must be noted that the above is only a guide: the detailed methods of exploration would depend on the type of construction and site involved.

Trial pits

This is the cheapest form of exploration in shallow depths (eg up to 3 m); above 3 metres deep the cost increases rapidly compared with boring. The main advantage is that soils and rocks can be exposed and examined insitu. This method shows changes in strata much more clearly than by borings. The pits are dug out either by local labour or by a small tractor-mounted excavator. The plan size of a pit depends on method of excavation, but approximately 1.2 × 1.2 m and should be dug at distances 20 metres apart in either direction. Holes should be kept well clear of the position of actual foundations, but should be in the vicinity of important structures such as heavily loaded walls or columns.

Problems occur in water-bearing soils, particularly sands, and therefore the economies of shoring and pumping pits may outweigh the savings gained against specialist borings. In dry conditions these pits are particularly valuable since they allow hand-cut samples to be taken, thereby minimising the disturbance of the sample and maximising the conditions for accurate testing. Deeper trial pits may be used for the investigation of rock fissures or to explore layers of weak rock which cannot be removed intact in normal boring operations. Such deep pits are costly to construct and would be used only in large scale exploration. Trial pits are the best method of exploring back-filled areas and sites overlain by variable natural deposits.

Borings

All borings should be spaced sufficiently close together to prevent false deductions concerning the uniformity of horizontal strata. The 'depth' of boring will be determined by the type of loading involved but account must be taken of any slope in the strata and variations in their thickness. In particular, when boring through glacial deposits, care must be exercised to ensure that boulders are not mistaken for bedrock.

There are three main factors which govern the depth of exploration:

- The depth to which the soil is to be significantly stressed.
- The depth to which weathering is likely to affect the soil.
- The depth at which impermeable strata occur.

The first factor depends on the type of structure, the intensity of loading and the shape and size of the foundation structure. Where large foundation areas are stressed the bulb pressures are far-reaching, therefore raft foundations may carry loads which will stress weak areas below normal good load-bearing strata (Figure 1.6).

It is necessary therefore to know the characteristics of soil at a depth of up to 1.5 times the breadth of foundation. For such works as roads and airfields the depth of exploration is normally taken as a minimum of 1.5 metres.

The second factor deals with seasonal changes in moisture content and this has been known to affect some soils down to depths of 1.5 metres. Frost is unlikely to affect soil more than 0.75 metres below ground level unless industrial equipment such as freezing plant is installed.

The third factor will apply to water-conserving structures such as reservoirs where an impermeable stratum is essential and must be located.

The seasonal presence of waterlogged ground may render the site unsuitable for the proposed works and the contours of the water table should therefore be plotted (Figure 1.7) in order to decide how the problem can be overcome. This type of survey will also be necessary as a preliminary to the design of intercepting lines of drainage to prevent the influx of ground water from higher levels.

Hand or mechanical auger borings are relatively cheap methods of sub-surface exploration for soils which will stand unsupported. Holes can be sunk to depths up to 3 metres provided there are no obstructions such as boulders. The diameter of the bore

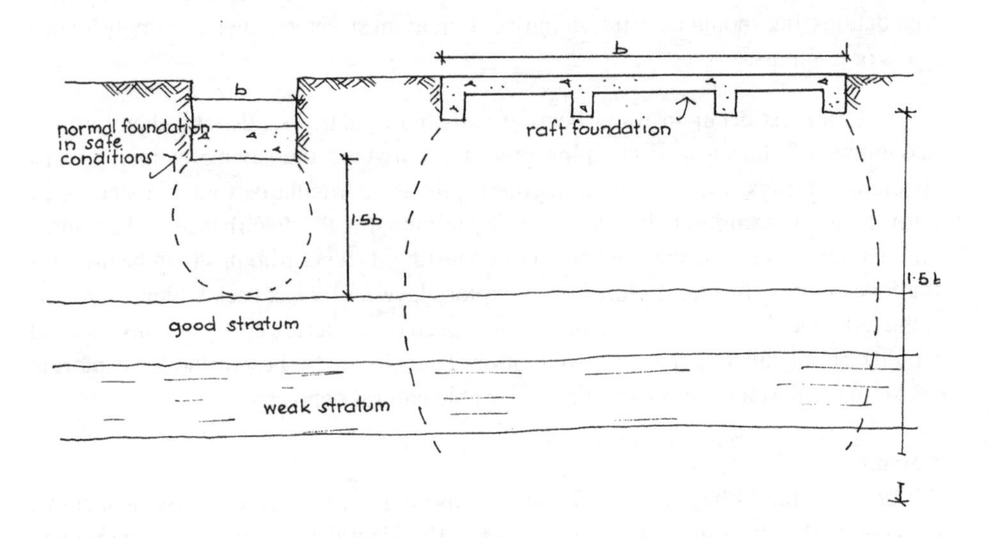

FIGURE 1.6
Showing bulb pressure below raft going down to weak stratum

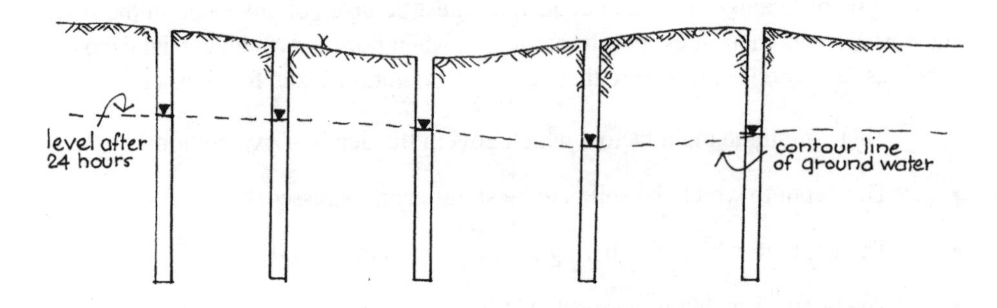

FIGURE 1.7
Ground water level in bore holes

hole is usually 150 or 200 mm: this allows soil sampling tubes to be used without difficulty. The mechanical auger is used in gravelly soil, which involves the use of a casing to prevent collapse of the boring.

Shell and auger boring is a method which can be carried out in all types of soils, because the bore hole is lined with a thick-walled steel casing. The boring is achieved by augers or open ended shells in cohesive soils and shells in the case of non-cohesive soils (Figure 1.8).

When boring through soils, the borehole is lined as the hole is bored and the section linings screwed together and driven as the hole deepens.

Rotary drilling is the traditional method of drilling hard stratum, such as rock, but may also be used in soft stratum to provide access for extracting soil samples. There are two types of rotary drilling:

- Core drilling.
- Open hole drilling.

In the first case, as the name suggests, a core of rock is produced by means of rotating a core barrel with an annular cutting bit at the end. The core of rock is retained in the barrel of the drill and lifted out to be placed in special core boxes for transportation to the laboratory. Open hole drilling is used in weak strata to allow soil samples to be removed at various levels, or to allow insitu tests to be carried out. In open drilling water or drilling mud is used to flush out the cut material and to lubricate the drilling bit.

Wash borings. The soil is loosened and removed from the bore hole by means of a strong jet of water or drilling mud. The liquid is jetted through a steel tube which is worked up and down the hole. The liquid disintegrates the soil and carries it up the annular space between the tube and casing. Wash boring has the advantage that the soil is not disturbed by blows of a tool or shell, but it is limited to soils which do not contain boulders or large gravels. Mud, such as bentonite, allows boring to be carried out without linings in non-cohesive soils. The soil in its settled-out state can be dried and used for identification purposes.

FIGURE 1.8
Boring rig
(Norwest Hoist Soil Engineering Limited)

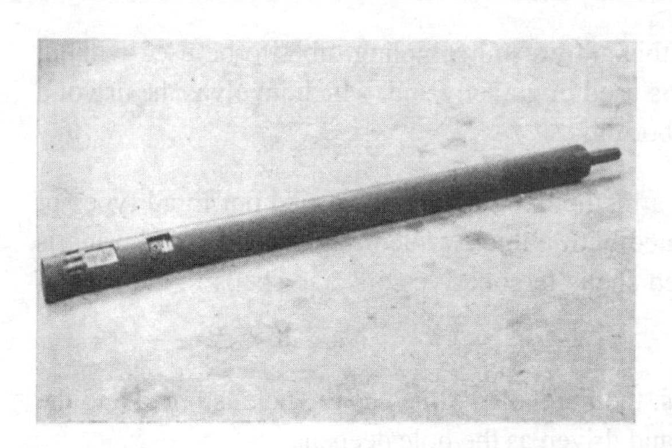

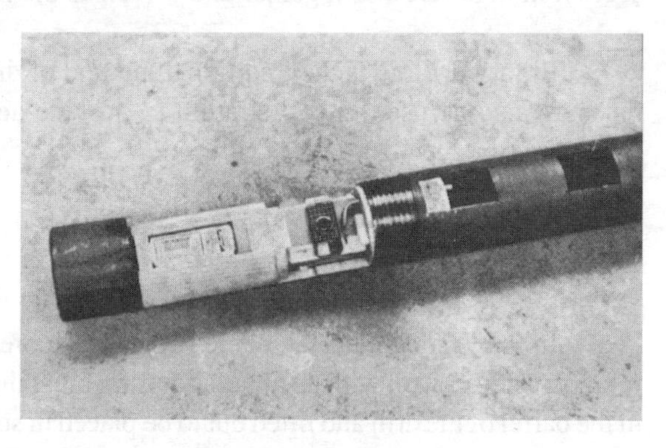

FIGURES 1.9(a) and 1.9(b)
The camera and flash-gun assembled in a 1.5 m length of HX core barrel (Ground Engineering)

FIGURE 1.9(c)
Borehole camera photograph of a 40 year old timbered heading to a sewer some 9 metres below a trunk road (Ground Engineering)

Headings and shafts

Headings are employed to explore steeply dipping strata. Headings are also used in the form of a pilot tunnel, in anticipation of the driving of larger tunnels later. They have an advantage over pits in that they can be drained easily and also allow easy removal of spoil. Shafts approximately one metre diameter can be bored using large power-driven augers. The sides of the shafts are supported to protect personnel engaged in soil inspection. This type of exploration allows the following to be achieved:

- removal of intact block soil samples;
- insitu penetration tests;
- loading tests;
- disturbed samples obtained by drilling.

Choice of method

The choice between these various methods of exploration will normally depend on the studies carried out in section 1.1.4. However, in general terms hilly country offers the choice between pits and headings whilst low-lying marshy areas are best explored by borings. Where rock or changeable ground conditions are expected, borings of various types should be used. Borings in fissured rock will allow photographs to be taken by a remote-controlled borehole camera. Cavities and underground can be investigated in the same way (Figure 1.9). In soils the normal method of exploration is by boring unless the loads expected are small. In such cases shallow pits will facilitate adequate investigation.

Types of samples

There are two types of samples, disturbed and undisturbed.

Disturbed samples are samples removed from boreholes with augers or other equipment which interfere with the natural structure of the material. Such samples are useful for visual grading and determining moisture content, and in some cases for laboratory testing. Samples are placed in airtight jars with identifying labels.

Undisturbed samples are samples removed by methods which preserve, so far as practicable, the natural structure and properties of the material. Samples in this category are easily obtained in rock and clay but difficult in certain other soils. Table 1.2 shows the method employed for obtaining samples. The graph in Figure 1.10 shows the comparative strength of various samples of soil from the same stratum.

TABLE 1.2 **Sampling methods**

Soil	Disturbed	Hand samples Auger samples Shell samples
	Undisturbed	Hand samples Core samples
Rocks	Disturbed	Sludge samples from percussion or rotary drills
	Undisturbed	Hand samples Cores

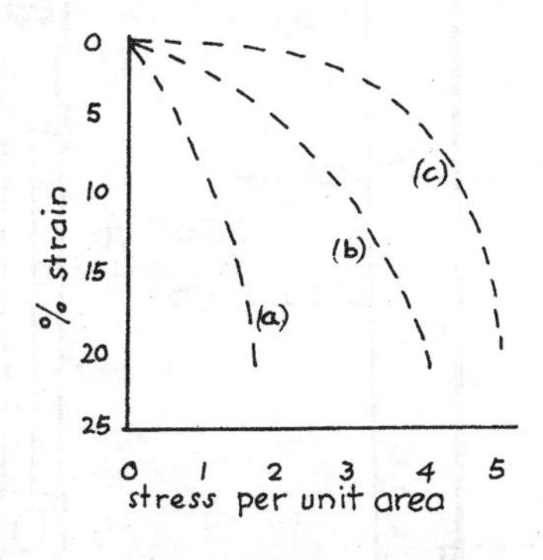

FIGURE 1.10
Comparative strengths of soil samples from the same stratum

The method used to obtain samples in cohesive soils is by 100 mm diameter sampling tubes. The tube, shown in Figure 1.11, is screwed on to a set of rods and pushed down or driven into the soil. The cutting edge of the tube is a detachable threaded ring in metal or plastic which cleaves the soil 1 mm under the internal tube diameter. This prevents consolidation of the soil during the overdrive. The tube, when removed, is sealed at both ends with wax or plastic caps and after being suitably marked is sent to the laboratory for the testing of its contents. On arrival at the laboratory the sample may be carefully extruded into a 100 mm plastic tube, thus reducing the cost of storing valuable steel sampling tubes.

1.1.7 Insitu testing

Tests to obtain the density or shear strength of soils insitu are very valuable since they can be carried out without disturbing the soil. Such tests are particularly valuable in sands and silts. The main tests are:

- Standard penetration test.
- Vane test.
- Plate bearing test.
- California bearing ratio (CBR).

Standard penetration test (BS 1377)

This test is particularly useful for testing the resistance of sands and gravels. As with all penetration tests this consists of measuring the resistance of the soil to penetration under static or dynamic loading. This particular test is made by driving a 35 mm (internal diameter) split-barrel sampler (Figure 1.12) into the soil at the bottom of a bore hole. The sampler, suspended on rods, is first driven 150 mm into the soil, or 25 blows, by a falling standard weight (65 kg falling through a distance of 760 mm). The sampler is then driven a further 300 mm and the number of blows required to effect each 75 mm of penetration is recorded. The test is used to establish the relative density of soils. Table 1.3 gives some indication of the results of such tests, but the data shown is only approximate and would require expert interpretation before being used.

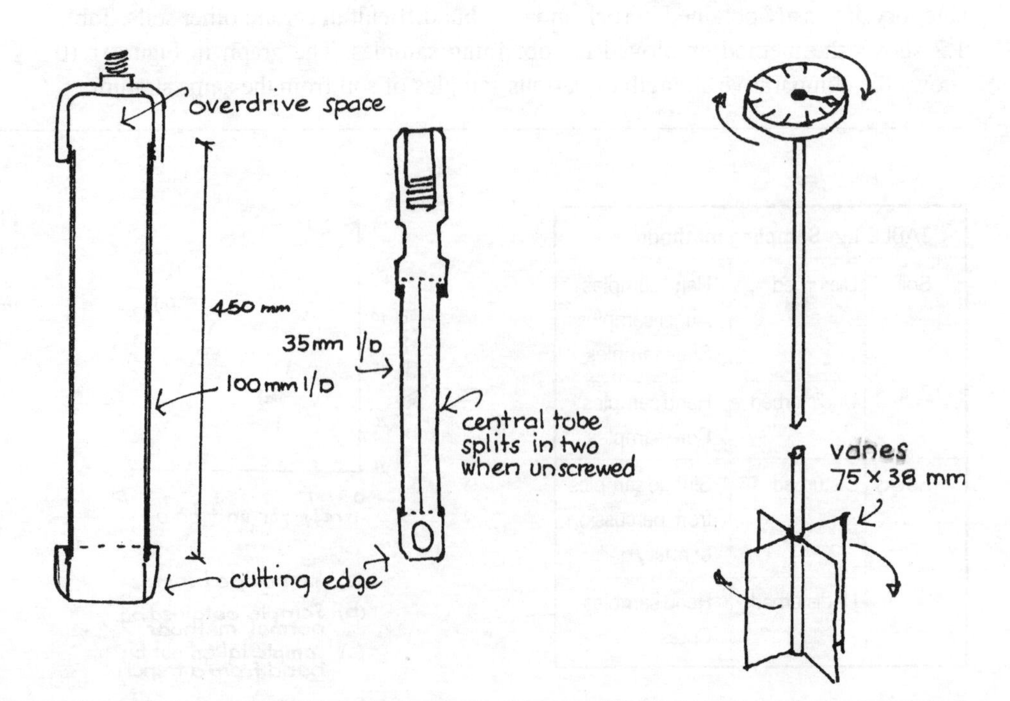

FIGURE 1.11 *Standard sampling tube*

FIGURE 1.12 *Split-barrel sampler*

FIGURE 1.13 *'Pilcon' hand vane test equipment*

Vane test (BS 1377)
This test measures the shear strength of soft cohesive soils insitu. The vane (Figure 1.13) is pushed into the soft clay and rotated by hand at a constant rate. The amount of torque necessary for rotation is measured, and the shear strength of the soil is calculated. The test can also be used to measure the remoulded strength of the soil. The test is normally used in uniform cohesive saturated clays. The advantage of this test is that soft soils can be tested without disturbing them.

Plate bearing test (BS 1377)
This type of test was once very popular and is still used on large engineering projects, as a means of providing insitu data on the behaviour of soils at foundation level. The procedure consists of excavating a pit to the level of the proposed foundation and then loading a steel plate, ranging from 300 mm to 1000 mm in diameter, at the bottom of the pit or borehole. The load can be applied in either of two ways; by loading it with increments of kentledge (concrete blocks or steel billets), or by means of a hydraulic jack bearing against a heavily loaded beam (Figure 1.14).

The load is applied in a controlled manner such that the chosen rate of penetration is both uniform and continuous. The test continues until the penetration reaches 15 percent of the plate width. Where there is no clear indication of failure prior to reaching a penetration of 15 percent of the plate width, the ultimate load may be defined as the load taken to produce a penetration of 15 percent of the plate width. The load is applied in increments and the settlement after each increment is recorded and plotted on a time graph. A typical increment of loading would be from approximately 1/5th of the proposed design load up to failure point.

TABLE 1.3

Sands			Clays		
No of blows	*Relative density*	*Notes*	*No of blows*	*Consistency*	*Compressive strength in kN/m²*
0–4	Very loose	Can behave	0–2	Very soft	0–24
4–10	Loose	like a liquid	2–4	Soft	24–49
10–30	Medium		4–8	Medium	49–98
30–50	Dense		8–15	Stiff	98–196
over 50	Very dense		15–30	Very stiff	196–392
			Over 30	Hard	Over 392

FIGURE 1.14
Loading test using hydraulic jack
(Norwest Hoist Soil Engineering Limited)

The safe load should be taken as one-third of that load which causes failure. One disadvantage of a test of this nature is inaccurate simulation: the bulb pressure from the test is usually far smaller than the bulb pressure from the actual foundation (Figure 1.15). This could lead to error in detecting settlement of a weak stratum.

California bearing ratio test *(CBR Test – BS 1377)*
This test is used in the design of flexible pavements and can be carried out on site. The test was designed by the California Division of Highways in the 1930s, as a result of an extensive investigation into the causes of pavement failure in 1928–9. The test shows the load-penetration of soils relative to a standard crushed stone sample. It is carried out using the weight of a vehicle to obtain the necessary reaction load through a hydraulic jack, and normally on soil at least 1 metre below ground level (ie below the level of any seasonal moisture fluctuation). A circular area, about 300 mm diameter, is trimmed flat to allow the plate to be seated before starting the test. The test can also be simulated in a soils laboratory by forcing a cylindrical plunger, of a known cross-sectional area, into soil at a given rate.

Other insitu tests include permeability tests, ground water level observations and ground water pressure. These tests are described in BS 5930. Geophysical testing is described in section 1.1.9.

1.1.8 Laboratory testing

Laboratory testing is undertaken to establish the following characteristics of soils:

- Identification and classification.
- Measurement of their engineering properties.
- Chemical content.

Identification and classification
This analysis involves a number of individual tests, such as:

- Visual examination.
- Moisture content.
- Liquid and plastic limits.
- Particle size distribution.

Visual examinations are made to note the colour, texture and consistency of disturbed and undisturbed samples, these being used later to describe the soil in the engineer's reports.

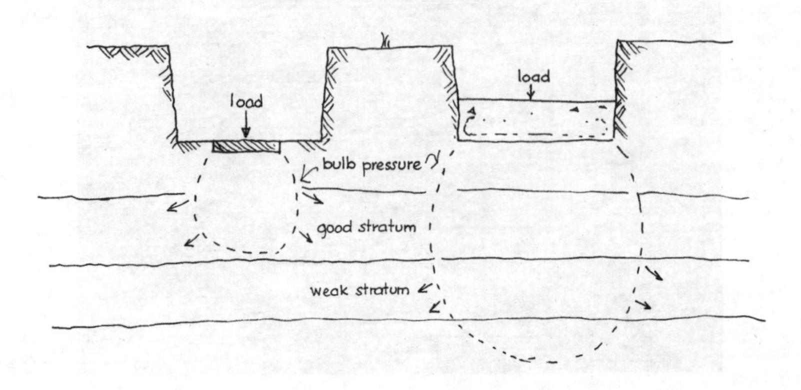

FIGURE 1.15
Showing difference in bulb pressure between plate test and actual foundation

The moisture content is important in all soil samples, since it helps to arrange a programme of testing (by relating samples to liquid and plastic limits) so that no doubtful sample will be overlooked. The higher the natural moisture content of the soil, the greater will be its compressibility. Clays in London have an average moisture content of 25 percent whilst those under Mexico City are about 250 percent, accounting for the excessive settlement in the city.

Liquid and plastic limit tests are made on cohesive soils for classification purposes and for assessing their compressibility. The liquid limit (LL) (BS 1377) is usually determined by using the cone penetrometer, in which a needle of standard shape and weight is applied to the surface of a soil sample for a period of five seconds; the penetration of the needle is then recorded. The process is repeated with different moisture contents. The moisture content for a 20 mm penetration is the LL. The plastic limit (PL) is determined by rolling out a 3 mm diameter thread of soil and noting the moisture content, which will allow the thread to be rolled out still further until it breaks up due to drying. When both LL and PL are known, the Plasticity Index can be established (Plasticity Index = LL – PL). A standard chart such as Figure 1.16 will indicate likely soil properties in further tests.

Particle size distribution is of particular importance when assessing problems of excavation in permeable soils below the water table. It is also useful for assessing the value of non-cohesive soils for use as aggregates. The first part of the test is achieved by sifting dried samples, to BS 410. In the case of cohesive soils a wet analysis is used, employing a hydrometer. The range of particle sizes is compared with a standard chart.

Measurement of their engineering properties

The foregoing tests give some indication of the engineering properties of a soil, but there are also specific tests which yield more definite information relating to, for example:

- Bulk density of soil.
- Shear strength of soil.
- Consolidation of soil.

The bulk density of material is the weight of that material per unit volume. It includes the weight of air or water in its voids. This information is essential in the design of retaining works, where the weight of a stratum is an important factor (eg stability of slopes, formation of earth dams, earth pressure of retaining walls etc). Dry density (weight of solids per unit volume) is used for the determination of optimum compaction in earth dams, embankments and other soil structures.

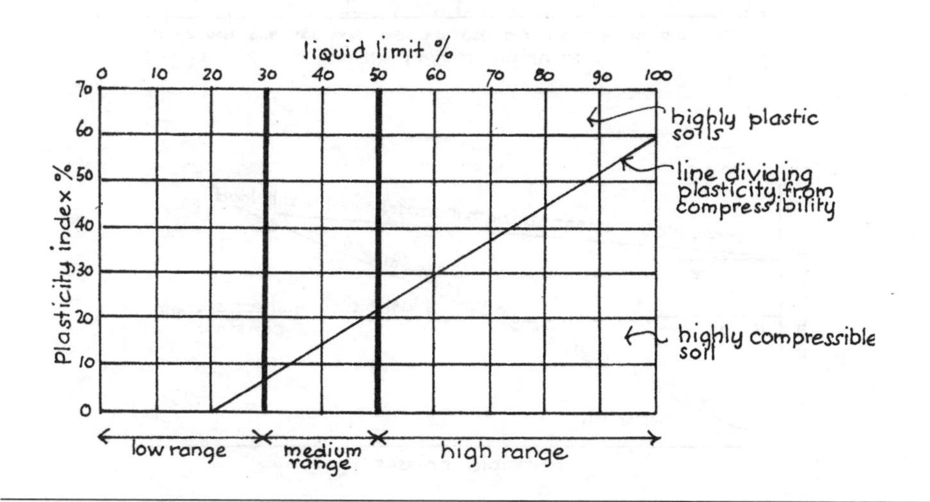

FIGURE 1.16
Plasticity chart

Shear strength of soil

The shear strength of a soil can be used directly to calculate its bearing capacity and also to calculate the pressure on supports in excavations. There are several tests available for ascertaining shear strength, but the most popular is the Triaxial Compression Test. This test can be carried out using any of three different methods:

- Undrained
- Consolidated undrained
- Drained.

In principle the test consists of subjecting a cylindrical sample of undisturbed soil (75 mm long × 38 mm diameter) to lateral hydraulic pressure in addition to a vertical load. This is achieved by placing the sample in a rubber sheath and capping it with porous end plates to receive the loading. The sample is supported within a plastic cylinder which is subsequently filled with water. Measurement of the forces needed to shear the sample is used in the calculation of bearing capacity. In the undrained triaxial test (often referred to as the quick test) the sample is capped with non-porous end plates to prevent the pore water escaping and allow axial loading of the ends. Three tests are carried out, one on each of three samples (all cut from the same large sample), each being subjected to a higher hydraulic pressure before axial loading is applied. The results are then plotted in the form of Mohr's circles (Figure 1.17). Stresses for various soils are indicated in Figure 1.18.

The consolidated undrained triaxial test allows the sample to drain while applying the hydraulic pressure, thereby allowing the sample to consolidate. After consolidation the sample is stressed without further drainage. In the drained test the axial load is applied so slowly that the pore water can drain off without building up any pressure in the sample; the drainage continues throughout the test and the amount of water drained off is measured. In both cases where drainage is achieved, the water

FIGURE 1.17
Test on stiff clay plotted on Mohr's circles

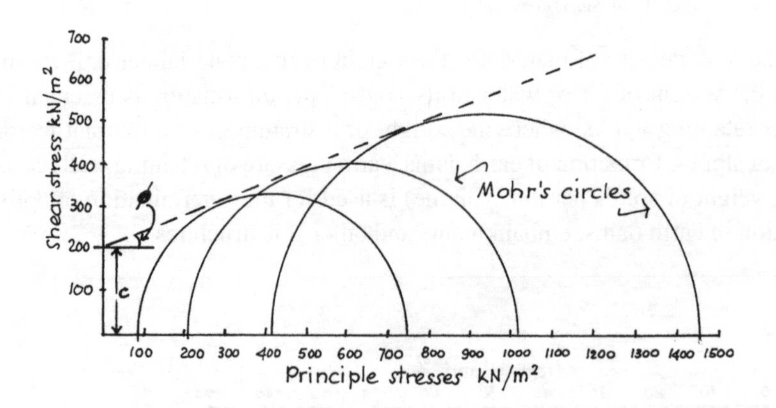

FIGURE 1.18
Showing results of triaxial tests for various types of soil

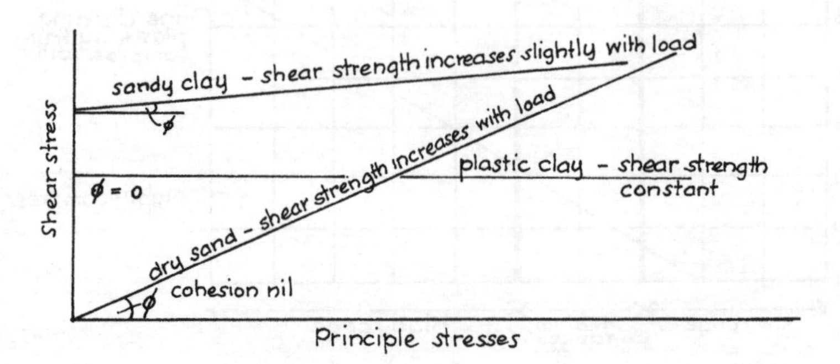

passes through porous discs at the ends of the sample and then through ducts in the apparatus. The consolidated undrained test and the drained test have particular application to the behaviour of soil in earth dams and embankments, and also to stability problems in general. Triaxial compression tests are suitable for cohesive soils only. Where non-cohesive soils have to be tested the Shear Box test is used. A sample of soil is subjected to a standard load, under which a horizontal force is applied to the lower half of the box until the sample shears.

Consolidation of soil

The consolidation test provides results which are used to calculate the magnitude and rate of consolidation of a particular soil. This is very important in calculating the movement of soil under foundations. The apparatus used is called an 'Oedometer'. The test consists of placing a cylindrical sample (75 mm diameter × 18 mm thick) in a metal ring and capped with porous discs. The sample is placed in a water-filled tray and subjected to load. The load is increased every 24 hours and a time-settlement curve is plotted.

Chemical content

A chemical analysis of soils and ground water is carried out to assess the effects, if any, which their composition might have on any materials to be used in the proposed works. The tests mainly cover sulphate content and pH value, although bacteriological analysis may also be required for works in tidal mud flats.

Figure 1.19 shows a typical layout for a testing laboratory.

1.1.9 Geophysical surveys and other techniques

This type of investigation is useful for extended sites where conditions are generally favourable. Favourable conditions include vast areas of land or water, beneath which a layer of rock may have to be surveyed for thickness and extent by a rapid and inexpensive method. The main contribution is to detect and locate any changes in material or stratification of material in the earth's crust so that the number of borings can be reduced to a minimum. Readings are taken between widely spaced borings (to ensure continuity or otherwise), allowing rapid coverage of large areas at very low cost.

FIGURE 1.19
Typical soil testing laboratory
(Norwest Hoist Soil Engineering Ltd)

The methods most commonly used are:

- Electrical resistivity
- Seismic refraction and reflection
- Magnetic.

Electrical resistivity is based on the difference in electrical resistance between one rock or soil type and another. The method uses four electrodes (an outer pair and an inner pair), a current being passed through the ground via the outer electrodes and the drop in potential being measured at the inner electrodes (Figure 1.20). The electrodes are spaced equally in a straight line and, by varying the centres of the electrodes, the depth of penetration can be varied. The results allow mathematical and graphical analysis to be made to establish the thicknesses and depths of the various sub-strata; the test is also useful for detecting swallow holes or other underground cavities. Problems can occur if unsuspected conductors are present, eg pipes and cables; such conductors can render the results unreliable.

Seismic methods involve setting up vibrations, normally by explosions, and measuring the time taken by the shock waves and the distance covered. The 'refraction' method is suitable for shallow exploration; in this method seismometers are spaced at increasing distances from the source of vibration and the shock waves are recorded and plotted graphically against a time scale. Vibration for this method can also be produced by a 6.5 kg hammer. The 'reflection' method is limited to depths of exploration greater than 150 metres. Both methods are particularly useful in surveying sites for tunnels, dams and harbour works. The test is used to determine the change in soil type, particularly the 'rock-head' level.

Magnetic methods of investigation are based on measuring the variations in intensity and direction of the earth's magnetic field. The most important application is that of tracing underground cavities such as mine shafts and swallow holes. The test is most valuable in detecting buried objects, such as cables and pipelines.

Geophysical surveys should be considered complementary to established methods of exploration rather than the sole means of obtaining accurate information on soil structure.

FIGURE 1.20
Showing electrodes in position for measuring resistivity

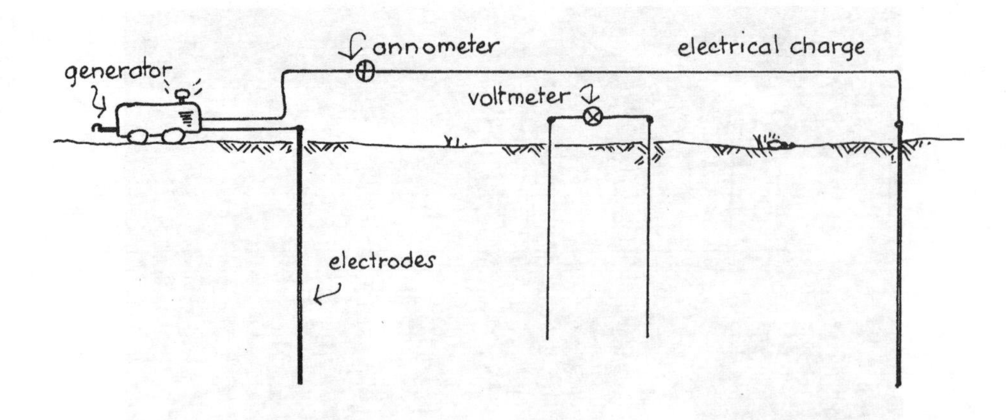

Pumping test

This is a test used to determine a permeability and storage coefficient of the soil where excavations are to be taken down below natural water level. The test involves the construction of a deep well or a ring of well points. A number of observation wells are dug around the pumping point and a check kept on the levels of water in these as pumping continues. The data thus obtained assists in the design of ground-water lowering systems as well as drainage and irrigation systems.

Off-shore investigation

In many cases investigation has to be made in seas and rivers for information concerning future tunnels, jetties, barrages and similar marine works. Drilling is often carried out from the deck of a floating pontoon (Figure 1.21), or alternatively from a jack-up or spudded pontoon, but perhaps the most modern and versatile drilling platform for off-shore investigation is the hover platform (Figure 1.22). The hover platform has many advantages over a conventional platform, including the ability to launch directly from a beach, and ease of working over mud trenches, sand banks and deep gullies.

In addition to the investigation for major engineering works, off-shore drilling is also used as a preparation for dredging. In this type of work the same platforms are used but with multi-drilling machines. The technique is known as 'drill and blast' in which rock is drilled and then blasted to a dredgeable size. The whole operation of drilling, placing of explosives and detonation is carried out without staff having to leave the platform. This work is discussed further in Chapter 3.

FIGURE 1.21
Floating pontoon with four drilling rigs
(Rock Fall Company Limited)

FIGURE 1.22
Hover platform
(J T Mackley & Company Limited)

The report

The site investigation report will include the following data:

- Preamble – giving client, job, terms of reference.
- General description of the site.
- General geology of the area
- Description of soil in bore holes – together with borehole sections
- Laboratory test results.
- Discussion of results commenting on type of foundation, type of cement, etc, if specified.
- Conclusions and recommendation, if specified.

Specification for site investigation

In preparing a specification for work to be undertaken in site investigation, the surveyor will have to consider two quite different situations.

- Onshore site investigation, which forms the greater proportion of all site investigation and is therefore the one with which the surveyor will have most contact.
- Intertidal and offshore investigation which, although not rare, is of a specialised nature and therefore dealt with by a small number of surveyors who are employed by very large organisations or consultants specialising in the field.

For further details on specifications for ground investigation the reader should consult Part 3 of *Site Investigation in Construction* (Thomas Telford 1993).

1.2 SITE ORGANISATION AND TEMPORARY SERVICES

1.2.1 Introduction

The nature of the site, its geology and subsoil conditions, the nature of the Works in relation to the site, considerations of access and strategic locations for stores, plant repair shops, supervisors' offices, mess rooms and the like, together with the Resident Engineer's accommodation, will all have been considered when preparing the tender. In jobs involving large volumes of earth transportation, such as airfields and roadworks, the broad plan for the movement of excavated material will also have been worked out. In many cases working drawings and diagrams will have been prepared for these preliminary planning considerations and in special cases even scale models may be constructed.

There are two types of models. The first type is experimental and might be used in the solution of complex structures, hydraulics or other analytical problems. The second type is visual and might be used to give an impression of the physical appearance of the work or to identify problems which may arise in construction. The latter is of great importance in site organisation. Such models would be used in a 'method study' of the work to be achieved, and the alternative solutions would be subject to technical and economic assessment. In addition, working drawings of contractors' temporary works and falsework will have to be prepared.

These considerations generally lead to the preparation of a Master Programme, which should take into account such matters as rate of construction, timing of operations, type of plant and equipment. In most cases this will take the form of a bar chart but in some cases, including those where the specification may require it, a 'critical path' network will be produced. All such programmes should include sufficient 'float time' to allow for circumstances likely to cause delay but for which the contractor will not be entitled to an extension of time.

1.2.2 General organisation and site clearance

On signing the contract, the contractor will use the data from the preliminary planning stage to establish and set up an appropriate site organisation. This organisation will include the layout of the site, which will involve the siting of access roads, storage areas, hutting, spoil heaps, plant areas etc. Access roads may be aligned on permanent access roads or hardstandings; this will allow grading to correct levels and the formation of a good base ready to receive final wearing coats later. Haulage roads, for taking materials within the site and off the site to tip and for bringing in fill material, must be planned and positioned to give economy and efficiency. In all cases, the roads will need to be considered in the light of the equipment and plant to be used, and the type of site being worked, ie compact or extended site.

Where the site is extensive, some distinction will be drawn between access roads and haulage roads. On such sites, permanent access roads will be positioned to serve both the main administration area and the internal movement of plant and materials. The rest of the site will be served by rough access roads which require constant maintenance during wet weather to prevent break-up of road surface and consequent damage to vehicles.

All the above organisation refers to normal sites on land, but some consideration must also be given to sites where access may be limited to water. This would include sites on or in a sea, lake or river. In such cases transport is normally undertaken by barges towed by diesel tugs, and organisation must allow for tides and bad weather. Temporary jetties will also be required for the loading of the barges. Other forms of transport which require a great deal of planning and organisation prior to installation are cableways, railways and conveyors.

Site clearance

Site clearance should be undertaken as soon as the contractor gets on site, so as to permit the initial siting of access roads and site accommodation. The method of working will again depend on the type of site.

In the case of compact sites, trees and other growth might be removed by a specialist subcontractor. This will be quickly followed by removal of the topsoil, which will be deposited in spoil heaps ready for re-use in finishing off verges and similar features immediately prior to completion. However, with an extensive site the problems are much greater. The clearance of such a site could involve diversion of sewers, drains, gas, power lines and telephone lines, and it might also involve phasing some or all of the work to facilitate harvesting of crops, all of which adds to costs and legal claims for compensation.

In particular, the problem of site drainage may be involved. When large-scale earth-moving is involved, it often interferes with natural drainage, which causes either a waterlogged site or a movement of mud into water courses. Both have their own disadvantages, the former making the site difficult to work and the latter bringing complaints from other landowners and river authorities because of contamination.

It may be an advantage to cut a perimeter or cut-off drain around major excavation works to intercept water during a rainy season. The same principle can be employed on large scale road works, where a carriageway can be cut to just above finished formation level during the dry months to intercept and channel water in the wet months. This will assist general earth-moving work in inclement weather. In some cases site clearance necessitates the installation and operation of dewatering plant: this technique is described and explained in Chapter 3.

In a great many instances site clearance will also involve large or small scale demolition of existing works and buildings.

1.2.3 Demolition *(Also see Chapter 8)*

Before any demolition work is commenced a detailed survey of the property should be made. Photographs should be taken of the works to be demolished, together with any adjoining property; this will serve as valuable evidence in the event of claims being made against the contractor. A check must be made of any legal rights which may affect the demolition; these may include easements, questions of ownership etc, as well as liability for damage and trespass. Detailed surveys of the buildings to be demolished should include the following:

- Framed buildings – will unbalanced thrusts occur when members are removed?
- Walls – load-bearing or non-load-bearing; stability of party walls if roof or floors removed.
- Support required for all elements where stability is to be reduced.
- Any cantilevered structure – nature of support.
- Basements and voids – method of filling or protecting.
- Underground storage tanks – position, size, contents of tanks.

- Supplies and services. These may include:
- Drainage, gas, electricity, water, telephone, television or radio lines; hydraulic pressure mains; district heating mains.
- Bench marks on buildings – which involves informing the Ordnance Survey Office.

Work on site

The demolition contractor is required to appoint a competent foreman to supervise the work and a programme of proposed sequence of operations should be prepared. Particular attention should be given to the following:

- Licences for hoarding, fencing, lighting etc.
- Scaffolding for access purposes.
- Closing of roads – special notice to LA.
- Access to site – may need special protection, planning permission will be required if access leads directly on to a highway.
- Services – may require specialist treatment; diversion.
- Health and Safety – workmen must be protected at all times
- Uncontrolled collapse – ascertain method of support shoring etc.

For methods of demolition, see Chapter 8.

1.2.4 Site offices and general accommodation

Accommodation for sites covers two types of structure: small mobile transportable units, and sectional prefabricated buildings. Both types are popular since they both fulfil specific requirements depending on site, space required etc. On small, short-duration contracts the mobile unit is economical and adaptable to site organisation and conditions. The sectional structure, however, is preferable on large contracts, allowing better coordination and control by providing many offices under one roof.

The main factors which govern the choice of site accommodation are location and duration of the contract concerned, together with the factor of space on compact sites. In some cases, where the contract is long term involving a large labour force, the accommodation may equate to a small township catering for living, sleeping, feeding and recreational facilities which may not exist locally. Where this is necessary a more permanent type of structure such as prefabricated concrete office buildings may be required. For most sites, however, the offices consist of prefabricated timber buildings, normally single storey, but adaptable for two storey layout.

Mobile units play a very important part in site accommodation and new developments have been introduced to satisfy site conditions. The older wheeled chassis unit has largely been superseded by 'skid mounted' and 'jack-leg' designs. Although the new designs are less manoeuvrable on site, they are just as quick to transport and are not so restricted in size as the wheeled chassis. The 'jack-leg' units have the advantage that they can be levelled easily on sloping sites and stacked on top of each other to form a double storey. These units can be delivered to site containing complete catering facilities or toilet facilities and therefore have much to commend them.

The amount of office accommodation needed will be determined by the type of organisation and amount of staff required for controlling the contract. The amount of general accommodation for the labour force should be calculated from a master programme which will show the labour force throughout the contract. Actual facilities, such as toilets, catering and first aid, must conform to the standards laid down in the current regulations. All hutments should be as centralised as possible and be near main access roads.

1.2.5 Material storage and compounds

Some form of hutting or more permanent buildings must be provided for storage of materials which are subject to pilfering, vandalism and deterioration from weather. Stores should be laid out in such a way that checking and issue of materials can be carried out easily.

Storage bins cast in concrete may be used for storage of bulk aggregates near central mixing plants, whilst heavy materials such as steel reinforcement, concrete pipes etc can be stored in the open. When materials are likely to be damaged by children or building operations, they should be stored in a locked compound for safe keeping. Compounds will also be necessary for the safe keeping of plant and equipment. The storage of liquids such as oil and petrol is subject to the specific regulations:

1.2.6 Temporary services

Every site, whether large or small, requires some temporary services. On the small site the requirements may be limited to electric power, water, telephone and sewer connections. On the larger construction site they will include compressed air, heat supply, forced ventilation, catering services and possibly a properly equipped medical centre. There is no set formula for establishing the type and range of services for a site; each site must be considered on its own merits and the services designed to fit the specific needs of the contract. Factors that affect the selection and design of temporary services are:

- Site location.
- Specific needs of the contract.
- Number of working shifts per day (eg whether continuous working would demand trouble-free equipment).
- Health and safety.
- Cost.

The following selection of temporary services does not describe in detail the design and installation of temporary services, which is a speciality, but rather lists some factors which merit consideration when planning the installation.

Electricity

Where electricity can be supplied by the local undertaking the following factors must be considered:

- Capacity of existing system.
- Demand of site including type of supply (single or three phase).
- Cost of installing transmission line to central sub-station.
- Alternative methods and cost of producing equivalent power supply.

Where the local undertaking cannot readily supply the site, then other additional factors must be considered:

- Design and position of generating house.
- Type or quantity of generating equipment.
- Method of earthing and distributing power on site.
- Health and safety.

One of the common failings when considering a supply is to underestimate the lighting requirement on site, particularly in winter working, when daylight fails quickly.

Water supply

The water supply to a large construction site may be divided into two types: industrial water and drinking water. The latter, which must be potable, is required for work areas, offices, mess rooms, shops etc. The former is required for rock drilling, feed water for boiler plant, moisture control of dam fill, transporting materials by sluice or dredge pipelines.

In both cases certain factors must be considered at the detailed planning stage. These include:

- Available sources.
- Quality and treatment.
- Water storage.
- Pumping requirements.
- Protection during winter period.
- Distribution.
- Fire fighting requirements.

Heating

Every site requires some form of heating for offices and mess rooms. The form of heating ranges from calor gas heaters to steam-generating plants which circulate steam through pipes to every work and rest area. High pressure steam lines can also be used for heating domestic water and defrosting material or steam curing of concrete. Factors for consideration include:

- Size of contract – amount of hutting.
- Availability of equipment.
- Availability of fuel supply.
- Economics of system.

Compressed air service

Most large sites are involved with the generation of compressed air, since it has become a power source for hand tools, rock drilling and hammers and a means of transporting materials in pipes, eg concrete. It is also used for sand-blasting, placing gunite, and pressure-grouting.

The multiplicity of uses is a very important factor to consider when planning the services requirements. Other factors include:

- Type of equipment (eg central or local).
- Method of generation (layout for power house).

- Distribution lines and air receivers.
- Air losses.
- Safety.

Ventilation

In certain work areas, such as tunnels, a system of forced ventilation may be necessary. The clearance of blasting by this method requires a great volume of air over a short period of time. This can be achieved by a system of fans placed at intervals along the tunnel route. Consideration must be given to the following:

- Size of fan.
- Time lapse requirement for achieving clean air conditions.
- Type of power required; for example, electrical demand is very high when using large fans and therefore consideration should be given to other forms of power.

Gas supply

In most cases gas supplies to sites are in the form of cylinders or main storage tanks on site. All such requirements must be subject to strict adherence to the regulations on storage and fire risk. Consideration must be given to:

- Type of gas; eg methane, propane, oxygen, acetylene.
- Quantities in store.
- Position of storage in relation to work areas.
- Fire fighting facilities.
- Safety in respect of explosion.

1.2.7 Safety

The change in attitude to safety measures in recent years has led to the development of complex solutions in the civil engineering field. The surveyor or engineer must be aware of the requirements which the contractor has to meet and for that reason this section is covered in some detail.

Site safety must be considered by the contractor during the early planning stages of the contract so that requirements and procedures can be established. The items for consideration include:

- Design and construction of temporary works.
- Site access and egress.
- Offices and compounds.
- Temporary power houses and services.
- Plant.
- Working in compressed air.
- Dangerous atmospheres.

Design and construction of temporary works

All such works must be designed and constructed by competent personnel. These structures require examination and maintenance on a regular basis so that the structure does not deteriorate and become a safety hazard. This is particularly applicable to support systems such as shoring, scaffolding, trench sheeting, gantries, falsework etc, where conditions may vary considerably during the contract period. In the case of falsework (formwork), special consideration must be given to live loads that are imposed during the pouring of the concrete. These loads are created by plant, labour force and flow of concrete.

Piling

Piling frames must be designed to withstand not only driving stresses but also wind loading. All platforms on the piling frame must be fully decked and ladders provided for the full height. The capacity of the piling winch should include a margin of safety above the weight of the maximum weight of pile. In addition, the weight of the driving hammer and helmet must be considered. Heavy concrete piles may require cranage and this will involve design of slings for inclined and eccentric loading.

Shafts

BS 5573 covers the safety aspects of large diameter boreholes for piling and other purposes. Access to and from the working surface for personnel may be achieved by means of a skip raised and lowered by a crane. The skip must be constructed for this use alone and include a device to prevent overturning. Where ladders are used, they must be screened from the operation of hoists or cranes and be well lit. Support material to be used during sinking should be available for securing the excavation, and one operative should be detailed to supervise the lowering of objects and materials into the shaft. Alternatively the shaft should be fenced or protected in some way.

Tunnels

Safety in tunnels is covered by BS 6164. Ventilation by artificial means is necessary in tunnelling by mechanical shields, where the rate of heat dispersal is high and where humidity is troublesome. Where explosives are used, some form of artificial ventilation will be necessary to reduce the fumes. For open tunnel working an air flow of 9 m^3/min per square metre of tunnel section is required.

For electrical supplies in tunnels, the supply should be either 110 V single phase, Centre Tapped to Earth (CTE), or three-phase and neutral (100/65) 50 hz, and the installation completely watertight. A 415 V, three-phase, 50 hz supply may be necessary for mining equipment. For lighting and hand-held portable tools the voltage to earth should not exceed 55 V, and portable tools must be double insulated. Circuit breakers and residual current devices (RCDs) should be used for earth leakage protection.

In compressed air tunnelling, where there is a risk of sudden flooding due to water breaking through the face, an elevated escape way must be provided which includes watertight plates across the upper sector of the tunnel to provide air pockets.

Site access and egress

The layout of site roads and haulage roads must be considered in the light of speed and volume of traffic. Where possible, a one-way system is advisable when moving large quantities of materials by wheeled transport. The layout should provide good visibility or, alternatively, incorporate warning and control systems to prevent collision. All junctions and hidden access points should be clearly signposted. Special consideration must be given to sites where public 'through ways' have to be maintained. This last factor may involve the use of hoardings and fans to protect the public, and permission to erect these structures may be obtained from the local authority.

Offices and compounds

Office buildings must be sited to minimise the fire risk and allow adequate means of escape. Where offices are raised above ground level, the area below must be enclosed to prevent storage of materials or accumulation of rubbish. Methods of heating and cooking in buildings must be carefully considered and naked flame heaters should be avoided. Full cylinders of gas and empty bottles should be stored in racks outside the buildings.

Where compounds contain workshops a number of factors should be considered:

- Hand operated electrically driven tools should be 110 V supply.
- Lifting devices must be tested.
- Machinery should be positioned to deal with long units.
- Fire extinguishers of the right type must be located in strategic positions.
- Lighting must be adequate for work being undertaken.
- First aid equipment must be readily available.
- Storage of inflammable materials such as timber must be kept clear of buildings, all stacks being separated by adequate fire breaks.

Temporary power houses and services

This section includes compressed air and electrical installations. All electrical installations must be laid out by an experienced engineer who understands the dangers of short circuits, and who can advise on the safest method of earthing the system. Where diesel installations are used for generation of power, there is some fire risk due to spillage of fuel on hot exhaust systems. In no case should large quantities of fuel be stored in the generating house. All cables must be protected and positioned so that working plant cannot foul them.

Compressed air and the generation of air are both controlled by regulations, but factors to be taken into consideration are:

- No unauthorised person allowed access to the compressor and receivers; one man to be responsible for the control and maintenance of the whole installation.
- The compressor should be contained in a building with a receiver of matching capacity.
- Where possible it is good policy to run a central air main to workshops.
- All joints in power lines should be flanged and spring-clipped rather than screwed.
- Water traps should be provided.
- Valves for blowing out water should be installed.

Services

Whilst services to the site include gas, electricity, water, telephones, sewers etc, care must be exercised in the diversion and installation of such work. The temporary service that requires greatest consideration is that of electricity.

The temporary supply will be provided by either the Electricity Company or, in the case of remote sites, the contractor's own generators. In either case, the main problem is a matter of earthing. Where the supply is provided by the Electricity Company by underground cable, it is possible to obtain earthing facilities by connection to the metallic sheath of the supply cable. This method is highly recommended.

Where the supply is brought in by overhead line or distributed from power houses over the site, then some form of earth electrode must be used.

Another method of earthing is one where the contractor provides his own earth electrode by burying suitable conductors in the ground. Such conductors require expert supervision and testing for earth resistance on installation and at intervals during use. All apparatus and equipment must be efficiently earthed.

Where it is difficult to obtain a low resistance earth path, use should be made of an earth-leakage circuit breaker unit. These units can be either voltage operated or current operated; both are available in single phase or three-phase forms. The voltage-operated device has its operating coil connected between earth and the frame of the apparatus. If the voltage in the apparatus rises higher than a certain level above earth, the current through the coil trips the circuit breaker and disconnects the supply. The time of operation for these circuit breakers is very low, thereby giving a very high degree of safety. In the current-operated circuit breaker, the incoming and outgoing currents are compared and any difference, which represents a leak, is used to operate a trip switch.

Whatever the supply, the main switchboard will normally provide for an outgoing 415 V three-phase supply for items of plant such as cranes, hoists, pumps, and concrete-batching plant. However, for most purposes provision should be made for a 32A 110 V three-phase and earth supply to step-down transformers which operate at 110 V for hand-operated tools and lighting equipment. This step-down transformer should be CTE, thereby reducing the risk of shock to a maximum of 55 V to earth. In some cases 240 V transformers will be used for offices and site lighting.

Where plant is provided with 415 V three-phase supply, the cable should be protected with metal armour, preferably of the steel wire type, and either buried or surface-laid on a clearly marked route. Overhead lines should follow the perimeter of the site and never hung across the path used by plant and high vehicles.

Travelling cranes requiring electrical supplies should have supplementary earthing though the wheels and rails. This means that rail joints must be bonded and rails effectively earthed. For moving plant an electronic device can be fitted to the machine, for example on a crane jib, which will detect a live supply line and give an audible alarm signal in the cab of the machine. These devices are particularly useful on cranage where the operator is concerned with watching the load.

Plant

Particular attention should be given to two very accident prone pieces of plant, namely hoists and cranes. The former may be used for goods or passengers and the regulations cover such points as:

- Enclosing the hoistway.
- Securing the hoist tower to the structure.
- Gates fitted at every landing.
- Hoist level indicators on goods hoists.
- All hoists to be inspected weekly.

In respect of cranes, the following factors should be considered:

- Proper calibration of radius and load indicators.
- Audible and visual indicators.

- Test certificate for the machine and equipment.
- Examination of machine and equipment.
- Systems of signalling.
- Lifting hooks must be tested to twice the maximum load.
- Warning lights for low flying aircraft.

Working in compressed air

The use of compressed air to achieve working conditions in water-bearing strata may cause men to suffer from Compressed Air illness. This illness is caused by incorrect decompression of the body which results in limb pains (known as 'the bends') or paralysis, or problems with sight, hearing and breathing (in some cases bone damage). The condition is created in the first place by the blood absorbing air proportionately to the pressure exerted on it and distributing it in solution to all parts of the anatomy. During the reverse process (decompression), the oxygen element in the inhaled air is used up in the tissues and nitrogen accumulates in the form of bubbles in the tissues of joints and muscles. These bubbles of nitrogen must be recompressed back into solution in the blood (where the illness occurs) and a method of decompression is employed to let the excess nitrogen leave the body via the lungs.

The principal means of curing compressed air illness is the medical air-lock, which allows medical attention without interrupting any recompression process. Prevention of decompression sickness depends upon two factors:

- Careful examination of each worker prior to admission to the pressurised atmosphere, rejecting the unfit.
- Strict observance of rates of decompression.

In the case of caisson sinking, an alternative method of decompression may be used, called 'decanting'. By this method persons are rapidly decompressed in the man-lock to atmospheric pressure, followed promptly by rapid recompression in a separate chamber and subsequent gradual decompression. In this method the total time spent between the start of the primary decompression in the man-lock and the completion of recompression in the separate chamber should not exceed five minutes.

Dangerous atmospheres

In certain works such as shafts, headings and underground works where machinery is being used, there is a risk of men being overcome by dangerous fumes.

Carbon dioxide (CO_2) is the asphyxiating gas most likely to be encountered in the course of well or shaft sinking. It is especially likely to occur in chalk, limestone or greensand and in all cases where acids or solid carbon dioxide are being employed. Since the gas is heavier than air, it tends to accumulate at low levels. Detection can be achieved by means of a flame, which will be extinguished by the pressure of the gas (where oxygen is reduced in the atmosphere by 20–25 percent).

Carbon monoxide (CO) is introduced into working areas by internal combustion engines, welding or brazier fumes. It is slightly lighter than air and is extremely dangerous; even 0.01 percent may be harmful. The capacity of this gas for combining with the blood is more than 200 times greater than that of oxygen; resulting in headaches, loss of breath, dizziness or vomiting, and collapse, followed by death in a matter of moments. Affected persons should be taken into the fresh air with urgency and kept as warm as possible with blankets and hot water bottles. If available, oxygen should be given freely.

Methane (firedamp) (CH_4) is found near the roof of headings. It is lighter than air and, although not poisonous, its presence in excess leads to a diminution of oxygen with a consequent danger of suffocation. Explosions can occur if as little as five percent of this gas is present, and this should be regarded as the principal risk. Detection is best achieved by means of a flame safety-lamp of approved design.

Nitrous fumes may occur during the process of well construction as a result of the use of explosives, particularly in consequence of incomplete detonation. Their effect on the body is usually delayed and symptoms of coughing and shortness of breath may not appear for some hours. The gas should be regarded as dangerous and medical help should be sought immediately. Other dangerous gases include:

- Nitrogen oxide.
- Hydrogen sulphide.
- Sulphur dioxide.
- Radon.

Attention must also be given to de-oxygenised air, fumes from welding and cutting, and gases which are stored on site for various purposes.

1.3 MATERIALS

1.3.1 Materials in general use

Since many of the basic materials in use in the construction industry are examined in detail in other books, this section is limited to a brief description of certain materials specifically used in civil engineering works.

Steel

Steel can be classified by the degree of carbon content as follows:

- Low carbon steel; up to 0.15 percent carbon, suitable for tin-plate and wire.
- Mild steel; 0.15 to 0.25 percent carbon, suitable for standard sections, eg universal beams, channels, angles.
- Medium carbon steel; 0.20 to 0.50 percent carbon, suitable for general engineering work; eg high tensile steel reinforcement and high tensile sections.
- High carbon steel; 0.50 to 1.50 percent carbon, suitable for tool making and heavy castings.

Weather-resistant steels comprise a group of high strength, low alloy steels containing elements which, under normal atmospheric conditions, give an enhanced resistance to rusting compared to ordinary carbon steels. The first example of this kind of steel was Cor-Ten, developed by the US Steel Corporation in the 1930s. The UK produces two main types under licence from the US:

- High phosphorus steel containing alloy additions – an example of which is Cor-Ten A.
- Low phosphorus steel which generally has higher carbon and magnesium content – an example of which is Cor-Ten B developed by the US Steel Corporation. This is a weldable steel suitable for bridges and other exposed structures.

Extras and allowances

In addition to having a basic knowledge of steel, the surveyor should also be aware of the extras and allowances that may be incurred by quality and specification. The carbon content has already been mentioned, the amount of carbon in the steel affecting the price. Other quality steels include manganese, re-sulphurised, niobium, copper, killed steels, and tensile and impact properties.

Cements and concrete

Cement has been produced since 1824 and uses clay and limestone as the main constituents. Little more needs to be said about this material in a text on civil engineering, apart from reminding the reader that certain cements are of particular value in the civil engineering field. One such cement is low-heat cement; of particular value in the construction of mass concrete bases and dam construction where hydration could create problems of expansion and cracking. However, the extent to which such cements are used is very limited when compared with ordinary Portland cement.

Concrete

BS 8110 covers the requirements of structural concrete. Cement is the most expensive ingredient in concrete and, although assumed to be the most reliable, since it is very carefully controlled during manufacture, should be sampled and tested.

Important qualities of cement include its strength, durability, and defined setting and hardening characteristics. These qualities are determined by the chemical composition, the manufacturing process and the fineness of grinding.

The relevant British Standards must be quoted when specifying the material. Cements are grouped according to their chemical composition and manufacturing processes, the characteristics of which determine the use of the material. It will be appreciated therefore that, although Ordinary Portland cement and Rapid-hardening Portland cement represent the bulk of all types of cement used, there are many special cements, such as low heat cement and sulphate-resisting cement, which have been developed for special purposes. In civil engineering the amount of cement used on a project may justify the production of one of these special cements rather than the use of other technical solutions.

Fine aggregates may be obtained from various sources. They must be clean, hard and chemically inert. Pit sand nearly always requires washing — hence the name 'washed sand' for concrete. Crushed stone may also be used provided that it is free from excessive dust and not too flaky. Crushed limestone and granulated slag should be carefully checked before using in reinforced concrete because it is often of low strength. Sea sand, suitably graded, may be used for concreting if taken from below high-water level, but tests should be carried out for organic impurities and salt content.

The sand, whatever the source, must be checked for correct grading (BS 410). The grading of fine aggregate is more important, when considering the strength of concrete, than the grading of coarse aggregate.

Coarse aggregates must also be clean, hard and chemically inert. They must be free of an undue quantity of elongated or flaky particles, organic matter, coatings of harmful chemicals or, if crushed, excessive dust. The material can be dug from pits, rivers and beaches, or formed from crushed rock from a quarry: large quantities of dredged marine aggregates are also used. Pit gravel requires more cleaning and screening than river gravel and may contain more large stones that require reducing by crusher. Beach gravel must be checked for organic impurities and salt content; again, materials taken below the high-water level tend to be acceptable. Gravel aggregates are slightly superior to crushed stone, owing to their lower volume of voids, and are therefore suitable for impermeable concrete. They are also more resistant to sulphate attack.

Crushed stone aggregates or angular aggregates can be produced from granite, flint or hard limestone. Sandstone should be avoided since it tends to be porous and therefore much weaker than the other stones. Whinstone chips make good coarse aggregate, but the combination of whinstone chips and whinstone sand or fines may produce a concrete which fails to harden; siliceous sand should be used with whinstone aggregate. Clinker and ashes must not be used as aggregates for structural concrete because of their low strength and possible high sulphate content; the exception to the rule is 'fly ash', which is produced by burning pulverised coal. Crushed blast furnace slag produces a high quality aggregate which is comparable with gravel for strength. It can be used to provide complete protection of reinforcement against corrosion; this is achieved by the dense concrete it produces.

Reinforcement

The main aspects of reinforcement for consideration include:

- Types of reinforcement.
- Methods of supply.
- Labours involved in preparation.
- Fixing.

The main types of reinforcement for concrete are:

- hot-rolled mild steel bars to BS 4449;
- hot-rolled and processed high tensile alloy bars to BS 4486;
- cold reduced steel wire to BS 4482;
- high tensile steel wire strand for prestressing, BS 5896;
- steel fabric to BS 4483.

All manufacturers or suppliers are in a position to supply mild steel bars to BS 4449, but few manufacturers are currently producing high tensile materials.

Generally, suppliers stock a normal range of 6 mm to 40 mm diameter, but some manufacturers will produce this range together with 50 mm diameter bars if requested. The maximum stock length of bar is 12 metres, but certain bars are available in 18 metre lengths.

All cold-worked bars and mild steel bars are produced from steel with a carbon content of below 0.25 percent the specification maximum.

Steel fabric (BS 4483) can be obtained in four different types, with preferred sizes in each type: square-mesh fabric; structural fabric; long-mesh fabric; and wrapping fabric. The fabric is made up of hard-drawn steel wire complying with BS 4482. It is formed by interweaving or electrically welding the wires so that the fabric will withstand normal handling. Square-mesh fabric has a mesh size of 200 mm × 200 mm and a weight range of between 1.54 kg and 6.16 kg per square metre. Structural fabric has a mesh size of 100 mm × 200 mm and a weight of between 3.05 kg to 10.9 kg per square metre.

Long-mesh fabric, which is used mainly in road slabs, has a mesh size of 100 mm × 400 mm and a weight ranging from 2.61 kg to 6.72 kg per square metre. While long-mesh fabric is suitable for Jointed Concrete Pavements (JCPs), it may not be suitable for Continuously Reinforced Concrete Pavements (CRCPs) due to the lack of steel in the main bars. However, it is possible to get a designed mesh fabricated for CRCPs to ensure that the design requirements are met.

Bricks

Numerous types of bricks are used in civil engineering work but the main varieties are:

- **Engineering bricks, Class A**. These are the densest, hardest and heaviest bricks used; they have a compressive strength of 70 N/mm^2. Trade names include Staffordshire Blues and Southwater Reds. They absorb no moisture of any consequence and can thus resist great extremes of weather. They are used in bridge abutments, boiler bases, culverts etc.

- **Engineering bricks, Class B**. These are slightly less dense than Class A, having a minimum strength of 50 N/mm^2 compared with 70 N/mm^2 for Class A. They are used in the construction of inspection chambers and in work below ground level.

- **'Calulon' bricks**: a proprietary brick, specifically made for highly stressed walls. These bricks are made in three strength grades, A10, B7.5 and C5, having strengths of 69 N/mm^2, 52 N/mm^2 and 34.5 N/mm^2 respectively. Tests show a 30 percent increase in the speed of laying and 40 percent reduction in the quantity of mortar used compared with solid brick walling. They are suitable for load-bearing structures above ground level.

- **Common bricks or Flettons**. This is a broad term applied to bricks which are suitable for internal work. 'Commons' may vary in different localities, so it is essential to state the type of common bricks required. They have a compressive strength ranging from 7 to 27 N/mm^2.
- **Facing bricks**. Although almost any brick can be used as a facing, the term is generally applied only to better quality bricks with an aesthetically pleasing texture. Engineering bricks can be used as facings, eg in retaining walls.

Timber

Timber used in civil engineering is mainly confined to temporary works such as strutting, formwork and gantries. With this in mind, the details given in this section will be limited to the strength aspect of timber.

Timber has a high strength both in tension and compression and is elastic. The strength of timber increases with density and varies greatly between species. BS 4978 deals with the structural use of timber and contains the basic stresses for specimen pieces of timber free from all visible defects.

The Code of Practice defines grades of timber in six categories; General Structural (GS); Special Structural (SS); Machine General Structural (MGS); Machine Special Structural (MSS); and two grades referred to as M75 and M50. These grades are determined by defects such as 'slope of grain', size of knot and wane in proportion to width of timber, size of fissures. It is now common to obtain machine-graded timber which is graded as it passes through a series of rollers under stress.

Plywood is graded in accordance with BS 6566, which covers eight grades. For good quality finishes to concrete it is usual to use a high grade Douglas fir ply, coated with an epoxy resin; this can be used many times before deterioration. For rough concreting and for formwork panels which require decorative finishes – normally achieved by a secondary lining – it is possible to obtain and use an inferior grade of ply known in the construction industry as 'shuttering quality'. This ply has all the normal strength qualities but has defects on one or both faces such as split laminates or missing knot infills. In all cases the plywood should be treated against the effects of contact with cement by coating with a vegetable oil or other releasing agent before every use.

Stones

Stones fall into three geological classes: Igneous, Sedimentary and Metamorphic.

Igneous stones, which contain crystalline silicates, are best known for the granite group of stones. Granites have a very high compressive strength as well as a highly decorative finish. They are highly resistant to chemical attack and are virtually impervious. Pink and grey granite can be obtained from Scotland and light grey from Cornwall.

Sedimentary stones include sandstones, consisting of fine or coarse particles of quartz with mica or feldspar bound together in a natural cement. Limestones consist mainly of calcium carbonate in the form of calcite, and vary in hardness. Limestone from the Pennines is very hard and therefore suitable for road works and concreting. Sandstone can be obtained from many parts of the country, but Yorkshire and Derbyshire stones are among the best. Limestones for walling purposes are mainly obtained in a belt from Dorset to the Wash, the best known examples being Bath and Portland Stones.

Metamorphic stones include slates and marbles which have been produced by tremendous heat and pressure. Slates come from Cornwall, the Lake District and Wales, and vary in colour with each area. Thin roofing slates are confined to the Welsh quarries and riven slate panels to the quarries in the Lake District.

Stone for road works, often called 'road-metal', can be obtained from numerous quarries in all counties, but wherever possible a local stone should be specified to avoid high transport costs.

1.3.2 Materials for bulk filling

Bulk filling materials in civil engineering include soil, rock and aggregates, and pulverised fuel ash (PFA).

Soil

This is used for the formation of embankments and other areas which require fill. Such soil may be brought to the site from other contracts but in the main forms part of the earth transportation on site. It is usual to 'cut' soil from the high points of the site and place the soil in layers in lower areas. This method, known as 'cut and fill', is the standard procedure for dealing with the mass transfer of soil, since it greatly reduces the cost of removal to spoil heaps with consequent additional spreading. However, soil used for bulk filling must be suitable for this purpose, and the Quantity Surveyor should specify, having first consulted a competent engineer, what is meant by 'unsuitable material'. Unsuitable material includes the following:

- Material from swamps, bogs and marshes.
- Perishable material such as peat, logs and stumps.
- Materials prone to spontaneous combustion.
- Frozen materials.
- Materials having a liquid limit exceeding 80 and/or a plasticity index exceeding 55.

Rock and aggregates

Rock fill consists of hard material of suitable size for compaction and may include crushed stone, hard brick, concrete or other hard inert material. The predominant supply of such hard fill comes from quarries where the rock is crushed to a suitable size. The crushed stone is transported either by road or rail; most quarries of any size have a direct link to the major railways.

Pulverised fuel ash (PFA)

This is a material obtained from coal-fired power stations. Coal is pulverised and blown into the combustion chamber by a jet of air. After burning, the resultant ash is extracted by cyclone collectors or electrostatic precipitators. This ash is known as pulverised fuel ash (PFA). It has a lower density but is stronger than natural granular materials; it requires very little consolidation; it has immediate strength from high value of apparent cohesion, and increased strength due to self-hardening: all of which results in stable conditions as a fill material.

PFA can be supplied in the following ways for use as a bulk fill.

- Conditioned PFA – material taken straight from the silos to which water is added to bring it to the correct moisture content for compaction.
- Lagoon PFA – which has been pumped hydraulically into lagoons to drain and dry out before delivery to site.
- Stock-piled PFA – previously conditioned ash which has been stock piled temporarily before delivery to site.

The material is transported to site in lorries or by rail, depending on accessibility. It is easily handled and compacted on site, is much lighter than conventional fill materials, and has a much higher shear strength than rock and soil fills. One of its greatest advantages is the absence of settlement when compacted correctly. To achieve good results, the following specification should be adopted:

- PFA must be delivered in covered trucks to prevent moisture loss.
- A thick drainage layer, 300 mm to 450 mm, should be provided below PFA bulk fill to prevent saturation of the lower levels.
- The material should be tipped at an agreed delivery point and transported and placed in layers not exceeding 225 mm, graded and compacted with a vibrating roller.
- If the material requires an increase in moisture content this should be achieved by spraying each layer before compaction.
- The site dry density after compaction should be 95 percent of the maximum dry density, with no results falling below 85 percent of the optimum.
- On completion the surface of the fill should be protected with 100 mm granular sub-base to prevent dusting or damage.

If PFA is to be used within 450 mm of the road finish, it must be stabilised against frost by using 10 percent (by weight) cement. Figure 1.23 shows PFA being compacted on site.

1.3.3 Explosives

The various types of blasting explosives can be divided into two groups: high explosives and low explosives. The former detonates with a rapid production of gas, while the latter deflagrates and produces gas slowly. Explosives can also be categorised into three main classes, gelatines, powders and slurries, although other types are available. Further details can be found in Chapter 8.

FIGURE 1.23
PFA bulk fill being compacted on site (National Power plc)

Gelatinous explosives have a good resistance to moisture both in use and in storage. The maximum storage life varies with conditions and climate but is in the order of one or two years. This type of high explosive is suitable for use in boreholes containing water. Gelatinous explosives contain a quantity of nitroglycerine, thickened with nitrocellulose to give the mixture a plastic consistency.

Powder explosives are of relatively simple design. They are made from ammonium nitrate, or sometimes sodium nitrate, and a combustible material, with nitroglycerine as a sensitiser. The nitroglycerine content is around 8—10 percent. Because of the low density of nitroglycerine powders, their bulk strengths are normally low. However, they can be used where economy of explosive is required. They are suitable for blasting medium and soft rock where there is no water problem.

Slurry explosives are composed of ammonium nitrate and other nitrates, both inorganic and organic, with the addition of fine aluminium powder and other light metals. They were developed to strengthen and waterproof ammonium nitrate. The ingredients are not in themselves explosive, and this is an advantage in the production process. They are very much less sensitive to accidental initiation by impact or friction than either gelatines or powder explosives. Operators do not suffer nitroglycerine headaches when handling slurry explosives. They are completely waterproof both in storage and use and can therefore be used in wet boreholes. The maximum storage life varies with conditions and climate but is generally at least six months.

Blasting accessories

Plain detonators for use with a safety fuse consist of an aluminium tube containing a base charge of high explosive and a priming charge of lead azide. This type of detonator is supplied in No 6 strength or No 8 strength. A No 6 strength detonator is adequate and gives efficient detonation with gelatines and powders. Slurry explosives cannot be reliably initiated by using only a plain detonator and safety fuse.

Electric detonators consist of thin-walled copper or aluminium tube sealed at one end and containing a high explosive base charge, a priming charge and a fuse head. The tube is sealed with a neoprene plug through which pass the leading wires of the fusehead assembly. The passing of an electric current through the wires heats the fuse-head wire, which heats up rapidly to fire the igniting composition. This in turn initiates the priming charge. They can be obtained in two forms: instantaneous and delayed action.

Handling and storage of explosives

The handling and conveyance of explosives is covered by various Acts as well as by Statutory Instruments. The Control of Explosives Regulations 1991 provides details on certificates, transfer of explosives, restrictions on prohibited persons, keeping explosives for private use, and the appointment of responsible persons for the security of explosives kept in factories or licensed magazines.

1.4 TEMPORARY WORKS

On many civil engineering projects the cost and design of temporary works forms a very high proportion of the total contract. Therefore care in design and planning is essential. Each temporary structure must be considered on its merits in relation to the importance of the contract and especially the consequences of failure. If under-design could lead to failure in operating conditions, then the cost of delay, together with loss of valuable plant and equipment (not to mention injury to persons) would far outweigh the saving in design. It is therefore important to design all structures to take the full working load envisaged. Allowance must be made for site conditions and human error in the erection of such structures in bad weather. Supervision in the erection, removal and maintenance of all these structures is paramount.

Where materials are used more than once, for example as in the case of falsework, they should be checked to ensure that they have not been weakened by their initial uses. Second-hand materials should be subject to careful scrutiny before being used in situations where the design was based on new materials.

Typical examples of temporary works are:

- Ground support, eg cofferdams, timbering, underpinning and shoring.
- Access bridges.
- Gantries and scaffolding.
- Trackwork for cranes and trains.
- Dewatering systems.

Specialised topics are dealt with in detail in subsequent chapters of the book.

CHAPTER 2

Contractors' Plant

2.1 MANAGEMENT OF PLANT

Plant is one of the most important resources in civil engineering; without it massive construction work and soil movements would be difficult to achieve in this country, which is not highly labour intensive. Plant will generally save time, manpower and cost, and produces a better finish to the work. It must be stressed, however, that costs are only reduced by good management of plant. Such management involves factors such as:

- Outputs.
- Method appraisal.
- Continuity of work.
- Training of operatives.
- Organisation of plant work.
- Maintenance and repair of plant.
- Economic selection of plant.

Once a contract has been signed, the contractors' plant department will be responsible for the provision of the requisite plant and equipment for the site. This will have to be brought from the owner's depot or from other contracts, bought in new, or hired from specialists. Alternatively, the contractor may sub-let certain works, which include heavy plant, eg earthworks, and be involved only in planning the continuity and progress of work.

The plant manager will assist in the preparation of programmes (Figure 2.1) and in the selection of plant for the work in hand. He will also prepare maintenance programmes and install a recording system for checking the performance of each piece of plant.

FIGURE 2.1 *Weekly plant programme*

Plant	*Mon*	*Tues*	*Wed*	*Thurs*	*Fri*	*Sat*
Crawler tractor No 1			Clear site			
Ditto No 2	Remove scrub		Form access road			
Scraper 13 m³ No 1			Reduced level work			
Scraper 13 m³ No 2			Reduced level work			
Articulated truck 12 m³			Haul spoil to tip			
Ditto			Haul spoil to tip			
Ditto			Haul spoil to tip			

When selecting plant and equipment the plant manager will consider the following factors:

- The work load to be undertaken.
- The time allowed in the construction programme for the work.
- The capabilities of the machine or equipment.
- The various tasks which any one piece of plant could accomplish.
- The transportation costs involved.
- Maintenance facilities.
- The cost, whether to hire or buy.

If plant is bought the economics will be established by the following:

- Capital cost.
- Residual value.
- Machine life.
- Capital interest.
- Maintenance costs.
- Terrain on which plant has to operate.
- Weather conditions.
- Availability of plant.
- Future utilisation potential.

As a general rule the large and expensive items of plant are best hired if required for short periods of time and high output of work. General items of plant which have a good utilisation factor are best owned. However, the duration of a contract and the amount of work involved might create a particular situation where it could be cheaper to buy all new plant, to be disposed of afterwards when the contract is complete.

The maintenance of plant is a major factor in large plant departments and all maintenance should be planned to ensure high productivity with low breakdown time. Maintenance costs depend upon:

- Age and condition of plant.
- Care of plant by the operator.
- Labour and overhead costs.
- Cost of spares.
- Cost of general maintenance, which includes lubricants, anti-freeze, batteries, tyres, tracks, hydraulics, replacement of ropes etc – all of which should be carefully planned.
- Availability of suitable facilities in relation to the contract operation.

2.2 EARTH-MOVING PLANT

2.2.1 Construction methods and selection of plant

Before commencing earth-moving, a plan should be produced showing all areas for excavation, tipping and filling. The quantities of excavated material and required fill should be shown on this plan, so as to facilitate the most economic movement of soil. When large quantities of soil have to be moved, a 'mass-haul' diagram should be prepared (Figure 2.2). Such a diagram shows the distances and direction of haul and gradients calculated to balance the cut and fill. If the excavated material is likely to be particularly variable, some indication should be given of its method of disposal, ie whether to haul to spoil heap rather than fill areas.These factors are considered further in Chapter 3 – Earthworks.

Construction methods for smaller sites, eg basements, deep pits, trenches etc, will depend to a great extent on the following factors:

- Type of soil or rock.
- Quantity of material to be moved.
- Presence of water.
- Depth of dig.
- Working space available.
- Whether excavated material can be left on site.

FIGURE 2.2
Mass haul diagram

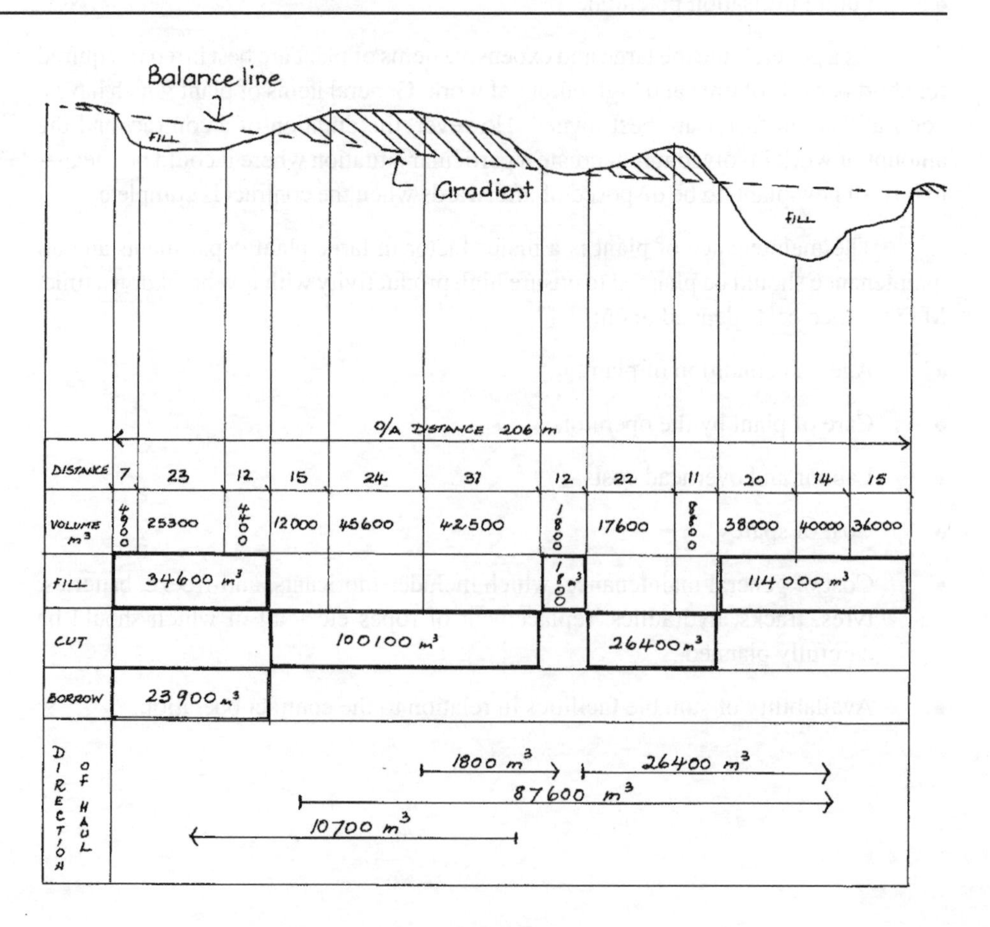

As a general rule, for large pits and basements, where space will allow, the excavation should be achieved by 'battering' the sides of the dig (Figure 2.3). This allows a clear working area, free from obstruction or shoring.

Selection of plant depends on the following factors:

- **Excavating plant**: quantities; type of soil; weather conditions; speed of removal; depth of dig; side cast or cart away.
- **Transporting plant**: quantities; distance to be moved; condition of site; conditions of tip; speed of excavating; size of excavating bucket; turn-round time.
- **Placing plant**: method of transporting and compacting; quantities involved; weather conditions; finish required.
- **Compacting plant**: nature of material; nature of contract; depth of fill; weather conditions.

Selection will also take into consideration availability of plant and costs per estimated unit of soil moved or placed.

Setting out for earth-moving plant

This differs somewhat from the normal setting out found in building and civil engineering works, since a different degree of accuracy is required. On some projects, the speed of earth-moving is more important than the achievement of precise final levels, eg road construction in forward areas. On the other hand, precision of final levels may be as important as speed of construction, eg as for airfield construction.

This difference in approach affects the setting-out procedure, which will vary from project to project. The normal procedure is to use timber pegs or stakes as guides and reference markers for earthwork operations. These stakes are used for centre lines, shoulder lines, batter and reference points. They are approximately 600 mm long × 75 × 25 mm in section and are marked according to purpose (see Figure 2.4).

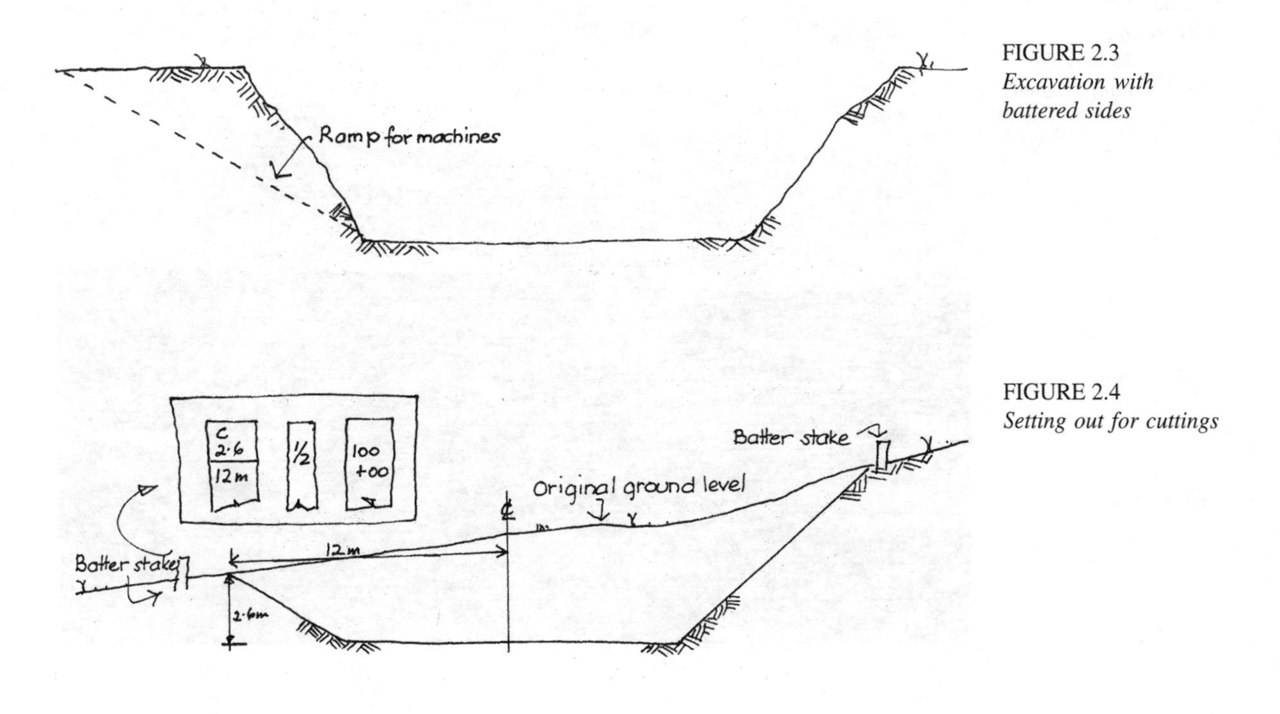

FIGURE 2.3
Excavation with battered sides

FIGURE 2.4
Setting out for cuttings

In Figure 2.4 the batter stake shows the depth of cut (C = 2.6 m) and its distance from the centre line 12 m. On the edge of the stake the slope (1 in 2) is marked as a fraction, and on the back face of the stake the distance along the centre line from the first station. After reducing the levels a series of final level pegs are positioned at the shoulder and centre line. This procedure can be used for almost any form of mass excavation.

2.2.2 Excavating plant

Hydraulic excavator (Figure 2.5)

This piece of plant is the best known item and probably the most commonly used in construction work. It can be fitted with different front or back acting equipment to carry out various jobs. The standard equipment is for backacter operations, in which trenches, deep pits etc may be excavated below ground level. The choice of plant depends on the depth of dig, the type of ground, the quantity to be moved and the speed of excavation required. A typical bucket size is 1.0 m^3 and the depth of dig 6 to 7 m. Machines can have extensions for greater depths.

Dragline equipment

Figure 2.6 shows the rope-controlled bucket filling the truck. It is suitable for a wide range of materials, but will not dig very hard ground or dig to fine limits. This equipment is suitable for large scale excavation, particularly in waterlogged ground or marine works. The long jib allows material to be dumped over a wide radius.

Hydraulic clamshell (Figure 2.7)

This is used for handling loose materials such as concreting aggregates or for digging pits which would prove difficult for the backacter. Hydraulic machines are being increasingly used as grabs and clamshells. This piece of equipment is very suitable for marine works such as caisson sinking and dredging.

FIGURE 2.5
Hydraulic excavator in operation
(Liebherr Great Britain Limited)

Trenching machines (Figure 2.8)
These are suitable for excavating extremely long trenches, as for pipelines etc. They vary in size but will normally dig trenches from 0.25 m up to 1.5 m wide and to depths of 3 metres. There are two basic types of trencher, one with buckets mounted on a wheel at the front of the machine and the other with buckets mounted on an endless chain which is carried by means of a jib. Both machines are self-emptying by means of conveyors which throw the soil clear on both sides of the machine. The wheel type is suitable for shallow (up to 3 m depth) trenches where high speed is required: the jib type is more suited for deeper trenches.

FIGURE 2.6
Dragline in operation (R-B International plc)

FIGURE 2.7
Long reach hydraulic clamshell (R-B International plc)

Crawler loader (Figure 2.9)
This is another common machine on civil engineering sites. It is suitable for stripping and loading the spoil on to lorries. It should be noted that various types of equipment can be attached to the rear of these machines, such as ripping tools.

The disadvantage of tracked equipment is that it cannot travel between sites under its own power. Every time the machine has to be moved, a low loader unit has to be employed. This increases the overall hire cost of the machine. However, it has an advantage over the wheeled-type machine in that much more manoeuvrability is achieved on tracks over wet or poor ground conditions.

FIGURE 2.8
Trenching machine excavating for a 0.9 m diameter pipeline
(Alfred McAlpine Services & Pipelines Limited)

FIGURE 2.9
Crawler loader with ripper attachment at rear
(Liebherr Great Britain Limited)

2.2.3 Transporting plant

This section deals with plant which is involved in moving earth some distance from the point of excavation. The items of plant under this section include rubber-tyred vehicles, tractor-drawn equipment, track-mounted equipment and belt conveyors.

Rubber-tyred vehicles

This form of transport, namely articulated trucks, dump trucks and lorries, form the largest section of earth-moving plant. They vary in capacity from a basic 5 m^3 lorry to a 30 m^3 dump truck. Such vehicles have heavily plated bodies which are hydraulically operated to ensure quick clean tipping. The lorry is designed for working on reasonably level ground and for carrying evenly distributed loads, but under rough site conditions they are subject to axle and spring damage. For this reason, where large quantities of earth have to be removed or taken across rough ground, the dump truck or articulated truck is the best piece of equipment to be employed (see Figure 2.10).

Tractor-drawn equipment

The most popular piece of equipment employed for moving large quantities of earth on extended sites is the scraper. The machine consists of a specially designed container on wheels which can be lowered at one end to pick up material while the machine is on the move. When full, the machine carries the earth to some other point and discharges its load by a similar principle.

There are three types of machine:

- Crawler-drawn scraper.
- Wheel tractor scraper.
- Elevating scraper.

The capacity of scrapers varies from 6 m^3 to 50 m^3 and the classification of a scraper is determined by its carrying capacity.

Crawler-drawn scrapers consist of a four-wheeled scraper unit being towed by a track-mounted machine. The scraper unit is connected to the crawler by a tow-bar and control cables. This type of equipment is slow in operation compared with other scrapers but is suitable for very rough ground. It is suitable for medium-sized sites.

FIGURE 2.10
Rigid dump truck
(Aveling Barford)

Wheel tractor scrapers consist of a tractor unit mounted on two wheels – which are large-diameter rubber-tyred wheels – and the scraper, which has its own power unit, also mounted on two wheels. The scraper is connected to the tractor and has hydraulically controlled linkages. These scrapers have a carrying capacity of 30 m^3 to 50 m^3. The arrangement includes a hydraulically actuated bail and cushion plate bolted to the front of the tractor and a hook on the rear of the scraper. This allows two push-pull scrapers to assist each other during loading (Figure 2.11). They are capable of high speeds over great distances and are therefore most suitable for contracts dealing with road-works.

Elevating scrapers are generally smaller scrapers, having a carrying capacity of around 12 m^3. They require less support equipment than other scrapers and can scrape and then elevate the bowl for rapid transportation across the site. They also have retracting floor and bulldozer ejector facilities to provide clean and quick material ejection.

The advantage of the elevating scraper is that it is very productive under favourable conditions – that is, loose, easy-to-excavate material. However, this makes the work highly weather-sensitive, which can result in a low degree of utilisation. Scrapers need a large working area in which to cut and turn; they also need space to dump the material if normal 'cut and fill' operations are not the norm for the site. They have high operating costs and require very skilled operators.

Bulldozers

Crawler tractors (Figure 2.12), commonly referred to as bulldozers, are suitable for stripping sites and for reduced level dig on medium to small sites. They are also useful for moving, placing and levelling soil.

They are used primarily for pushing earth from one point to another. It is quite normal for a bulldozer to excavate to a depth of 200 mm and then push the material 50–100 metres for a spread and level operation. Angle and tilt dozers are another variation in this plant group where the blade is set at an angle to the machine for the purpose of pushing soil to one side instead of carrying it forward for levelling. Dozers can be fitted with winches for hauling and rippers for breaking up ground.

FIGURE 2.11
Wheel tractor scraper being assisted by another scraper
(Finning Limited)

Graders

This piece of plant is neither truly earth-excavating nor earth-moving, in the sense of bulk movement of earth. It is used to finish earth formations, such as roads and embankments, to a fine limit. The machine (Figure 2.13) has six wheels, two at the front and four at the rear. It is similar to a dozer, in that it pushes the earth or other fill material into position, but it does so in a unique manner to remove undulations. This is achieved by a long slender blade, which can be adjusted hydraulically, hanging from a frame suspended between the cab and the front wheels.

There are three main types of motor grader, based on their steering mechanisms:

- Straight frame.
- Articulated turn.
- Crab steering.

The **straight frame** grader has its main frame centred and only the front wheels are used for steering: this is the best type for long-pass grading. The **articulated turn** machine can pivot from the rear wheel section. This results in easier manoeuvring in close quarters, quicker turn-round at the end of the pass, plus the ability to carry a full blade load around a curve. The **crab steering** machine also pivots from the rear wheels but also has a mechanism that compensates for side drift when turning and increases stability for side-slope work.

FIGURE 2.12
Crawler tractor with electronic steering control
(Liebherr Great Britain Limited)

FIGURE 2.13
Motor grader
(Aveling Barford)

Conveyors

Belt conveyors can be used for earth-moving, as well as for moving other material. They are often used in conjunction with earth-excavating plant, such as trenching machines. They are available in various forms; the most important ones are:

- Independent units for the carriage of materials over considerable distances, eg sand and gravel.
- Units for moving materials in confined spaces.
- Track or wheel-mounted units – better known as portable units.
- Integral units, which form part of a large machine such as elevating graders or multi-bucket trenchers.

The advantages of a belt conveyor are:

- They deliver a high output with a steady flow.
- They can deliver horizontally or on a limited incline.
- Movement is not affected by bad ground conditions.
- They can be used in situations where access for vehicles would create problems.
- They require minimal maintenance.
- Efficiency is high and power used is low.

The disadvantages are:

- They are not as flexible in movement and rearrangement as other forms of plant.
- The belts are subject to damage by sharp irregular materials, which if large may fall off.

2.2.4 Compaction plant

This section is limited to plant used on large earth-moving contracts and therefore does not cover items such as small vibrators.

Smooth-wheeled rollers such as those used in road works may be used for producing a smooth surface on top of embankments to facilitate the shedding of water, but other than that have a limited application to earthworks.

Sheepsfoot rollers consist of a hollow steel cylinder on which are mounted rows of projecting studs (called feet). There are two types of foot, the taper foot and the club foot. For extra weight the roller is filled with water or sand; alternatively ballast boxes may be fitted to the roller. The sheepsfoot roller is towed by a tracked vehicle across a fill area until the roller has fully consolidated the material. On reaching consolidation the roller 'walks out', meaning that the feet no longer dig into the top of the fully compacted material.

Vibrating tamping rollers (Figure 2.14) are the latest development in the consolidating of fill material. This type of plant is capable of optimising the consolidation of various types of material and are most suited for compaction.

Pneumatic-tyred rollers are designed to give a kneading action to the material. The roller has two lateral rows of wheels, the front row being out of line with the back row, so that no gaps are left in compacting. The wheels are mounted in pairs on separate axles, allowing each pair to follow the irregularities of the ground. This has given the name of 'wobbly-wheels' to the rollers. The load is applied by means of kentledge or sand and the roller is towed by a tractor or alternatively can be self-propelled.

2.2.5 Performance and outputs of earthmoving plant

The actual performance of earthmoving plant is very difficult to establish because no two sites are the same. The only way of arriving at a performance figure is by one of two methods:

- Reference to output charts and tables, or
- Detailed calculation.

In the first case output charts are produced either by the manufacturers, and must be treated with caution, or they are produced by a contractor's estimating department from information received from actual sites. The latter source is much more reliable since weather, quantities, time of year etc can all be considered.

With detailed calculations it is necessary to assess the initial soil-moving operations by means of a method study. This will take into consideration the actual site conditions and allow site management to base their decisions on accurate information.

It must be mentioned, however, that time for prime operations does not represent the optimum time for all operations, since on every job there is a 'learning curve' and a job familiarisation factor to be considered.

If the plant has much moving to do during the task, the efficiency factor is greatly reduced. Table 2.1 shows typical efficiency factors for excavators.

For example, a backacter trenching with a 1 m^3 bucket will excavate and load 60 buckets per hour. 60 m^3 × Efficiency factor of 0.8 = 48 m^3 actual performance (allowance should be made for bulking).

These performance figures must be multiplied by the operator's efficiency factor, which is normally 75 percent. This allows for personal needs etc during the working day. Taking the above example of 48 m^3 actual performance of the backacter times, the operator's efficiency factor gives the following overall performance figure for the machine: 48 m^3 × 0.75 = 36 m^3 per hour.

It can be seen from the above factors that outputs from manufacturers' catalogues cannot be relied on.

FIGURE 2.14
Vibrating tamping roller under test at TRL
(The Department of Transport – Transport Research Laboratory)

The task efficiency for moving plant will depend on the length of each working 'pass' and speed of machine. Typical outputs are shown in Tables 2.2 and 2.3.

The output of earth-moving plant will depend on the haul distance and should be calculated as follows:

- Estimate loading and discharge time.
- Calculate travelling time to and from discharge point.

These figures will give the cycle time:

Cycle time = Loading + discharge time + travelling time.

$$\text{Output per hour} = \frac{\text{Capacity of machine} \times 60 \text{ minutes}}{\text{Cycle time}}$$

Example for scraper (30 m^3 capacity):

Loading time = 2 mins.

Discharge time = 1 min.

Travelling time = 15 mins.

Therefore, cycle time = 3 + 15

= 18 mins.

$$\text{Maximum output per hour} = \frac{30 \text{ m}^3 \times 60 \text{ mins}}{18}$$

= 100 m^3 per hour

(Note: Allowance should be made for bulking.)

TABLE 2.1 **Task efficiency factors for excavators**

Equipment/machine	*Task*	*Task efficiency factors*	
		When loading into vehicle	*When side-casting*
Backacter	Trenching:		
	Bucket width	0.8	1.0
	More than bucket width	0.6	0.7
Dragline	Bulk excavation	0.8	1.0
	Wide open ditches	0.7	0.9
Trenching M/c	Continuous trenching	—	1.0
Face shovel	Excavating above track level:		
	Little movement	0.65	0.8
	Much movement	0.45	0.6
	Moving blasted rock	0.40	0.5
Scrapers and bulldozers	Cutting, hauling or pushing soil	—	1.0
Tractor shovel	Clearing site	0.8	1.0
	Reduced level dig	0.8	0.9
Grab (hydraulic)	Excavating deep bases:		
	Min obstruction	0.7	0.8
	Max obstruction (timbering)	0.5	0.6

Note: The above table must be used in conjunction with standard outputs for clayey soils, which are shown in Table 2.2.

The output must be multiplied by the operator's efficiency factor, ie 0.75, giving a basic output of $100 \times 0.75 = 75$ m³ per hour. (Note: The task efficiency factor for a scraper is unity.)

Soil types and site conditions affect the loading and spreading time as well as travelling time.

Where a number of vehicles are being filled by a loader, the number of haulage vehicles required can be calculated as follows:

$$\text{Haulage vehicles required} = 1 + \frac{\text{Cycle time per vehicle}}{\text{Loading time of vehicle}}$$

TABLE 2.2 **Plant outputs for clayey soil**

Plant	*Remarks*	*Output*
Backacter	In confined conditions	Allow 30 buckets/hour (ie 30 × bucket size)
Backacter	In normal trenching	Allow 60 buckets/hour
Dragline	In open excavation	Allow 80 buckets/hour
Dragline	In restricted area	Allow 35 buckets/hour
Face shovel	In normal excavation	Allow 80 buckets/hour
Trenching machine	In normal trenching up to 1.5 m deep	Allow 2 metres/minute
Crawler tractor	Reduce level dig	Allow 50 buckets/hour
Bulldozer	Haul = 30 m	Allow 60 m³/hour
	Haul = 60 m	Allow 20 m³/hour
Scraper 15 m³	Haul = 300 m	Allow 60 m³/hour
	Haul = 200 m	Allow 80 m³/hour
	Haul = 100 m	Allow 110 m³/hour
Grab	Normal conditions (bases)	Allow 40 buckets/hour

Note: In the majority of cases the above table shows outputs by number of buckets of earth which can be moved under normal conditions. The actual output, in cubic metres, can be calculated by multiplying the bucket size by the figure given in the output column.

TABLE 2.3 **Task efficiency factors for moving plant on large contracts**

Machine	*Task*	*Task efficiency factor*				
		Length of pass in metres				
		50	*100*	*200*	*500*	*Over 500*
Grader	Grading roads	0.4	0.6	0.8	0.9	1.0
	Spreading and shaping fill	0.4	0.5	0.7	0.8	—
Haulage plant	Carrying spoil to or from cut and fill areas	0.7	0.75	0.8	0.9	1.0
	Carrying spoil over poor terrain (undulating)	0.5	0.6	0.7	0.75	0.8
Towed rollers	Rolling fill areas	0.5	0.7	0.9	0.9	1.0
Rooters and scarifiers	Breaking up ground	0.5	0.7	0.8	1.0	1.0
Mix in place stabilisers	Road works	0.5	0.6	0.7	0.7	0.7

2.3 CONCRETING PLANT

2.3.1 Methods and selection of plant

A concrete structure involves four main production stages:

- assembly of formwork;
- mixing of concrete;
- distribution of concrete;
- placing concrete.

Each of these stages may be carried out in different ways and the best way for each operation must be determined for each site. The following factors must be taken into account when making the plant selection:

- Topography of the site – boundaries, restrictions, noise, contours of land, soil conditions.
- The total volume of concrete required.
- The maximum amount of concrete required at any point at any one time.
- Availability of plant.
- Time of year in which concreting is to be carried out.
- Amount of space available for setting up plant.
- Quality of concrete required, ie specification, varying mixes.
- Cost of producing concrete by various methods.

From the above it can be seen that every site will have a set of conditions which require individual solutions for all stages of production. The best solution will depend on careful selection within each stage and on the interrelationship of each stage in terms of speed and efficiency.

2.3.2 Concrete mixing plant

There are four main types of mixing plant:

- Tilting-drum mixers (T).
- Non-tilting drum mixers (NT).
- Reverse drum mixers (R).
- Paddle mixers (P).

The first type, **tilting drum**, is used mainly for mixing very small amounts of material on site and is normally used by a builder rather than by a civil engineer. It consists of a conical drum rotating on a movable axis. When the materials are mixed the drum is then tilted to discharge the mix. However, larger output mixers are available for central mixing plants and these can produce up to 3 cubic metres per mix. The tilting action is controlled by hydraulic rams and batching is achieved by overhead weigh-batch hoppers.

The second type, **non-tilting drum**, is suitable for larger outputs still of, say, 10 cubic metres per hour. It consists of a circular drum with a side outlet for loading (usually by means of a hopper), and an outlet on the opposite side for discharge, which occurs when the chute is inclined into the drum. The concrete falls on to the chute from the top of the drum. During mixing the chute is inclined to face the bottom of the drum, thus preventing concrete from spilling out.

The third type, **reverse drum mixer**, is similar to the non-tilting mixer but it mixes when rotating in one direction and discharges in the reverse direction. Special baffles retain the concrete until the drum is reversed.

Paddle mixers consist of a stationary pan with rotating paddles. The paddles may be fixed or may themselves rotate as they go epicyclically round in the pan. This form of mixing unit gives very consistent mixes and is used for high-grade concrete. Since this type of mixer is not as portable as the others, it tends to be used at a central mixing point or at locations where pre-cast units are made.

All the types mentioned come in various sizes, giving outputs from as low as 200 litres to 4 cubic metres per cycle. In some cases a high level discharge is required and this can be achieved by setting a non-tilting mixer up on a steel frame. This allows discharge direct into large dumpers. Alternatively a skip-lift high discharger may be used.

The selection of suitable plant for mixing will depend upon the following:

- Amount of concrete required at any one time.
- Quality of mix required.
- Availability of plant.
- Amount of room for setting-up plant.
- Type of distribution plant.

In some cases it may be more economic to supplement the site output with ready-mixed concrete rather than set up a second machine which may be under-employed most of the time.

Ready-mixed concrete has become increasingly popular over the last decade for the following reasons:

- Congested city sites make it difficult to set up mixing plant.
- High-strength concrete is easily obtained, thereby reducing expenditure on sophisticated mixing plant.
- Small quantities can be obtained.

However, there are disadvantages, such as:

- Delivery times are unreliable, due to traffic problems.
- Prices are considerably higher than the cost of site mixing.
- Large loads create difficulties in handling the bulk quickly on site.
- Access to site must be fit to carry the combined loads of delivery – vehicle and concrete.
- Washing down on site may create problems.

2.3.3 Concrete distribution plant

There are many ways in which concrete can be distributed on site; the most common of these are:

- Tipping barrows.
- Dumpers.
- Mono-rail.
- Hoists.
- Pumps.
- Placer units.
- Cranes with skips.
- Cableways.
- Conveyors.
- Tremie pipes, elephant trunking and chutes.

Hand-operated **tipping barrows** (Dobbin barrows) have two wheels and are easy to handle and to discharge quickly. Powered barrows can be used to eliminate manual work; with this type of equipment the operator walks behind the machine controlling it by clutch and brakes.

Dumpers vary from small (500 litre) bowls to large (3 cubic metre) bowls. They can be either two-wheel or four-wheel drive and may be tipped by hand or by hydraulic mechanisms. In the hydraulic range there are various options, such as discharge to both sides as well as the front, and a turn unit which allows discharge through an arc of 180 degrees.

Mono-rail equipment consists of a power unit mounted on a single rail. The power unit has a side-tipping skip attached to it and it can also tow another unpowered truck behind it. The power unit can travel at 90 metres per minute. Trip-mounted rails provide automatic stopping and therefore eliminate the need for a driver. The power unit is capable of climbing slopes with a gradient of up to 1 in 8, and the capacities of the skips may vary from 300 to 500 litres. This method is particularly useful when working in areas of bad ground condition congested sites, heavily reinforced slabs etc.

Hoists have been developed for handling concrete economically. One such development is the 'tip skip' hoist in which a manually operated tipping skip replaces the platform of a normal hoist. The skip carries the concrete vertically and discharges it into receiving hoppers which are placed at each working level. The hoppers then feed barrows or other plant at each working level.

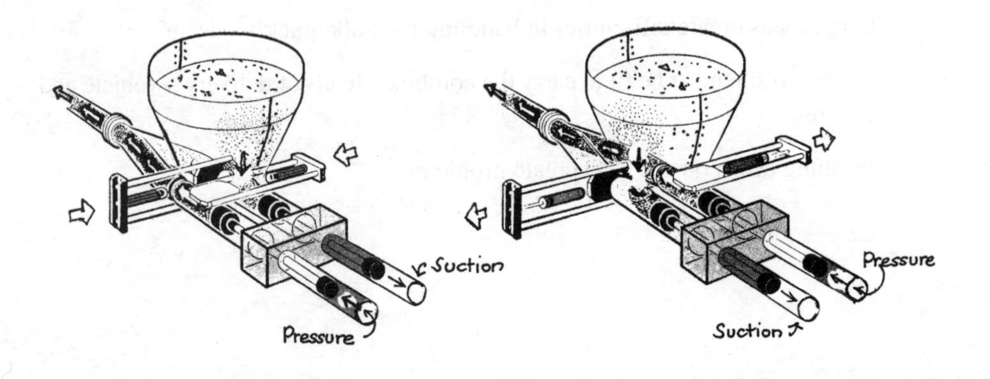

FIGURE 2.15
Hydraulic concrete pump

Another development is the static skip which discharges the concrete via a bottom door which is open by pre-determined trip points. For large outputs twin skips can be used, one for filling while one is travelling. The skips travel on either side of a central hoist mast.

There are two types of concrete **pumps** available: mechanical and hydraulic. With the mechanical pump the concrete is forced along the pipes by a piston which is driven by a diesel engine. This type of equipment requires careful consideration when being set up because of its semi-permanent nature.

Hydraulic pumps using either water or oil as a working fluid are more commonly used because there is much less work in setting up (Figure 2.15). They are self-contained in operation and are available as trailer units which can be towed into position on site and set up very quickly with the aid of jacking legs. Lorry-mounted hydraulic pumps with hydraulic booms to carry the delivery pipe to high levels are often used. This particular piece of equipment is often linked with ready-mixed concrete (see Figure 2.18).

Both pumps are capable of straight line pumping for a distance of 300 metres horizontally and 30 metres vertically. Pipe lines are normally 110 or 150 mm in diameter.

Placer units. In this method of distribution, concrete is fed into a hopper and then into an airtight cylinder. Compressed air from an air receiver is admitted to the cylinder, causing the concrete to be driven along the pipe line at a predetermined speed. A special cone device prevents air from blowing a core through the batch of concrete, thus ensuring an even flow. The concrete is discharged at the delivery point through a special discharge box incorporating a vent for the compressed air.

Cranes are often employed in the distribution of concrete and little is required to explain their use. The type of crane varies with the type of work job being undertaken. The main development in this form of distribution is the type of skip or bucket used. The two main types of skip are:

- Roll-over.
- Constant attitude

The former is best suited to discharge into formwork, but the latter occupies less space under the crane hook. In both cases, where the bucket is capable of carrying large amounts of concrete, the method of controlling their discharge needs consideration. Discharge can be controlled by geared-gate mechanism, by air-operated mechanism or by hydraulic operation.

Special buckets are available for placing concrete under water, in which the concrete is protected by an enclosed skip whilst being lowered through the water to the point of discharge.

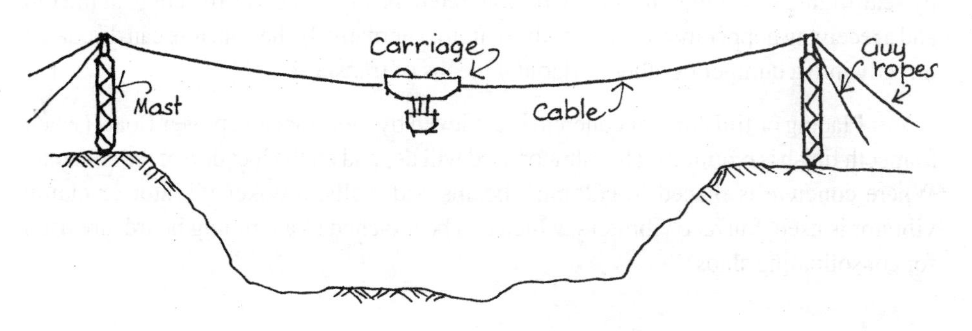

FIGURE 2.16
Cableway

Cableways (Figure 2.16) are used mainly for large civil engineering work, such as dams, where the concrete has to travel great distances over inaccessible ground. Cableways can be constructed in various ways such as:

- Travelling.
- Radial travelling.
- Luffing.
- Fixed.

Travelling cableways have masts mounted on tracks, thus enabling two-dimensional coverage. Radial travelling cableways have one fixed mast and one travelling mast, thereby allowing the concrete to be distributed over a segmental area.

Luffing cableways have both masts mounted on pivots to allow some sideways movement from the base.

Fixed cableways, as the name suggests, have no tolerance in movement and these are therefore used only when concrete is to be placed in a straight line.

The output of cableways is very high, usually in the region of 8 to 60 cubic metres per hour. Travelling speeds of up to 350 m per minute are possible and spans can range from 300 to 600 metres.

Conveyors consist of a narrow continuous belt which runs over a series of rollers and carries a constant stream of concrete. The belt can travel at speeds of up to 150 m per minute, giving delivery of between 50 and 100 cubic metres of concrete per hour, depending on belt width. The advantage of this type of equipment is one of full coverage, the conveyors being capable of easy removal and repositioning for large slab areas (Figure 2.17). One type of conveyor – the bridge conveyor – is suitable for spanning over reinforcement and other obstacles. This particular conveyor can span distances of upwards of 30 m. Elevating units are also available for moving concrete to a higher level, the angle of elevation being restricted to approximately 25 degrees.

Tremie pipes, **elephant trunking** and **chutes** are all pieces of non-mechanical equipment which are used for placing concrete in position below ground level. Such positions occur in piling, basements, diaphragm walling etc, where segregation of the mix must be prevented.

Tremie pipes consist of rigid tubes of metal or plastic, with a feed hopper at the head of the tube. As the concrete is placed the tube is shortened.

Elephant trunking consists of PVC flexible tubing supported at intervals as it hangs into the area for concreting. The concrete is fed into the tubing by means of a hopper.

The chute method of placing concrete has been used for many years and consists of an open metal channel or large diameter plastic tube (200 mm diameter), supported by scaffolding down into the area to be concreted. A hopper serves the chute at the top and a receiving hopper may be used at the bottom; alternatively the concrete can discharge directly into a dumper or other horizontal means of transport.

Placing or finishing of concrete is achieved by vibrators and power floats (when a smooth finish is required). The vibrator used will depend on the location of the concrete. Where concrete is placed in columns, beams and walls, a poker vibrator or clamp vibrator is used. Surface vibrators, which can be attached to a tamping board, are used for consolidating slabs.

FIGURE 2.17
Placing concrete by chute and conveyor belt into a tremie pipe

FIGURE 2.18
Placing concrete by mobile pump

2.4 PILE-DRIVING PLANT

2.4.1 General considerations

When selecting plant for piling operations it is necessary to establish the type of pile to be used. In the case of displacement piles, in which some form of pile or tubular casing is driven into the ground, consideration must be given to the support of the unit being driven. This normally takes the form of a pile frame or crane and leader, although latest developments use hydraulically operated telescopic back struts in lieu of a crane jib. In the case of replacement piles a hole is formed in the ground and then filled with reinforced concrete, the plant used varying with the size of hole formed. A tripod rig is used for small diameter holes (up to 600 mm). This rig is equipped with a winch for raising the cutting auger and the whole equipment weighs approximately 1.5 tonnes. For larger diameter holes large rotary augering equipment is used; these are shown in Chapter 4. Casings have to be sunk into the borehole to protect against collapse, which is normally done by the use of a vibrator unit.

In each case consideration must be given to the height and manoeuvrability available on site, and whether vibration and noise would create problems to either adjoining buildings or residents. Further considerations include:

- Type of sub-soil.
- Surface conditions, eg slope of site.
- Surface drainage, eg waterlogged conditions.
- Obstructions, eg old basements, existing services.

2.4.2 Methods of driving piles

Pile frames and leaders are used to locate and guide a pile during the initial stages of penetration as well as guiding and supporting the hammer. The leaders for guiding the hammer and pile extend the full height of the frame and consist of steel channels set 150 mm apart. Where drop hammers are used the space between the leaders accommodates and facilitates the sliding of the hammer. A winch is used to lift the hammer into position and may also be used for positioning the pile.

Pile frames may be vertical or raking, and vary in height from 10 metres to 25 metres, adjustment for a raking frame being made by hydraulic rams. Stability is achieved by guy ropes from the head of the frame.

Cranes and leaders (Figure 2.19) are commonly used instead of specialised piling frames. This equipment consists of a standard crane with a purpose-built leader unit attached to the crane at both top and bottom. Leaders are used on sites when normal piling frames would otherwise prove cumbersome. The crane can lift the leader unit and move it easily across the site to a new position.

The actual 'driving' mechanism will depend on the type of pile, as set out below.

- **Driven steel casing**. Driven from the top usually by compressed air, diesel or steam hammer; by driving a mandrel and casing; or by driving a shoe or plug of material by internal drop hammer. (See Chapter 4 for details.)
- **Pre-cast piles**. All types of hammer are suitable but the hammer should weigh a minimum of half the weight of the pile being driven. The head of the pile must be protected against spalling; this is achieved by using a special helmet attachment.

- **Special pre-formed steel piles.** Any type of hammer, but the heavier the hammer with reduced drop, the less damage is done to the pile head.
- **Screw piles** require a crane for pitching and some form of guide frame to hold the pile during its screwing operation.
- **Timber piles**: drop hammers and single- or double-acting hammers are suitable. Where hard driving is anticipated, the weight of the hammer should equal the weight of the pile.

2.4.3 Pile-driving hammers

Various types of hammer are used in pile driving. The main types are:

- Diesel hammer.
- Air hammers – single-acting hammer (compressed air).
- Double-acting hammer (compressed air).
- Hydraulic and vibratory (see Chapter 4 – Piling)

FIGURE 2.19
Crane with leader to support pile
(BSP International Foundations Limited)

Diesel hammers (Figure 2.20) are self-contained units which can be suspended from a crane or slide in leaders. The driving action is achieved by a falling cylinder of steel within the unit, which is raised by the explosion of gas. This cycle of operation is repeated continuously, giving 60 blows per minute for a single-acting diesel hammer. The fuel is injected automatically, giving rise to an explosion for the continuous action. The weight of these hammers varies from 2 to 6 tonnes. Single-acting and double-acting hammers are available.

Air hammers
Single-acting hammers are semi-automatic in action. The hammer is guided by leaders or suspended from a crane. The blow is provided by a heavy falling cylinder or ram, which is raised by means of compressed air. The cylinder slides up and down a fixed piston. Compressed air is admitted by a valve to raise the ram and exhausted through another valve to allow the ram to fall. Normal working speeds are 36 blows per minute up to a maximum of 45 blows per minute. Hammers of up to 6 tonnes are available. These hammers are suitable for driving trench sheeting and bearing piles.

Double-acting hammers (Figures 2.21 and 2.22) are powered for both the upward and downward strokes, giving a rapid pattern of blows which keep the pile moving. This type of hammer is suitable for all driving, with the exception of long, heavy piles. The force is less than that of a single-acting hammer, but the number of blows per minute is greatly increased, varying from 95 to 500 blows per minute depending on the weight of the hammer. The cast iron cylinder remains stationary and the blow is achieved by a piston moving at high speed. The double-acting hammer is suitable for use in situations where reduced head-room prohibits the use of a conventional hammer.

These hammers can be used for underwater driving controlled by compressed air and for driving steel sheet piling. The latter use can be achieved by suspending the hammer from a crane without the use of leaders, providing the piles themselves are in a guide frame.

FIGURE 2.20 *(Left)* *Single acting diesel hammer* *(BPS International Foundations Limited)*

FIGURE 2.22 *(Centre)* *Hydraulic hammer* *(BSP International Foundations Limited)*

FIGURE 2.23 *(Right)* *Extractor for sheet piling* *(BPS International Foundations Limited)*

2.4.4 Pile extractors

Pile extractors can be either specially designed for the sole purpose of extraction or in the form of a double-acting hammer with extracting gear.

In the former type, a rapid number of blows in the upward direction causes the extracting jaws to grip and lift piles out of the ground (Figure 2.23). The specially designed equipment, which is similar in action to an inverted double-acting hammer, is more efficient and more compact than the double-acting hammer. The double-acting hammer has to be inverted for use and the extractor works on the principle of hitting the pile out of the ground. The extracting gear is fitted around the inverted hammer and a winch is used to exert an upward pull of 1 to 2 tonnes in excess of the weight of the extracting gear. The specially designed extractor can work at speeds of 350–500 strokes per minute.

2.4.5 Winches and other equipment

Winches are mounted on a base frame of steel construction which is long enough to accommodate the winch, a boiler and an engine or motor. The whole unit is usually mounted on wheels. The winch can be powered by steam, diesel, compressed air or electricity. They may be single, double or multiple drums and can have reversing gear if necessary. The normal winch will be either single or double drums with a slip-clutch mechanism to give positive release of the rope for the drop hammer. Winches vary in weight from 1.5t to 4t for single-drum and 2t to 6t for double-drum types. Each winch can be selected for a particular type of frame or driving situation.

Further specialist equipment used in piling is discussed in Chapter 4 to which reference should be made.

FIGURE 2.21
Double-acting air hammer
(BPS International Foundations Limited)

2.5 CRANES AND HOISTS

2.5.1 General considerations

When selecting lifting plant there are many factors to be considered, such as:

- Weight and size of load involved.
- Height of lift.
- Utilisation factor.
- Whether lifting operation can be static.

Such factors will help to decide whether to use a crane or a hoist. When selecting the type of crane, further consideration must be given to the following:

- Access – type of ground over which the crane may travel.
- Radius of swing.
- Amount of lateral movement.
- Whether 'luffing' will be required. (A luffing jib is a hinged jib which facilitates lateral movement without moving the complete jib.)
- Type of plant being used in conjunction with craneage, eg concreting plant.

To achieve the most economic cost of craneage, a programme showing the planned sequence of working is necessary. This can be checked against alternative methods of hoisting and selection is thereby made on an economic basis.

2.5.2 Mobile cranes

Mobile cranes can be divided into four distinct groups:

- Mobile wheeled cranes.
- Truck-mounted cranes.
- Track-mounted cranes.
- Gantry cranes.

Mobile wheeled cranes consist of a simple crane on motorised wheels. Lifting capacities vary from 3 to 50 tonnes, but most standard mobiles have a capacity of up to 10 tonnes only. They need a hard level surface on which to run, and are used mainly in plant and goods yards for lifting moderately heavy loads. Telescopic jib cranes, the jib length of which can be altered at will, are tending to replace the conventional mobile crane. The development increases the mobility of the crane and provides a solution for lifting in difficult situations, eg where a fixed jib could not be raised.

Truck-mounted cranes consist of a crane mounted on a lorry or truck which has been specially designed to carry an increased load (Figure 2.24). The truck has its own conventional engine and controls for normal driving, plus an extra engine, cab and controls for the crane. A vertical frame extending above the driving cab carries the jib when the machine is travelling from one place to another. Once on site extra jib lengths can be achieved by bolting jib sections together; these sections are carried from job to job by separate transport. The capacity of these cranes is between 5 and 20 tonnes, but this can be increased by means of outriggers. The outriggers are incorporated in the crane chassis and jack down to timber bearers at ground level. Truck-mounted cranes are capable of travelling on the highway at 30 mph and are therefore suitable for short hire periods.

Track-mounted cranes (Figures 2.25 and 2.26) are basically the universal excavator (rope controlled) fitted with a long lattice mast and additional lifting ropes. The mast often incorporates a fly jib to obtain better coverage of the site, especially where there are physical restrictions such as scaffolding. The advantage of this type of crane is its multi-purpose utilisation – the basic machine can be used as a dragline or backacter or other similar item of plant and then quickly adapted for use as a crane.

The track-mounted crane is the only type of crane capable of operating on bad ground conditions and hire rates are much lower than for lorry-mounted and mobile types. They have a lifting capacity of between 5 and 30 tonnes, although normal lifting capacities are in the region of only 10 tonnes.

Gantry cranes, sometimes named portal cranes, are becoming increasingly popular in medium-rise as well as low-rise buildings. A portal-frame type of structure straddles the work area and moves along two sets of rails, one on either side of the work area. The lifting gear is suspended from a horizontal frame which is capable of moving the full width of the portal.

2.5.3 Derrick cranes

The derrick crane is widely used in civil engineering work because of its lifting capacity and wide reach. Such cranes are capable of lifting up to 30 tonnes and providing jib reaches of up to 45 m. The size of the crane and the fact that lengthy site assembly is necessary for each crane make this type of crane suitable only for heavy lifting or specialised construction, where the crane will be fully utilised over a long period of time. There are three types of derrick crane:

- Scotch or stiff-leg derrick.
- Guy derrick.
- Mono-towers.

FIGURE 2.24
Truck-mounted crane lifting bridge sections (R M Douglas Construction Limited)

The **Scotch derrick** (Figure 2.27) is the most commonly used derrick on civil engineering projects. It has a short vertical centre post, two rigid stays and a long jib (usually twice as long as the centre post). The crane is normally powered by an electric motor but diesel-driven models can be used in isolated locations where electric power is not available. The jib is fixed to the base of the centre post and, when slewing, the jib and centre post rotate together. Anchorage is provided by the three legs, on to which heavy kentledge, ie weights of iron or concrete, is loaded to prevent overturning. Alternatively, the legs may be bolted down to temporary concrete bases.

FIGURE 2.25 *(Left)* *Track-mounted crane used in driving sheet piling* *(Liebherr Great Britain Limited)*

FIGURE 2.26 *(Right)* *Track-mounted crane involved in pile formation operations* *(Liebherr Great Britain Limited)*

FIGURE 2.27 *Scotch derricks (rail mounted) working on a new lock entrance* *(Edmund Nuttall Limited)*

This particular type of derrick can be either stationary or mobile. The travelling derrick can take one of two forms: either rail-mounted for work on shore; or floating, in which the derrick is mounted on some form of raft.

Scotch derricks can be elevated by placing them on lattice steel towers (known as Gabbards). Such towers are braced together and either ballasted as stationary units or mounted on rails to give greater coverage. In all cases the Scotch derrick is limited in slewing to an arc of 2700°, owing to the position of the rigid stays. Derricks are unwieldy to move horizontally and so are moved only as the construction advances across the site; a crawler tractor is used to tow the derrick along its rails.

Guy derricks are used mainly in connection with steel erection. They are easily and quickly erected and dismantled and removed from one required position to another; this facility is often required on large, widespread steel frame erection, hence their popularity for such work. With this form of crane the vertical centre post is longer than the jib and is kept vertical by four wire ropes (guys), connected to a top plate. The jib and centre post both revolve on a base plate and, because the jib is shorter than the centre post, it can slew through the full 360°, clearing the guys by passing beneath them.

Mono-towers have developed from the elevated Scotch derrick. The mono-tower consists of a braced tower up to 60 m high surmounted by a derrick crane. The centre mast of the derrick is pivoted well down inside the tower and the jib fixed near the centre of the mast. The whole mast, jib and operator's cab slews on a bearing plate on top of the tower. The jib is counterbalanced by a ballast counterweight. The derricking can either be controlled by ropes or hydraulic rams. Such cranes are capable of lifting 2 tonnes at a reach of 30 m.

2.5.4 Tower cranes

These may be of several types:

- Static – self-supporting.
- Static – attached to the building framework.
- Climbing cranes.
- Rail-mounted or travelling tower.

Static self-supporting cranes have their masts firmly anchored at ground level to massive concrete bases. Special base mast sections are provided which are cast into the foundation ready to receive the rest of the crane. Where poor ground conditions exist, it may be necessary to use piles beneath the crane base; the pile foundation may then be utilised as part of the structure when lifting operations are complete. The crane should be positioned in front or to one side of the building under construction, thus preventing delay in completion of lower parts of the structure, such as might otherwise occur if the crane is situated within the building. These cranes are best used where the site does not allow room for a travelling crane.

Static, building-supported cranes, are similar in construction to self-supporting cranes, but are used for lifting to much greater heights. The crane is fixed to the building structure at intervals during construction of the building. Since this type of crane induces stresses in the supporting structure, there may be extra cost incurred in designing and strengthening the structure.

Both the above cranes have static bases and static towers or masts, but it is possible to have a self-supporting crane with static base and slewing tower. In such a case the base of the mast is in two sections to permit slewing. A luffing jib would normally be used with the revolving mast.

Climbing cranes (Figure 2.28) are the best form of craneage for very high structures. The crane climbs with the structure, having a relatively short mast (20 m) and therefore occupying little space within the building. The crane is normally sited so that the jib has full coverage of the building, but should not be positioned in lift shafts since this could prevent preliminary lift installation work being carried out. The best position is one where little work is required to complete the structure after the crane has climbed beyond that point. Initially the crane is constructed on a foundation at ground level; when the first four floors of the building are in position, two special frames or collars are fixed to the floor around the mast to transfer its load to the floors. A winch or jacking equipment is used to raise the mast to its new position and the mast is then bolted to the collar. This operation is repeated as necessary until the structure is completed, and the crane is finally dismantled into sections and lowered from roof level by means of a winch. The decision to use a climbing crane can only be made when the structure is capable of supporting the loads involved. The frames or collars must be designed to suit the particular structure.

Rail-mounted or travelling cranes are used on sites which cannot be served by static cranes. The crane is mounted on heavy wheeled bogeys travelling on a wide gauge rail track. The track is laid on sleepers and ballast, requiring great accuracy in level and gauge. In particular, the level of the track must be checked frequently for settlement, since a slight movement of the track could render the tower unstable. Any gradient for travelling cranes should be less than 1 in 200. Corners may be negotiated by radius rails or special turntables. Most travelling cranes are operated by an electric motor, and to prevent the cable trailing and becoming subject to damage the crane is equipped with a spring-loaded cable drum. This drum allows the cable to be paid out when the crane travels in one direction and draws it in when travelling in the reverse direction. The crane can also be used as a static crane by using stays, whenever loads and radius require greater stability.

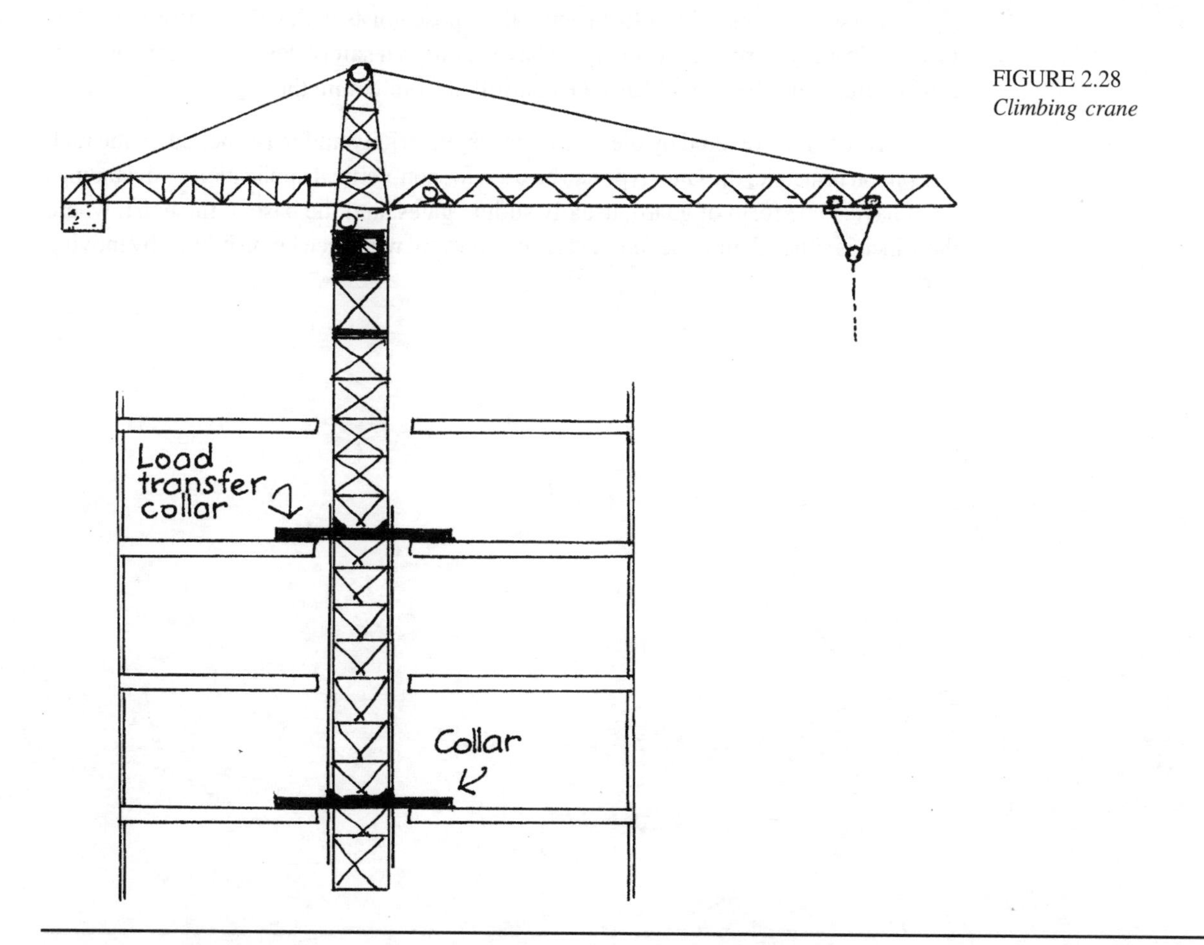

FIGURE 2.28
Climbing crane

Jibs. Tower cranes use one of two types of jib, namely luffing jib or horizontal jib (sometimes called saddle jib). The luffing jib has a larger minimum working radius than horizontal jibs, while the horizontal jib has a smaller maximum lifting capacity. The horizontal jib gives better and faster control in three directions. The masts of all horizontal jibs remain stationary while the jib revolves, this being either fitted on top of the mast or built on to a collar around the mast. The luffing jib has the advantage of reaching into corners which would be out of the reach of the horizontal jib.

2.5.5 Hoists for personnel and materials

The types of hoist vary from the small jib hoist, which can be fixed to tubular scaffolding, to large passenger hoists which are used for transporting personnel to heights of 450 m.

Mobile hoists are the most common type used and are capable of lifting light loads to a height of 30 m. The mast and winch unit is mounted on a wheeled chassis, the whole unit being adjusted for verticality by screw jacks. For movement between sites, the unit is lowered to a horizontal position and towed by a vehicle. Where such hoists are extended for use up to 30 m the mast is secured to the scaffolding or frame of the building.

Tower hoists are normally used for hoisting general materials to any height. The tower consists of a rectangular mast made in sections suitable for transportation. The mast is secured to the building or scaffolding throughout its full height and is provided with landing stages within the scaffolding at each floor or access level.

The platform for material hoists takes the form of a steel frame, covered with some form of durable decking such as plywood. The platform is normally interchangeable with hoppers or special carrying jigs to accommodate the lifting of special materials. One such jig might carry a monorail truck, thus allowing the truck to be lifted vertically to an elevated layout of track.

Passenger hoists differ from general purpose hoists in that the platform must be fully enclosed to form a cage and must have additional safety devices to prevent over-run and free fall. They should also be controlled from within the cage.

Hoists are powered by diesel engines or electricity and may operate on the rack and pinion principle; the latter type can be used for non-vertical applications. All landings must have some form of guard, usually sliding gates, and the base of the hoist, where the winch is situated, must be fully guarded to prevent workmen being injured by moving parts.

2.6 COMPRESSED AIR PLANT

2.6.1 General considerations

Pneumatic tools are used extensively on civil engineering projects because the power source can be very mobile and the range of tools can easily be handled. By comparison with other forms of power the efficiency of compressed air is low, but the mobility of the power far outweighs this disadvantage. This mobility of power allows work to be carried out rapidly on isolated sites or in situations where access or work space may be confined. These advantages have resulted in the development of a wide range of tools for concrete breaking, demolition, rock drilling, steel sharpening, woodworking, riveting, grinding, tamping and vibrating concrete, pumping, pile driving, winching and spraying, blowing fine debris out of formwork before concrete is poured.

The type of compressed air system will depend on the nature of the work involved, but will be either a local or a centralised system. In the case of local systems the compressor unit, with air receiving tank, is mounted on a four-wheeled chassis for towing around the site or on public highways. In some cases the compressor is lorry-mounted. These mobile units are powered by diesel, petrol or electric motors, the most common being diesel.

The centralised supply of compressed air involves the installation of a larger semi-permanent plant. A central compressor house is then built to accommodate the plant and a network of air pipes is taken from the compressor house to a variety of outlet points. The range of such a supply will cope with most extensive sites and is more reliable than a number of mobile units. Pipes between the supply and the tools should be calculated to give a minimum pressure reduction and have gradients to prevent the collection of water. Low points in the supply system should be provided with traps for collecting condensate. This type of supply is more convenient for the running of workshops and plant areas and can be used for supplying temporary needs to any area of the site at little additional cost.

2.6.2 Air compressors

There are two main types of compressor, the reciprocating or piston type and the rotary impeller type.

The **reciprocating compressor** may be either single stage or multi-stage in action. In the single stage type the air is drawn into a cylinder and compressed by one stroke of a piston. With the multi-stage type the air passes through a series of cylinders, each cylinder containing air at increasing pressures. With a two-stage compressor, the air is compressed into a large cylinder at an intermediate pressure then further compressed into a smaller cylinder at the required final pressure. Two-stage compressors give greater efficiency, up to 20 percent more air being compressed than a single stage compressor of the same power. In addition, the air in the two-stage plant is cooled between the first and second stage, thereby eliminating valve trouble which can be caused by high temperatures in a single stage compressor. The multi-stage compressor is suitable for high pressures and high volumes of air. This type of compressor will be used in a centralised plant, where the amount of air required can be as high as 130 m^3 per minute.

Rotary compressors consist of a high speed cylindrical rotor mounted essentially in a cylinder. The rotor is slotted to receive a number of blades which are free to move radically in the slots. This compressor, known as the sliding vane compressor, draws air from the intake port by centrifugal force and delivers it through the delivery port when the desired pressure is reached. The rotary compressor is quieter than the reciprocating unit and maintenance is less because of the smaller number of moving parts.

Air receivers are used to store the compressed air as it leaves the high pressure cylinder and provide a constant source of pressure for the air lines. The main functions of an air receiver are:

- To minimise pressure fluctuations, created by the compressor.
- To minimise frictional losses due to pulsation of air through distributing lines.
- To store energy for a short period and so enable an unloading device to be used.
- To reduce the temperature of the air and remove moisture before the air reaches the supply lines.

2.6.3 Pneumatic tools and equipment

Pneumatic tools can be divided into four basic types:

- Percussion type, in which a piston reciprocates and delivers rapid blows to tools.
- Percussion type, requiring rotation of the tools.
- Rotary type, which use the power of air motors.
- Miscellaneous equipment.

Percussion type tools

The most common tool used is the concrete-breaker (jack hammer) which is designed for breaking up road surfaces. They vary in weight from 15 to 40 kg and have an air consumption of between 10 and 20 m^3/min. These tools can be fitted with a variety of heads including a small pile driving head for driving trench sheeting.

Pneumatic picks are similar to jack hammers but much lighter in construction. They weigh 9 to 14 kg and have a consumption of air of 6 m^3/min. Chisels and clay spades are normally fitted for use in hard excavation.

Backfill rammers, used for backfilling trenches and similar work, weigh 23 kg and have a consumption of 10 m^3/min.

Chipping, caulking and riveting hammers are all small, lightweight tools, having a low consumption of air (0.5 m^3/min or less), and especially suitable for specialised work at low cost.

Pneumatic tools with rotary action

Hammer drills are the most common type, and are used for drilling rock or concrete. The piston which produces the hammer action also produces a rotary motion of the return stroke. A quick-release retaining device allows rapid changing of drills. This tool can be used for wet or dry drilling, weighs 9 to 32 kg and has an air consumption of from 5 to 10 m^3/min.

Rock drills are similar in design to hammer drills but are larger and heavier, weighing 45 to 90 kg. They are used in tunnelling and quarry drilling, with either a dry or wet process. Because of their weight, the drills are mounted on some form of frame. This frame supports the drill and applies pressure to the bit by automatic feed. In the case of quarry drilling, the drill is mounted in a cradle on a small pneumatic-tyred chassis; the cradle can be pivoted to drill at any angle; air consumption varies from 15 to 24 m^3/min.

Rotary type tools
These include tools which are powered by air motors with rotary sliding vanes. They include chain saws weighing up to 31 kg, with an air consumption of 10 m^3/min, circular saws, drilling, and general forming tools, including wood-boring machines.

Miscellaneous tools
These include concrete vibrators and compactors for road construction, external and internal vibrators used with shuttering, pile-driving equipment, sump pumps, paint-spraying equipment, concrete-placing and spraying equipment.

The biggest problem with air tools, especially concrete-breakers and drills, is noise. This can be greatly reduced by using jacket-type silencers, which do not impair the efficiency of the tool.

2.7 BITUMINOUS MIXING AND LAYING PLANT

2.7.1 General considerations

When considering plant for bituminous paving, two stages of work must be considered, namely mixing and laying. In some cases both operations are carried out by the one piece of plant, while in other cases many types of plant carry out specific operations, each forming only part of either mixing or laying.

The plant used will depend on the type of material to be handled, and this requires careful specification since terminology may get confused if compared with practice in the USA. The binder used in bituminous-bound materials is either bitumen, which is produced by refining crude oil and some natural deposits, or road tar, which is obtained from crude tar produced from coal or coke. Where aggregate is mixed with bitumen an artificial asphalt is produced. Lake asphalt, such as Trinidad Lake Asphalt, consists of bitumen and mineral matter found in large natural deposits, and is normally refined to remove vegetable matter before use. Rock asphalt is another natural deposit, in which a limestone or sandstone contains a high level of bitumen.

All the materials mentioned above may be used in the process of producing good quality road surfaces. Other materials, of a low viscosity, may be used for treatments of road surfaces. These include cut-back bitumen, which is bitumen reduced in viscosity by the addition of a suitable thinning agent, and emulsions, which consist of the suspension of one liquid (usually bitumen) dispersed minutely through another liquid in which it is not soluble. The type of plant used will depend on the process selected, together with the type and quantity of aggregates and binder specified. The greatest requirement is road maintenance, and special techniques have been developed to deal with this requirement. In the UK and wider Europe the Surface Dressing process is now common (known as Chip Seal in the US). The plant for this process is discussed below.

A process that is currently under development in France, and being demonstrated in the UK, is the bitumen emulsion sealing coat. This process uses a bitumen emulsion at a density slightly less than used in a surface dressing, which is followed by screeding a coated mix over the top. Specialist machines have been developed for the process; however, contractors involved in traditional asphalt-laying are modifying paving finishers with spray bars. The surface laid is thicker than surface dressing and, by utilising a screed, the uneven nature of the existing surface can be dealt with.

2.7.2 Heaters and boilers

There are many types of bitumen heaters and boilers ranging from small mobile units to large permanent plants. The most common form of boiler is the mobile boiler; it is mounted on a steel chassis having four wheels and a tow bar. The furnace at the base of the boiler is either gas or oil fired. Mobile boilers have capacities ranging from 350 to 9000 litres.

Static heating and storage tanks are available for high outputs, having a capacity ranging from 9000 to 18000 litres. These units are heated by oil-fired burners. Models are strengthened for transportation and are provided with four lifting eyes to allow the whole unit be transported on a low loader. Mobile heating and storage units in the form of truck or trailer-mounted tankers can be used for either supply or bulk tank distribution. The bitumen is supplied to the mobile unit in a pre-heated form and oil heaters are incorporated in the unit to keep the material at the required temperature.

The choice of boiler will depend on the type of work involved. Repair and resurfacing work requires constant mobile supplies and a central mixing plant requires large static boilers or a battery of heating and storage units. Mobile units have a storage capacity of up to 27000 litres.

2.7.3 Binder distributors

There are two methods of distributing binder material, namely bulk tank sprayers and trailer-mounted units, but the bulk tanker is the most common one in use (Figure 2.29). Bulk tank sprayers are heat-insulated and fitted with heating units to maintain the binder at an adequate temperature. The binder is fed into the bulk tanker from a static heating unit or bulk supply unit. Bulk tank sprayers are truck-mounted with sealed tanks having a capacity ranging from 9000 to 20000 litres. A power-driven pump sprays the binder over the surface of the road through a horizontal spray bar mounted transversely at the rear of the unit. This spray bar is variable in width and a road can be covered in two passes. Trailer-mounted units vary in size from 540 to 4500 litres capacity, mounted on a trailer chassis with pneumatic road wheels. The spraying unit can either take the form of a transverse-mounted bar or power-operated lances.

2.7.4 High speed road-surfacing units

The most common unit in use is one which lays a surface dressing as a low-cost wearing coat. This form of wearing coat, in which the road is sprayed with hot binder and coated with grit, is used for a majority of resurfacing projects. The machine, approximately 10 m in length, carries out three operations as it moves forward at a speed of 15 mph. The three operations are:

- Road cleaning.
- Spreading of bituminous sealing membrane.
- Grit spreading.

FIGURE 2.29
Bitumen distributor with variable width spray-bar
(The Phoenix Engineering Company Limited)

The alternative method of re-surfacing is to use the bitumen distributor, detailed above, and follow through with a chip-spreading machine (Figure 2.30). A lorry keeps the chip-spreader supplied with chippings as the work proceeds. The chippings are fed into a rear hopper and transferred to the spreader via a conveyor.

Road planers and heaters

Road planers are used to shave off the uneven surface of bituminous roads by heating the surface till a softened state is achieved. It then removes the humps with revolving blades. The machine has a long wheel base which prevents too much undulation when planing the surface. The heating burners are positioned below the chassis and protected by hoods. The revolving blades are hydraulically adjusted and on cutting leave a smooth surface ready for further surface treatment.

Road heaters are designed for drying, heating and burning off the road surface ready to receive further treatment. This form of plant is used on road surfaces which have become too smooth, as a result of movement of the binder material in hot weather.

2.7.5 Asphalt and bitumen mixing plant

Asphalt and bitumen mixing plants range from permanent plant installations to small mobile units which are towed from site to site. These units carry out a complete sequence of operations, namely:

- drying;
- heating and mixing of aggregates;
- coating aggregates with binder;
- delivery ready for laying.

Drying units consist of a long cylindrical steel drum rotating at a slight angle. The unit is heated internally by the passage of hot gases, through which the aggregate passes. The purpose of the drier unit is to ensure that the material is thoroughly dry prior to coating with a binder. The unit may also be used for drying sand for hot rolled asphalt.

FIGURE 2.30
Chipping spreader being fed by lorry (The Phoenix Engineering Company Limited)

Batch-mixing plants have the following individual sections:

- aggregate feeder – normally conveyor;
- drier unit;
- storage bins for heated aggregate;
- binder heater with tank storage;
- weighing plant;
- mechanical mixer.

The **mixing** units consist, in the main, of some form of paddle mixer. For batch-mixing the paddle mixer will be the twin-shafted type, the paddles rotating in opposite directions to give a thorough mix in a very short time. They have a capacity of 750 kg to 5000 kg and an output of 60 to 400 tonnes per hour.

2.7.6 Spreading and finishing machines

The method of spreading macadam will depend on whether the material is cold-mixed or hot-mixed. Cold-mixed macadam can be either spread by hand, using forks and rakes, or, when the volume of material is large, by a small spreading machine. Hot mixes can now be stored in hot asphalt containers, such as the Phoenix Unistor (Figure 2.31). The Unistor can be mounted on a lorry and is ideal for repair work on roads.

Larger machines are available for spreading and finishing the surface coating in a single pass (ie only passing over the area once). The larger model is capable of laying pavements up to 150 mm thick and 3 to 4.25 m wide. It has a laying speed of 2.4 to 27 m per minute and has a 10-tonne hopper capacity. The screed unit comprises an oscillating tamper, thickness control, screeding heater and screed plate, all of which are carried on a long cranked arm pivoted from the tractor unit. The machine is filled by tipper trucks which back on to the hopper and unload while the paver is in motion. Rollers on the front of the hopper connect with the lorry tyres and push the lorry forward as it is unloading. The thickness of the surface finish is controlled by hydraulic adjustment of the screed plate which is automatically levelled as the machine moves forward (Figure 2.32).

FIGURE 2.31
Hot asphalt container – lorry mounted
(The Phoenix Engineering Company Limited)

2.8 PUMPS AND DEWATERING EQUIPMENT

2.8.1 General considerations

One of the most common and yet most important pieces of plant in civil engineering is the pump. It is a piece of equipment that must be reliable in adverse conditions and able to cope with varying factors, such as the total pumping head and various types of water. Since these factors may vary as a contract proceeds, it may be good policy to use an all-purpose pump rather than a series of pumps for specialist operations. This choice will depend on the complexity of the project and the amount of liquid to be moved in any particular period of time. Although pumps may be used for moving other liquids and materials such as sewage and concrete, this section deals specifically with the movements of water and sludge.

When selecting pumps for a site the use of the pumps must be considered. The following operations may be involved:

- Keeping foundations, pits etc free from water.
- Lowering of water table to below the level of excavation.
- Pumping out cofferdams or other large quantities of water.
- Supplying water for jetting and sluicing.
- Supplying water for general purpose use.

Having decided on the operations, there are further factors which will have to be considered before selecting the pump. These include:

- The rate at which the water is to be pumped.
- Height of suction lift (distance from water to pump).

FIGURE 2.32 *Spreading and finishing machine (Tarmac Quarry Products Limited)*

- Height of discharge (distance from pump to discharge – vertical).
- Altitude of the project.
- Loss due to friction.
- Size of pipe to be used.
- Pressure required at head if pressure is required.

Size of piping is very important since the loss of pumping power due to friction increases rapidly as the pipe diameter decreases. An example of two pipe sizes in steel will show the difference. Water to be pumped = 4500 litres per min. Pipe diameter (internal) 150 mm and 100 mm. Then loss due to friction, shown as extra head, is as follows:

150 mm pipe = 1.88 m per 30 m of pipe
100 mm pipe = 15.30 m per 30 m of pipe
} obtained from standard tables

This indicates that it may be advantageous to increase the diameter of suction and delivery pipes to reduce friction loss.

2.8.2 Pumps – types

Pumps for general use can be grouped under the following heads:

- Centrifugal – normal, self-priming, air-operated.
- Displacement – reciprocating, diaphragm.
- Submersible.
- Air lift.

Centrifugal pumps

Normal type centrifugal pumps contain a rotating impeller which revolves at such a speed as to cause the water to flow at considerable pressure. The speed of the impeller creates a vortex which sucks air out of the hose and atmospheric pressure causes the water to rise to the pump to commence the pumping action. This pump will require priming every time it runs dry.

Self-priming centrifugal pumps work on the same principle as the normal pump but have a reserve supply of water in the impeller chamber. When the pump is started this supply of water produces a seal against which the pump can draw air from the suction pipe. When the pump is stopped it retains the supply of priming water indefinitely.

Air-operated centrifugal pumps, sometimes named 'sump pumps', are particularly useful in tunnels, foundation pits etc, and will handle sewage, oil or sludge. The pump consists of a small centrifugal pump rigidly fixed to an air motor. The whole unit is enclosed in one tubular casing.

Sludge pumps are available, using the centrifugal principle, and are classified as:

- Unchokeable or fullway pumps.
- Open impeller pumps.

Unchokeable pumps are capable of passing large solid objects through the impeller. This is achieved by having widely spaced vanes on the impeller for the objects to pass through. This design of impeller has an effect on the efficiency of the pump and therefore limits the suction lift. The open impeller pump is similar to the normal centrifugal pump but is capable of moving objects up to 20 mm in diameter before clogging. A strainer is fitted to the suction hose intake to prevent larger objects from entering the pipe.

Displacement pumps

Reciprocating pumps work by the action of a piston or ram moving in a cylinder. When the piston moves in one direction the water is drawn into the cylinder in front of the piston and pushed out at the rear of the piston. This is classified as a double-acting pump; where the water moves in one direction only with the movement of the piston it is classified as single action. Larger pumps may have two or three cylinders and are classified as duplex, single- or double-acting, and triplex, single- or double-acting. These pumps have the following advantages:

- Ability to pump at a uniform rate against varying heads.
- High reliability.
- High efficiency regardless of the head and speed.

They have certain disadvantages however, which include:

- Better for low-flow conditions.
- Water is delivered in pulsations.
- They cannot handle water containing solids.

Diaphragm pumps work on the principle of raising and lowering a flexible diaphragm within a cylinder by means of a pump rod which is connected to an engine crank. When the diaphragm is raised, water is drawn into the cylinder through a valve on the suction side of the pump. The downward movement of the diaphragm closes the suction valve and pushes the water out through the delivery pipe. This pump, sometimes called a 'lift and force pump', can have two diaphragms to increase the efficiency. The pump is very popular since it can handle liquids and mud containing 10 to 15 percent solids. It is suitable for work where the flow of water varies greatly, because it is self-priming.

Submersible pumps

Submersible pumps are required for ground-water lowering when using the 'deep well system', or for removing water from any form of deep sump. The pump unit, normally powered by electricity, is suspended from the rising main or from a wire cable if flexible hose is being used. The pump consists of a centrifugal unit and motor mounted in a single cylindrical unit having an annular space between pump and casing. The space allows the water to move upwards to the rising main. For civil engineering work these pumps are specially designed for heavy duty work that involves lifting gritty water. They have a capacity of between 200 to 1200 litres per minute against a head of 10 metres.

Air lift pumps

Air lift pumps differ from the other air-operated pumps in that no moving parts are used in the lifting operation. This type of pump consists only of a long vertical pipe to the lower end of which is connected a supply of compressed air. The air carries the water up the vertical pipe to the discharge area. This pump is useful for moving silt from the bottom of a cofferdam.

2.8.3 Dewatering equipment

The techniques of dewatering are fully described in Chapter 3, and are not therefore repeated in this section. The main operations which involve the use of pumps are:

- Jetting the riser pipes.
- Dewatering the header pipes.

The first operation is undertaken by using a 'Jetting pump' (Figure 2.33). This is a pump which is capable of delivering a considerable amount of water at high pressure. The pressure required at the nozzle of the well point depends on the nature of the ground, but will be up to 10 bar. The pump normally used for this work will be of the centrifugal type, having an output of 140 to 1100 m^3 per hour. The jetting pipe is normally 75 mm diameter and the suction hose or dewatering hose 100 mm or 150 mm diameter, although diameters up to 300 mm are possible; both hoses are heavy quality armoured hoses or steel tubing. The jetting hose is connected to the top of a jetting tube, which is driven into the ground by water pressure. On completion the wellpoint is connected to the riser pipe and lowered into the tube, the jetting tube is then withdrawn.

The 'dewatering' operation has two stages, first to create a vacuum in the ring main - this is done by a vacuum pump; secondly, to dispose of the water that rises up the wellpoints to fill the vacuum in the header pipe. This requires a heavy duty pump having an output to deal with the number of well points being used. This means that the pump for dewatering will vary from job to job based on the number of dewatering points required, as well as the volume of water to be moved. The dewatering pump consists of an air separator valve at the intake, an open impeller type centrifugal pump, a vacuum pump to create the vacuum in the ring main, and a power unit to drive both pumps.

The water passes through the air separator to the centrifugal pump, then it is pumped through a valve to the discharge point. Water is prevented from entering the vacuum pump by means of float-operated containers. The equipment is mounted on a four wheels (Figure 2.34). The most common-sized pump used is a 150 mm diameter suction and delivery unit which has an output up to 400 m^3 per hour and a vacuum lift of 9.5 metres. It is usual to provide a duplicate pump on standby during pumping operations, to deal with problems of breakdown. The standby pump will be connected to the main header pipe and should be ready for immediate use.

FIGURE 2.33 *(Left)*
Jetting pump
(Millars International)

FIGURE 2.34 *(Right)*
Dewatering pump
(Millars International)

CHAPTER 3

Earthworks

3.1 GENERAL CONSIDERATIONS AND PLANNING

3.1.1 Site considerations

Having examined the information contained in site investigation reports, the engineer will then consider the following factors which affect the practical planning and costs of earthwork operations:

- Nature and extent of excavation.
- Available work area.
- Disposal of soil.
- Existing services and structures.

Nature and extent of excavation

The nature of the excavation, whether for reduced levels, road works, trenches, basements or pits, will have to be considered in the light of the time available for excavation and the sequence of completed earthworks. In the case of large cut and fill operations, a detailed plan of the movement of spoil and plant will need to be prepared (see Section 2.2.1). Areas of fill may require some form of retaining wall or drainage, prior to depositing the spoil; this must be taken into account when planning the sequence of operations. Extended work such as trenches for pipelines can be divided into sections and may progress simultaneously. Deep excavations for basements and 'cut-off' walls must be progressed to suit the work area and site conditions; in some cases cut-off walls are completed well in advance of other earthworks.

Work area

'Work area' can be defined as the total space available for the manipulation of plant and storage of materials. It does not include areas for site administration or accommodation, since such areas are often outside the perimeter of the work area.

The total work area should be indicated on a site plan so that movement of plant and materials can be efficiently planned. On certain sites the work area is sufficient to allow the sides of deep excavations to be battered to a safe angle of repose, thereby giving a work area free from obstructions such as timbering etc. On confined sites the plant may have to work from a position outside the actual work area, such as on a gantry or public highway. Alternatively, plant may have to work its way into the excavation and be lifted out by crane on completion. It is essential therefore that a study be made of the space available and of the effect progress will have on this. The selection of plant for earthworks is discussed in Section 2.2, where further details are considered.

Disposal of soil

The disposal of soil is achieved by one of the following methods:

- Immediate use as backfilling elsewhere on the site.
- Storage in spoil heaps, for use later or removal at a later stage.
- Immediate removal from site to tip or other destination.

When the material is used for backfilling it must be suitable for the particular operation in hand. This should involve separating the cut material into two categories, one suitable for filling and the other for removal or other earthworks. Topsoil will be one such material that is separated into spoil heaps for finishing off embankments and general areas to be grassed.

Consideration should be given to the position of such spoil heaps in order to reduce the amount of handling they receive. They should be so positioned that they do not interfere with access to work areas or become a danger or nuisance because of slumping in bad weather. Consideration must also be given to the stresses induced by spoil heaps on structures or services below ground or adjacent to the spoil heap.

The immediate removal of spoil from the site may involve the control of lorries to allow maximum utilisation of earthmoving plant. The cycle time for removing spoil and returning lorries to site can be calculated to allow for traffic delays and plant 'turn round' time. Some consideration must also be given to spillage of soil when leaving the site: public highways must be kept clean at all times.

Existing services and structures

Care should be taken to establish the position of all pipes, cables and underground services, which should be clearly marked before earthworks commence. Where excavation involves the disturbance of services, they should be carefully unearthed and supported to prevent damage by movement or vibration; nevertheless, breakages almost invariably occur. Pumping may interrupt the natural flow of underground water, and sheet piling may be required around the excavation to prevent any adverse effect. (See Section 4.1).

Existing structures adjacent to excavation areas will require support during excavations: this may take any of the following forms:

- Cut-off walling.
- Strutting and shoring, using sheet piles.
- Underpinning.

These techniques are described in subsequent chapters. In very loose or wet conditions the adjacent structures may be subjected to movement by heavy pumping or grabbing.

Whatever the condition of the site, it is expedient to establish a record, both photographic and written, of the state of the existing structures. This will involve a record of levels against a known fixed datum and measurements which delineate the position of the structures. If there are signs of movement in or around the structures, it will be necessary to fix 'tell-tale' markers in strategic positions on the structures to allow rapid checks of horizontal and vertical movement to be made.

3.1.2 Ground conditions

Chapter 1 dealt with the reasons for establishing a clear picture of ground conditions before excavation commences. The information obtained will assist in establishing the following factors:

- Ground support required during excavation.
- Best method of keeping the excavation free from water.
- Type of plant to be used.

The ground support required will depend on the strength and stability of the soil, the depth of excavation and the length of time that the excavation remains open. Consideration must be given to the change in soil stability due to adverse weather conditions. Some clays are particularly susceptible to softening and movement during periods of heavy rain. The movement of clay will also increase the pressure on shoring and this must be taken into consideration when calculating sizes of supports.

In conditions where there are frequent changes in ground strata, attention must be given to the strata at the bottoms of the excavations; these strata may be heavily stressed and subject to slip movement.

The best method of keeping the excavation free from water will depend on the type of soil and the source of water. Surface water can be intercepted by some form of ditch, and field drains can be diverted. Sumps or pits can be dug to receive surface water which finds its way into the excavation during heavy rainfall. Ground water can be dealt with in many ways, depending on the permeability of the soil and the methods of control (the latter is discussed fully in Section 3.3).

3.1.3 Contract duration and weather prospects

Earthworks are subject to two forms of change due to weather. The first is a superficial change due to immediate weather conditions, eg heavy rainfall; the second is a change due to seasonal conditions, by which soils may be affected for a longer period of time. With this in mind, consideration should be given to the duration of the earthwork programme and the time of year in which the work is to commence. In a rainy period, certain parts of the earthworks may be carried out well in advance of a programme appropriate for a dry period. This would involve excavating cuttings which could be used to channel water away from the site. One carriageway of a motorway could be excavated to within say 600 mm of formation level to drain the other carriageway during inclement weather. Where these methods are not practicable, it may be necessary to cut drainage channels during the dry season and to discharge them into ditches or streams in the wet season.

Embankments should be cambered so that water cannot lie at the centre of the surface and create instability by percolating through the material. Hardcore roads should be formed to facilitate the movement of plant during heavy rainfalls, pneumatic-tyred vehicles being particularly hindered by wet site conditions.

The compaction of soil

Compaction of soil is affected by changes in moisture content, and wet conditions may render satisfactory compaction impossible. Conversely, very dry or hot periods may involve much more wetting and rolling to achieve satisfactory results. It follows that extremes of weather may pose particular problems in the formation of embankments and fill areas. These problems can be eased by leaving fill material in position, ie unexcavated, until immediately before it is required for use; this will protect it from excess moisture or from drying out

Earthworks carried out in winter may be affected by frost and therefore uneconomical to excavate. Frost action causes the moist material to expand and lose density, involving extra problems or work in compaction.

3.1.4 Economic aspects of earthwork design

The economic aspects of earthwork design take into account the following points:

- Nature or type of work.
- Availability of suitable plant.
- Availability of suitable fill material, if required.
- The time of year.

The **nature of the work** can be placed in the classification of sites in Chapter 1, namely, extended sites or confined sites. The extended site, which would include roads, railways and pipelines, may require flow charts or mass haul diagrams (see Chapter 2) to indicate distances and volumes of spoil involved, so that the most economic procedures can be determined. The mass haul diagram will assist in establishing the most economic gradients by comparing the cost of cut and fill areas. The confined site will involve the economic selection of excavation method, ground support and spoil disposal.

The availability of suitable plant for a project is a problem that often faces contractors who have a large amount of plant tied up on sites. It may involve buying new plant or hiring plant, the latter proving less economical on very large earthmoving contracts. Availability of suitable fill materials may involve importing material from outside sources or the excavation of 'borrow pits' (ie areas made available by the employer from which suitable soil may be excavated).

Borrow pits can be refilled with the excavated material from any other parts of the site. The availability of suitable material will affect the degree of slope on embankment work and the subsequent maintenance of such work. The nature of the fill material will affect the cost of compaction and general soil movement costs.

3.2 EXCAVATIONS

3.2.1 Bulk excavation

Bulk excavation may include the following operations:

- Cuttings.
- Cut and fill areas.
- Basements and large pits.
- Hand excavations.

Cuttings

Cuttings include large excavation cuts for roads, canals and similar forms of construction where the excavated material is usually moved to some other part of the site either for bulk fill or for general 'spread and level' operations. With road works the amount of excavation is normally balanced against the fill areas, and normal cut-and-fill operations take place. In other forms of construction where the spoil may be partially waterlogged, such as dock areas, the excavation may require different types of plant or some means of dewatering to allow bulk excavating plant to operate.

The type of plant used will depend on the quantity of soil to be transported and the distance to the disposal point. Bulldozers can be used efficiently if the pushing distance does not exceed 100 m, whereas scrapers would probably be used for distances of over 100 metres: the type of scraper would depend on the haul distance (see Chapter 2). In some cases the section of the cut will be too narrow for the scraper to turn or too deep for the scraper to ascend the ramps. This may involve the use of drag lines or a combination of plant, eg track-mounted excavators, dozers; and loading equipment, eg dragline or loading shovels.

Where the excavation is of restricted length and depth and has suitable road ramps for lorries, it can be done by a hydraulic excavator. The angle of repose for the cutting will depend on the type of ground, but can be as low as 20° for some clays and 90° for stable rocks. This angle should be calculated from laboratory tests.

Shallow cut and fill

Shallow cut and fill operations occur mainly in road works and airfield construction. The work is normally carried out by scrapers, or dozers, which first strip the topsoil for re-use and then reduce the level of the site to the required formation level. This type of work can be hindered by wet weather because the limited depth of dig does not allow economic two-stage excavation. Two-stage excavation is suitable for deep cut and fill areas where the first stage is taken out to within 300–600 mm of the final formation level and then completed with a second cut when the weather is suitable or when some form of protection to the formation can be placed. In shallow excavation the plant may have to stop work to prevent damage of the formation level, or temporary roads may have to be provided over large areas, either way resulting in extra cost. Areas of cut and fill can be adequately drained by temporary trenches, which could be incorporated into the final sub-grade drainage. The formation level may be protected against water, and the drying-out action of wind and sun, by some form of waterproof dressing.

Basements and pits

For construction purposes the term 'deep pit', which applies to excavations over 4.5 m deep, is synonymous with the term 'basement'. Shallow pits, which are 1.5 m deep, present little or no problem in terms of excavation or ground support. They are normally excavated with a small backacting machine or by hand. Medium pits, classified as 1.5 to 4.5 m deep, require careful selection of plant and ground support. Small machines may be limited in depth of digging, normally up to 4 m deep, larger pieces of equipment may have difficulties in manoeuvrability between such pits, and since pits of this nature are usually small in superficial area, the excavation must be completed from ground level. The type and amount of ground support will depend on the nature of the soil, the amount of water present and the length of time the excavation will be open. These factors are discussed in detail in Section 3.2.4 below.

Methods of excavation of basements and deep pits vary. The following are methods commonly used by contractors:

- Unshored excavations.
- Shored excavations.
- Dumpling method.
- Cut-off walling method.

All these methods involve some form of support to the ground and as such are dealt with in detail in Section 3.2.4; but they also have an effect on the choice of plant to be used. With unshored excavations it can be assumed that there is ample working space around the excavation to allow battering of the excavation. This will, to a great extent, allow more freedom of choice in excavating plant than the other methods. The work may be executed by drag line, backacter or grab, depending on the superficial area of the pit. Other factors to be considered are access to formation level for lorries or dumper, the amount of water to be encountered, and the permeability of the soil. With shored excavations, the shoring can be made watertight by sealing the joints of sheet piling, thereby eliminating the free flow of water. The method of excavation in this case would depend on the nature of the shoring: the soil could be moved by grab, backacter or small-tracked excavator.

The dumpling method (see Figure 3.10) and cut-off walling method (see Figure 3.11) – the latter being in the form of contiguous piling – have this in common: they both involve the construction of a retaining wall, usually concrete, around the excavation area prior to bulk excavation taking place. In the former case some shoring is necessary during the construction of the retaining wall: in the latter case shoring may be necessary as the excavation proceeds. Both methods lend themselves to the use of hydraulic excavators, grabs and draglines.

Hand excavations

It may be necessary in many cases when excavating deep basements and pits to use pneumatic tools such as clay-spades and picks. This will occur in excavations which are heavily supported, leaving little room for mechanical excavation; it will also be necessary in the vicinity of services which would be subject to damage by machine, or in the case of removing obstacles such as boulders, logs or other projecting objects. Spoil from such excavations would be put into skips for periodic removal.

3.2.2 Rock excavation

The methods of breaking and excavating rock or other hard material will vary according to the type of material, quantity involved, conditions on site and equipment available. Such methods include:

- Use of pneumatic breaker.
- Breaking by hand with hammer and wedges.
- Drilling with pneumatic machines and breaking by driving plugs or freezing liquid in the holes.
- Drilling with pneumatic machines and breaking by blasting.

The first three methods are suitable where any of the following conditions prevail:

- The noise of blasting would cause annoyance.
- Adjacent buildings may be subject to damage.
- Blasting may cause inconvenience or stoppage of traffic.
- Landslides or rock falls might result.
- Accurate cutting is necessary and excessive 'overbrake' would be uneconomical.

However, modern methods of control allow very accurate vibration-controlled blasting in limited spaces. One example involved the excavation of a large basement in very hard rock, some of which was within 30 metres of an important computer installation. These innovations have made drilling and blasting the most effective and economical method of hard rock excavation.

There are two basic methods of drilling and blasting rock in excavations:

- Benching.
- Wellhole blasting.

Benching is used for most foundation and trench excavation in which the face of the rock is taken back in steps or 'benches'. Each bench or step is used as a platform for loading the loosened rock. The height of each bench varies with the type of rock but can be as shallow as 1.5 m or as deep as 6 m (Figure 3.1). Holes are drilled by the use of pneumatic hand tools or rig-mounted drills to a depth of 600 mm below the required excavation level. This allows the blast to clear the rock to or slightly below the required formation. The diameter of the holes is important: increased hole diameters require an increase in the charge required, thus leading to greater vibration. Up to 3.5 m deep the holes should be 38 mm dia; between 3.5 m and 9 m the diameters can be increased to between 50 and 75 mm.

Wellhole blasting involves the drilling of large holes, 150 mm to 250 mm diameter, at large intervals, the spacing of which should be equivalent to the 'burden' or depth of

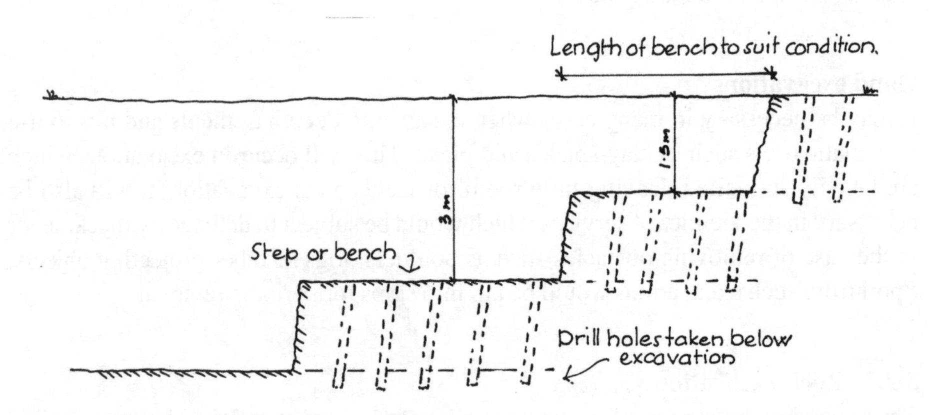

FIGURE 3.1
Drilling and blasting by benching method

face to be moved. To be economical, the depth of boring should be at least 9 m and up to 24 m.

The holes are machine drilled to the required depth and loaded with suitable explosives; this form of blasting produces good fragmentation and thereby reduces the amount of secondary blasting. Since the technique is only economical for deep blasting, it has a use restricted to very large foundation excavations.

Explosives

Explosives are discussed Chapters 1 and 8 and therefore only those related to excavation are considered here.

Gelatine, gelignite and the range of slurries are the main explosives recommended for most types of excavation. They have good water resistance and reliable performance in wet conditions, and give good fragmentation. In the case of very hard rock, gelatine or one of the high-powered slurries should be used; in the case of soft rock a lower-powered explosive will give satisfactory results at lower cost. The amount of explosive used varies with the type of rock; as low as 0.25 kg per m^3 for soft rock and up to 0.50 kg per m^3 for hard rock.

When the charge has been placed, the holes are 'stemmed'. This is the term applied to the consolidation of the back filling of the hole prior to blasting. Stemming is best achieved with damp sand or a mixture of clay and sand; large holes can be partly stemmed with rock chippings.

Explosives for secondary blasting. In some cases the main blasting operation may produce stones which are too large for crushing or for loading on to lorries. When this occurs there are two methods of dealing with the problem: 'pop shooting' and 'plaster shooting'.

With 'pop shooting' a hole is drilled into the centre of the boulder and fitted with a charge of gelatine. The charge is fired by means of a safety fuse and plain detonators or by electric shot firing.

'Plaster shooting' provides a means of breaking large boulders where drilling is difficult or expensive. A charge is primed with a detonator and safety fuse, laid on the surface of the boulder, covered with an amount of plastic clay and pressed into position by hand – a high-strength gelatine type explosive is most suitable for this type of work. This method of breaking boulders is limited in use. It should not be used within 400 m of a building or structure.

Underwater rock excavation

Underwater rock excavation by blasting is similar to surface operations in that similar techniques are employed in drilling and blasting. Drills are mounted on either platforms or barges and holes are drilled to a pre-determined grid. Factors needing special consideration for underwater work are:

- Fragmentation must be sufficient to enable the rock to be dredged easily.
- Explosive charges need to be heavier.
- Drilling must be deeper than normally required in 'on-shore' methods, preferably as far below the required blasting depth as the spacing between the holes.
- Shot holes must be closer together.

The last three considerations take into account the hydrostatic pressure on the face of the rock.

There are two main methods of achieving the breakdown of rock under water; drill and blast, which involves normal drilling methods and the use of a suitable explosive such as a high density, high powered explosive. Plaster shooting, which involves the placing of charges on the rock bed in a grid pattern. The technique requires sufficient head of water to confine the energy produced in the blast – usually a minimum head of 8 m. It is also suitable for clearing peaks of rock to allow drilling methods to be used.

The method used will depend upon depth of water, depth of rock to be stripped and the effects of vibrations or shock waves on property in the vicinity. Such work is used for underwater trenching, removing sandbanks, deepening and widening channels, deepening harbours, demolition of wrecks and obstacles, cutting piles, etc.

Methods of removing rock from excavations

The choice of method will depend on the position and size of the rock pieces which may be removed by either of the following methods:

- By hand loading on to flat bottomed rock skips.
- By mechanical excavators, such as draglines, and loaders (a machine specially designed for rock loading).

In both cases the material will be finally loaded into trucks, dumpers or railway wagons.

Backfilling

Backfilling with rock must be undertaken with great care to ensure consolidation of the excavation; this will limit the size of the rock fill to that which allows maximum consolidation. In the case of rock fill to embankments, the size should be limited so that it can be compacted in layers not exceeding 450 mm. Most rock bulks in volume when excavated, owing to the voids, and is never returned to an excavation without some surplus; but there is one exception to this rule that the surveyor must be aware of: namely, the excavation of chalk. When backfilling with chalk the material readily compacts into a volume smaller than that of its natural state, owing to the percentage of voids in the material. This can lead to the need to import extra fill material to complete the backfilling, and some allowance must be made for this in the specification or measurement of the work.

3.2.3 Trench excavation

The choice of method of excavating, supporting and backfilling trenches depends on the following factors:

- Purpose for which the trench is being excavated.
- The nature of the ground.
- The time scale of the work.
- Ground water conditions.
- The location of the trench.
- Number of obstructions.

The first three factors greatly influence the choice of plant. Some trenches can be dug and backfilled with a single pass of the machine, a flexible pipe having been laid as an integral operation. Where the trench has to be left open for pipe-laying or other work for some hours or days, consideration must be given to ground support. Where possible, the sides of the trenches can be battered to obviate the use of shoring, which would hamper foundation work or pipe-laying. This must be given economic consideration, since the extra excavation and fill may not offset the cost of shoring and decreased production caused by supports. Where supports are to be used, the method of support must be decided having regard to the ground-water conditions and method of ground-water control.

The location of the trench, whether across open land or along public highways, will affect the selection of plant and in some cases the method of ground support. Location may also involve organisational problems such as diversion and control of traffic, fencing, and lighting during the hours of darkness. In built-up areas there is a greater incidence of underground services that may have to be negotiated, and this will affect the speed of the operations.

Excavation methods

The methods of excavating trenches are as follows:

- Full depth, full length excavation.
- Full depth, successive stages of excavation.
- Stage depth, successive stages of excavation.

The first method is suitable for long narrow trenches of shallow depth in which the machine completes the trench non-stop ahead of any other operation. This method is suitable for pipelines and sewers.

The second method is suitable for deep trenches where several operations of work can proceed in sequence; this would prevent stretches of trench from being left open too long and thereby being subject to collapse. It also reduces the amount of support and protection employed at any one time.

The third method of excavation is suitable for very deep trenches in confined areas or adjacent to existing property. It involves the support of the trench as the work proceeds and is most suited for operations such as deep foundations and underpinning.

The first method is also suitable for trenches with battered sides, and would be most suited to works requiring freedom from struts, eg cast-in-place foundations, walls and culverts. The plant used for battered trenches ranges from special trenching machines which are capable of producing the required batter in a single pass, to standard equipment such as hydraulic excavators.

Special equipment is available for the formation of narrow trenches; these may include mechanical trenchers with a continuous chain action. Alternatively, some form of 'trenchless' technology might be considered. This includes a range of moles and other boring equipment that can be steered underground. Such equipment is ideal for services that pass under roads, canals and other areas that may prove difficult to excavate. These techniques are discussed further in Chapter 5.

3.2.4 Support of excavations

Support of excavations is governed by the following factors:

- Type of soil.
- Ground water condition.
- Depth and width of excavation.

The **soil types** for discussion are as follows:

- Loose sand, gravel and silts.

- Compact sands and stiff clays.
- Rocks.

In addition to support to the above soils, special consideration is given to large excavations.

Support for excavation in loose sand, gravel and silts requires some form of continuous support, which may consist of trench sheeting (Figure 3.2) – lightweight pressed steel sheets – and steel sheet piling (Figure 3.3). Since such soils are likely to slump quickly, the support must be placed immediately after excavation takes place. This means that deep excavations will have to be dug in stages; the first stage by

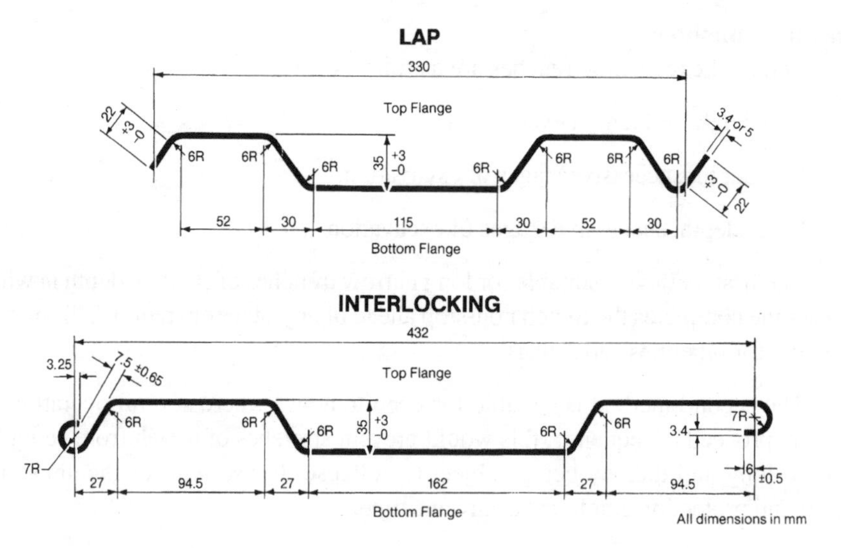

FIGURE 3.2
Types of trench sheeting for ground support
(British Steel plc, Narrow Strip, Ayrton Goddins)

FIGURE 3.3
Use of Larssen steel sheet piling in ground support
(British Steel plc)

machine, and, after supporting the first stage, any subsequent stages by hand or by grab. Hand excavation is very expensive and therefore the use of driven sheet piles may be more suitable for deep trenching (also see Section 4.1).

Trench sheeting is available in two main types, overlapping and interlocking. The overlapping sheets are also available in different thicknesses, 3.4 mm and 5.0 mm. All types can be supplied galvanised.

Table 3.1 may be used as a guide when establishing the support required. Support to excavation in compact sands and stiff clays can be satisfactorily achieved by the use of open shoring or timbering. This involves the use of poling boards or trench sheets at intervals of approximately 1 m; the boards are supported by continuous walings and trench jacks at 2 m centres. If the soil is subject to drying out and crumbling, the spacing of the poling boards can be reduced accordingly. Dry clays which have become fissured by wind and sun are likely in wet weather to take up rainwater and expand: this produces extra stress in the struts and allowance should be made for this at the design stage.

TABLE 3.1 **Support required for excavations with vertical sides in uniform ground**

A indicates that no support is required
B indicates that open sheeting should be employed
C indicates that close sheeting or sheet piling should be employed

Type of soil	Depth of excavation		
	Up to 1.5 m (shallow)	*1.5 to 4.5 m (medium)*	*Over 4.5 m (deep)*
Soft peat	C	C	C
Firm peat	A	C	C
Soft clay and silt	C	C	C
Firm and stiff clay	A*	A*	C
Loose gravel and sand	C	C	C
Cemented gravels and sands	A	B	C
Compact gravels and sands	A	B	C
Gravel and sands below water table	C	C	C
Fissured and jointed rock	A*	A*	B
Sound rock	A	A	A

*Open or close sheeting or sheet piling may be required if site conditions are unfavourable.

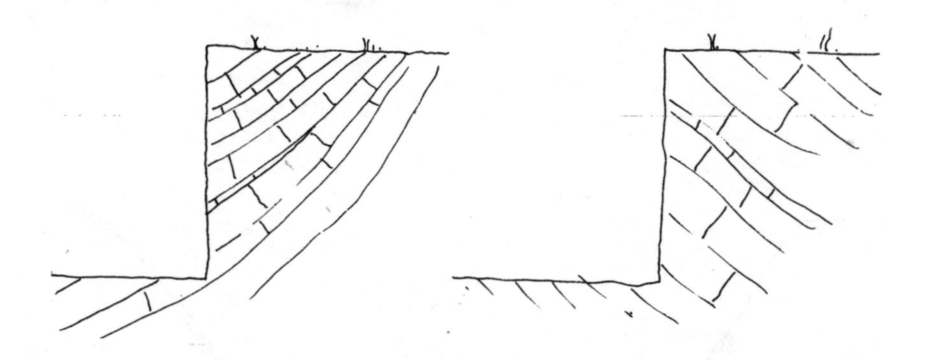

FIGURE 3.4
(a) (Left)
Unstable conditions in rock excavation
(b) (Right)
Stable conditions in rock excavation

Support to rock excavation depends to a large extent on the type of rock and the slope of rock strata. Figure 3.4 shows two different situations likely to be encountered in rock excavation. In the case of unstable rock faces, open timbering should be used to prevent slip.

Alternatively, where the depth of excavation is excessive, the rock face may be stabilised by rock bolting (Figure 3.5). Rock bolting techniques are discussed further in Section 4.5.2, but in general terms consist of solid steel rods or cables which are fixed in deep drill holes by means of wedges, anchors or grouting processes. Light steel sections or steel plates are used to support the rock face through which the rods are threaded. Some shales and chalks are subject to movement through weathering, and support of these materials may be expensive; it may be more economical to cut such materials back to a safe angle of repose if space allows such treatment.

Support of large excavations

Large, deep excavations can be supported by a number of different methods, depending on the type of soil, proximity of structures and depth of excavation.

Unshored excavations (Figure 3.6) involves the battering of the sides of the excavation to a safe angle of repose; for many soils this can be taken to be an angle of 45°. In some cases the site may not be large enough to cope with the extra spoil and this may incur extra transporting costs. However, as this method of ground support leaves the area free from cumbersome struts, it allows greater productivity and is the preferred system.

Shoring to deep excavations by traditional methods can be undertaken in two ways, the first by internal support and the second by external support. Internal support

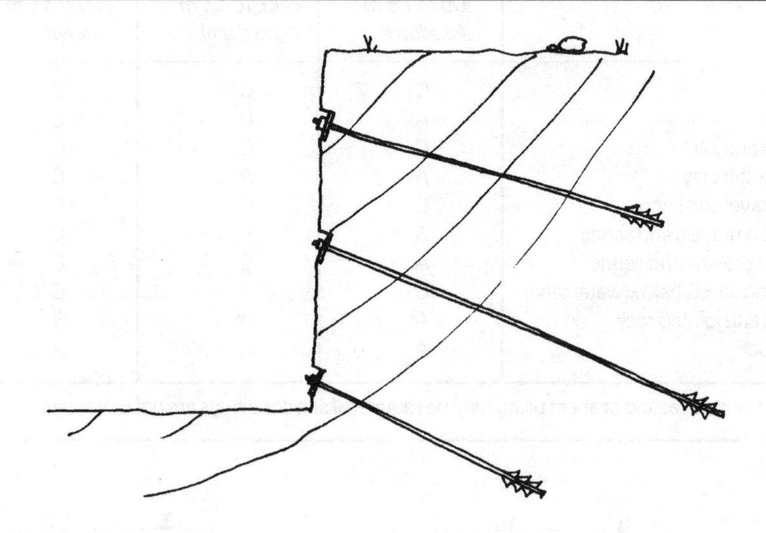

FIGURE 3.5
Rock bolting in unstable conditions

FIGURE 3.6
Battered sides to excavation

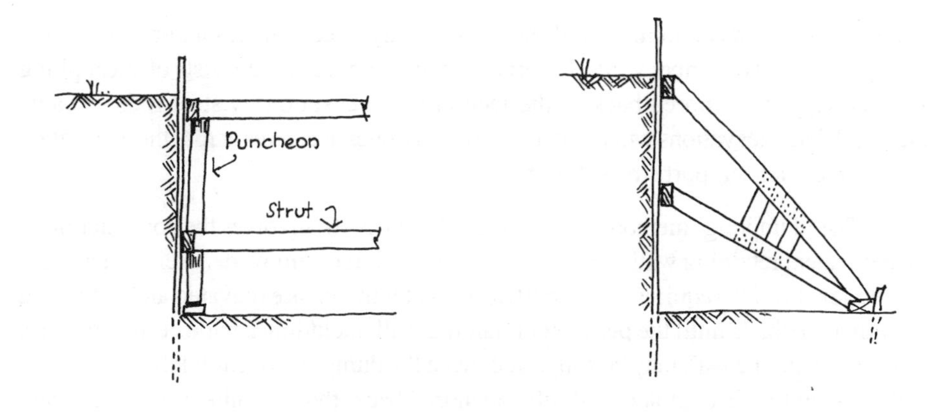

FIGURE 3.7
Internal support to excavation

FIGURE 3.8
External support to excavations by cable anchors
(Platipus Anchors Limited)

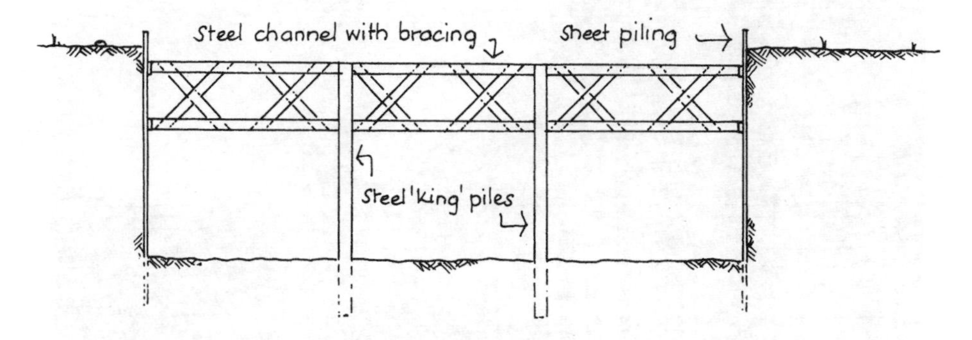

FIGURE 3.9
Support to wide excavation using H-piles

(Figure 3.7) consists of poling boards or, more usually, sheet piling, supported by heavy walings and steel or timber struts. External support (Figure 3.8) consists of sheet piling or other support anchored back to the face of the excavation by steel rods or cable anchors. Wide excavations will involve the use of H-piles at centres across the excavation to carry the main support struts (Figure 3.9).

The 'dumpling' method of excavation (Figure 3.10) involves the construction of the permanent retaining walls before the bulk of the soil is removed. After reducing the site to a level which requires no support, a perimeter trench is excavated and supported by normal methods until the perimeter retaining wall, including a wide perimeter strip, is constructed. The wall may be supported, from the dumpling of earth left in the centre of the site, until it is complete and fully anchored back; the dumpling is then removed by dragline or other suitable excavator.

Cut-off walling methods of support are formed by steel sheet piling or concrete walls. The concrete is cast insitu in the form of deep narrow walls or contiguous bored pile walls (Figure 3.11). They are of particular value when supporting deep excavations

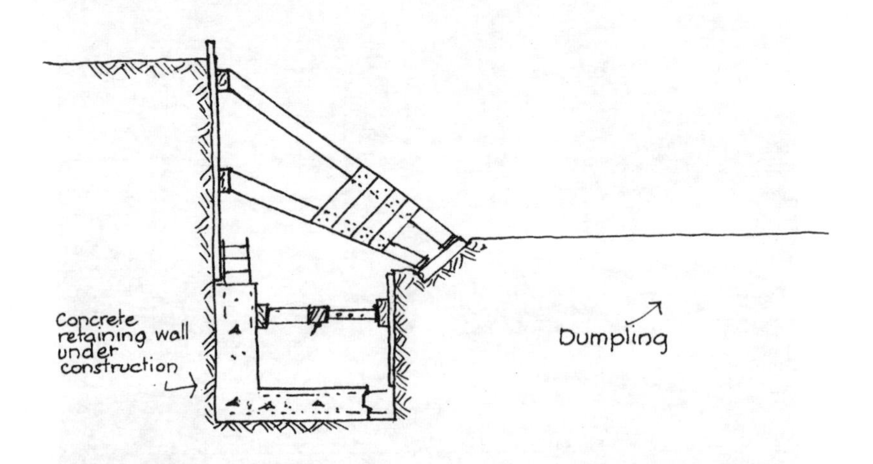

FIGURE 3.10
Dumpling method of support

FIGURE 3.11
Contiguous bored piles forming cut-off wall and support to excavation
(Westpile Limited)

which are adjacent to heavily loaded foundations of other structures. They prevent any movement of the existing foundations and allow greater freedom of movement for mechanical excavation. This is achieved by one of two methods: either by taking the wall deep enough to support the ground by cantilever action, or by anchoring the walls back into the soil using ground anchors (Figure 3.12).

3.2.5 Embankments

The construction of embankments and the design profile of the sloping sides will depend on a number of factors, such as:

- The purpose for which the embankment is constructed, e.g. the loads involved.
- The consolidation of the fill in the embankment under the proposed loads.
- The stability of the ground on which the embankment is to be constructed.
- The extent to which the strength properties of the fill may be affected by the method of construction.
- The cost of obtaining suitable fill material.
- The difficulties in construction during adverse weather, when using clays and fine sands.

Embankment construction – general

The method of constructing an embankment will depend upon the extent of the works, the type of fill material being used and the nature of the site.

The site must be stripped of all vegetable matter which would readily consolidate under the heavy load of fill material. Fill material should be tipped and spread in layers of such a thickness that it can be compacted to the required density – which will be established by laboratory testing. Where large volumes of fill are involved, the density factor obtained in the laboratory may not be achieved on site. This is due to practical

FIGURE 3.12
Diaphragm cut-off walling tied back with ground anchors
(Cementation Ground Engineering Limited)

difficulties in varying the moisture content of large volumes of earth between the two stages of excavating and filling. This discrepancy can be minimised by the correct selection of compaction plant: the plant most suited will depend on the soil type and its working moisture content. Multi-wheeled rubber-tyred vehicles (wobbly-wheel rollers) are very effective for compacting fine fill materials such as Pulverised Fuel Ash (PFA) and sands; very heavy vibrating rollers, towed by bulldozers, are also suitable for compacting PFA and coarser granular materials.

Construction in water

Where it is uneconomical to drain the water from the site, attention should be given to the maximum and minimum water levels and to the type of soil beneath these levels. All soft materials, such as mud and peat, should be removed and replaced with coarse granular material, such as rock waste. This material should be placed carefully where the foundation is subject to slip movement through uneven loading. In other conditions the fill can be tipped direct from lorries; or, in the case of deep water, bottom-opening barges may be used. The shape of the embankment below the water line must be monitored to prevent wastage of fill material. In some cases, where appreciable settlement can be tolerated, material which would normally be classified as unsuitable can be used as the centre core of the embankment, provided that it is protected by other suitable material.

Where wave action is likely, the slopes of the embankment must be protected against erosion; and where water pressure affects only one side of the embankment, precautions must be taken to avoid seepage underneath. Further consideration concerning fill materials in water is given in Chapter 6.

Construction on soft ground

When constructing earthworks on ground containing soft material, the soft material should preferably be removed to a depth which will enable the earthworks to be formed on a firm stratum. This may be achieved by normal methods of excavation or by displacement methods. Displacement methods include:

- **Bog blasting** – in which the soft material is displaced by explosives. This can be achieved in one of three ways:
- **Underfill blasting** – a technique used when the soft stratum exceeds 10 m in depth. The embankment is formed on the soft foundation and the soft material is blasted out, thereby allowing the embankment to subside into the cavity.
- **Trench shooting** – a technique used for blasting out the soft material to form an open trench into which the fill material is tipped.
- **Toe shooting** – a technique used in conditions where the sides of the trench would collapse if blasting was undertaken using the trench shooting technique. The site is loaded with fill material and explosives placed under the toe of the bank. As the explosives move the soft material, the fill flows into the toe cavity, thus extending the width of the bank.

Alternatives to the above are overloading and jetting.

Overloading is a technique in which the bank is formed to a considerable height above the finished level, thus overloading the soft ground which flows out from the toe. Surplus material is removed from the embankment when settlement is complete.

Jetting is a technique in which holes are jetted through the bank to a level just above the required formation level. Water is then jetted down through the bank, causing the soft material to flow laterally; the bank moves downwards as the soft material moves out. This technique can be used on soft deposits up to 15 m thick. When peat is to be displaced the embankment should be formed with fine granular materials to prevent impedance of the jetting operation.

Construction on sloping ground

Where the slope is gradual it is necessary only to maintain a good foundation and to provide drainage for the flow of water beneath the embankment. Where the slope is steep, it may be necessary to form 'benches' or horizontal steps in the side of the slope, to provide an adequate key for the new earthworks. Drainage must be carefully designed to catch water from the upper side of the embankment and to channel it safely under or through the embankment, thus eliminating any danger of instability due to water pressure.

When the natural slope is 1 in 5 or steeper, it will be necessary to investigate the possibility of soil slip and to take preventive action where necessary.

Embankment slopes

The safe angle of any embankment slope will depend on the nature of the fill material used and the height of the bank. The safe angle will range from as much as 45° for rock-waste fill down to as little as 20° for some clays. An average range for most rock fills can be taken as 33° to 42°, but the behaviour of existing embankments in materials similar to those being used is the best guide to the slope to be adopted. Slopes may also be stated as ratios and percentages: eg a slope of 1:3 or 33%; a slope of 1:5 or 20%. Where existing works are not available the safe slope may be determined from the following:

- For coarse-grained materials the slope can be taken as the angle of repose for the material, adjusted to give a margin of safety.
- For materials such as coarse and medium sands the slope should take account of surface erosion.

Very fine non-cohesive materials may be subject to instability due to pore water-pressure which could cause the slope to flow. The angle would therefore have to be lower than the angle of repose.

Cohesive soils such as silts and clays are subject to factors which do not affect the design of slopes in non-cohesive soils: therefore the angle of repose method cannot be used. The safe angle of these soils can only be determined from laboratory tests which take into consideration such factors as shear strength of soil under adverse conditions, height of bank, development of pore water-pressure during construction, etc.

3.3 CONTROL OF GROUND WATER

There are many ways in which ground water may be controlled during the construction period; some methods deal with water lowering and others with water exclusion. The various methods of ground water control and ground conditions in which these may be best suited are shown in Table 3.2, and are described in detail below.

3.3.1 Pumping systems

The control of ground water by pumping is the cheapest and commonest form of control: the various systems of pumping include:

- Pumping from sumps
- Pumping from wells
- Pumping from wellpoints.

Pumping from sumps

Pumping from sumps is the most widely used method of ground water control, since it can be applied to all types of ground conditions and is economical to install and maintain. The only problem is one of soil movement due to settlement: the ground is likely to move as the water flows towards the sump area. There is also a risk of instability at the formation level in supported excavations, owing to the upward movement of water. These problems can be partially overcome by positioning the sump at a corner of the excavation at a level below the formation level (Figure 3.13).

For excavations which are likely to be open for long periods of time, a peripheral drain filled with gravel can be dug to intercept water at formation level and channel it to the sump, so giving a drier and more stable work area. When the excavation is taken through permeable soil and continues in impermeable soil, it is better to form a drain at the line where the two soils meet. This type of drainage channel, known as a 'Garland' drain, prevents the impermeable soil being softened by the flow of water (Figure 3.14), and carries the ground water to a sump at one corner. The types of pumps used are discussed in Chapter 2, but it should be mentioned that the suction lift of most pumps is limited to a depth of 7.5 m, although some manufacturers claim a maximum lift of 9 m. This may affect the position of the pumps: for deep excavations where the depth exceeds 9 m the pump will have to be placed on a level suitable for the suction lift.

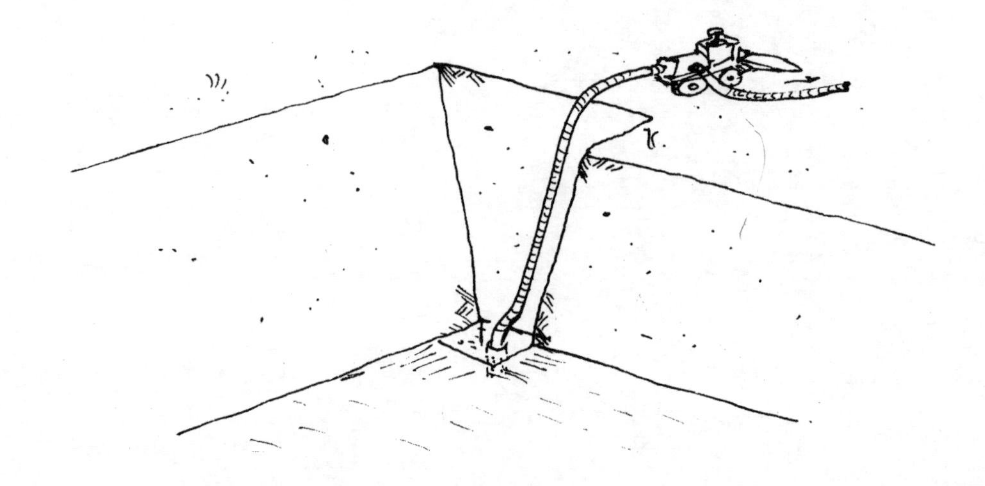

FIGURE 3.13
Sump below formation level in corner of excavation

TABLE 3.2 **Ground water control processes**

Method	*Soils suitable for treatment*	*Uses*
Group 1 – Diaphragm, exclusion of ground water		
1 Sheet piling	All types of soil (except boulder beds and rock)	Practically unrestricted
2 Diaphragm walls (structural concrete)	All soil types including those containing boulders (rotary percussion drilling suitable for penetrating rocks and boulders by reverse circulation using bentonite slurry)	Deep basements Shafts
3 Slurry trench cut-off	Silts, sands, gravels and cobbles	Practically unrestricted Extensive curtain walls round open excavations
4 Thin grouted membrane	Silts and sands	As for 3
5 Contiguous bored pile walls	All soil types but penetration through boulders may be difficult and costly	As for 2
6 Cement grouts	Fissured and jointed rocks	Filling fissures to stop water flow (filler added for major voids)
Grouted cut-offs		
7 Clay/cement grouts	Sands and gravels	Filling voids to exclude water To form relatively impermeable barriers – vertical or horizontal, also suitable for conditions where long-term flexibility is desirable, eg cores of dams
8 Silicates	Medium and coarse sand and gravels	As for 7 but non-flexible Joosten, Guttman and other processes
9 Resin grouts	Silty fine sands	As for 7 but only some flexibility
Freezing		
10 Ammonium/brine refrigeration	All types of saturated soils and rocks	Formation of ice in the voids stops water flow
11 Liquid nitrogen refrigerant	As for 10	As for 10
Group 2 – Water Lowering		
12 Sump pumping	Clean gravels and coarse sands	Open shallow excavations
13 Wellpoint systems with suction pumps (Vertical and horizontal)	Sandy gravels down to fine sands (with proper control can be also used in silty sands)	Open excavations including rolling pipe trench excavations
14 Bored shallow wells with suction pumps	Sandy gravels to silty fine sands and water bearing rocks	Similar to wellpoint pumping More appropriate for installations in silty soils where correct filtering is important
15 Deep bored filter wells with electrical submersible pumps	Gravels to silty fine sands, and water bearing rocks	Deep excavations in, through, or above waste-bearing formations
16 Electro-osmosis	Silts, silty clays and some peats	Deep excavations in soils to speed dissipation of construction pore pressures
17 Drainage galleries tunnelling	Any water-bearing strata underlain by low permeable strata suitable for tunnelling	Removal of large quantities of permeability strata suitable for dam abutment cut-offs etc
18 Jet educator system using high pressure water to create a vacuum as well as lift the water	Sands (with proper control can also be used in silty sands and sandy silts)	Deep excavations in space so confined that multi-stage wellpointing cannot be used. Usually more appropriate to low permeability soils

Courtesy: Ground Engineering

An alternative method of preventing soil movement due to open sump pumping is the use of jetted sumps (Figure 3.15). The sump is formed by jetting a metal tube in the ground by means of water pressure. A disposable wellpoint, consisting of disposable hose and intake strainer, is lowered into the tube and a sand media placed around it. The metal tube is then withdrawn and the flexible suction pipe is connected to a pump. Table 3.2 shows the various processes for ground water control.

Pumping from wells

Since normal pumping methods limit the depth of suction lift to a maximum of 9 m and conditions may not allow the pump to be placed in the excavation, well pumping may be employed. The main use is the lowering of ground water to a considerable depth below 9 m or where the ground may not be suitable for the wellpoint system. The well is formed by sinking a lined borehole, of a diameter between 300 and 600 mm to the required depth. Into this borehole another tube is placed, known as the inner well lining

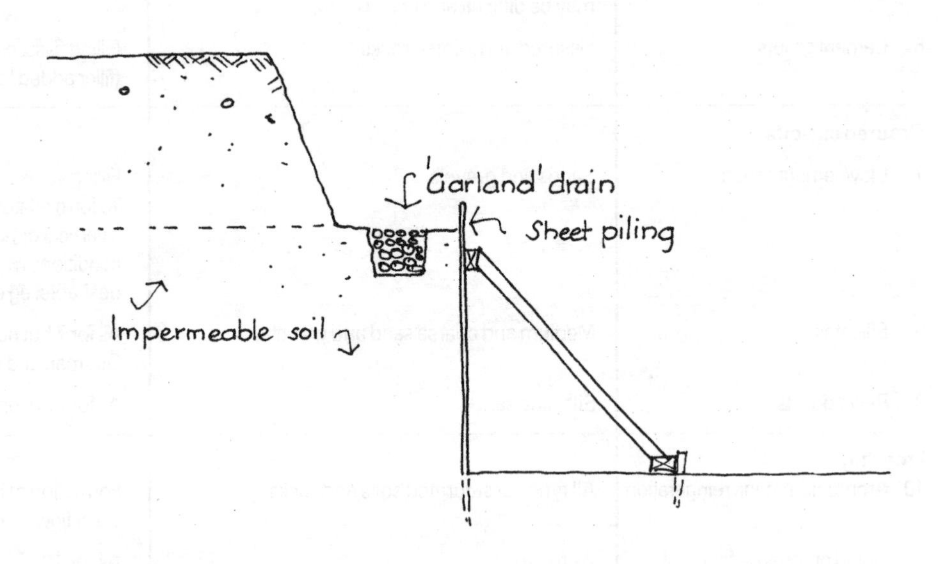

FIGURE 3.14 *Garland drain intercepting water at the impermeable level*

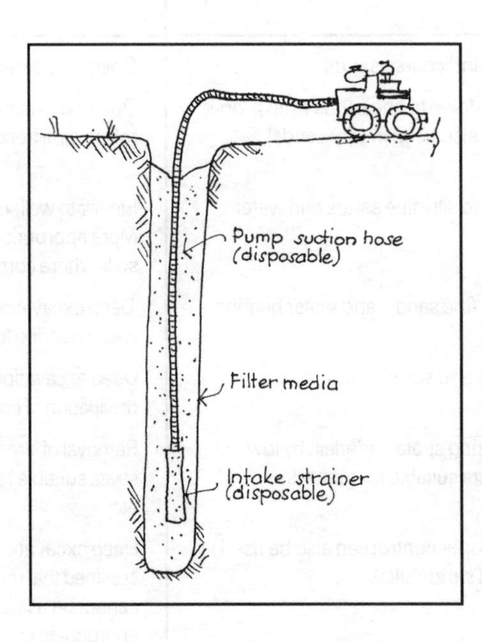

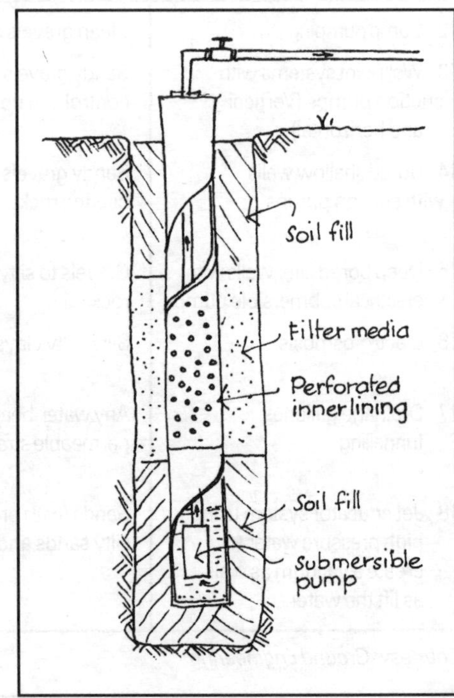

FIGURE 3.15 *(Left) Jetted sump*

FIGURE 3.16 *(Right) Section through tube well*

(Figure 3.16), which is provided with a perforated screen for the section for which dewatering is required. The lower end of the inner lining is unperforated and acts as a sump for the settlement of fine material.

The annular space between the two linings is filled with filter material over the length of the perforated section: the remainder of the borehole is backfilled with any suitable material. Depending on the depth of the well, the outer lining is withdrawn as the annular space is filled. Before the pump is placed in position, the water in the well is 'surged' by some form of plunger to promote flow through the filter and wash out unwanted 'fines'. The pump used is the submersible type as described in Section 2.8.2. The spacing of the wells is determined by the type of soil being dewatered. The depth of the well depends on the depth of the impermeable stratum: where this stratum is well below the excavation formation level the spacing between the wells can be increased until the draw-down curve is just below the formation level (Figure 3.17).

Horizontal borings are suitable when the formation level is at or just slightly within an impermeable stratum in which the draw-down curve in vertical wells would not lower the water to the required level. A well is sunk outside the excavation area to a level below the proposed formation, and horizontal borings are made in a radial pattern from the vertical well. This allows water to drain from the upper surface of the impermeable stratum into the large well and then to be pumped out by submersible pump.

Another horizontal system of ground water control can be used. This consists of laying a PVC perforated pipe at a depth of up to 6 m around the excavation area and connecting it to a pump (Figure 3.18). The suction pipe can be covered with a nylon filter sleeve which prevents particles of soil from entering the pipe: it can be laid over sites at a speed of up to 160 metres per hour by a special placing machine (Figure 3.19) which digs the trench, lays the pipe and backfills the trench in one operation. The length of pipe handled by one pump will depend on the soil conditions and the size and type of pump to be employed. The laying of the pipe automatically breaks up the ground and forms a drainage channel to the pipe. For continuous drainage of pipelines an overlap between the ends of the pipe is required to effect a continuous draw-down.

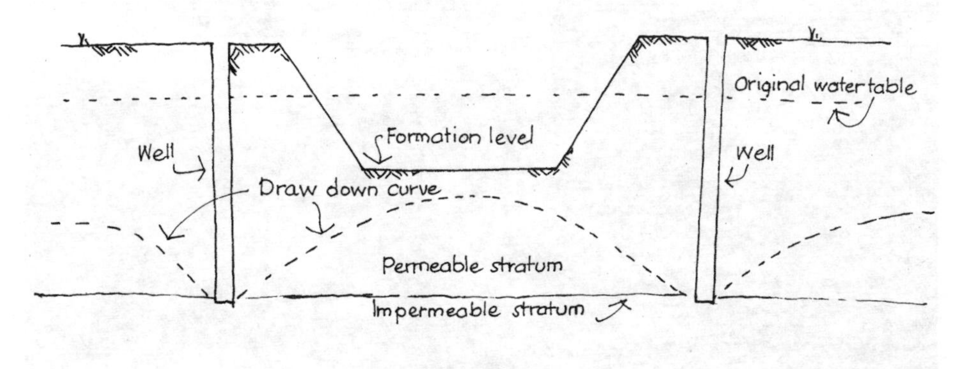

FIGURE 3.17
Wells used to lower water table to a level below formation of excavation

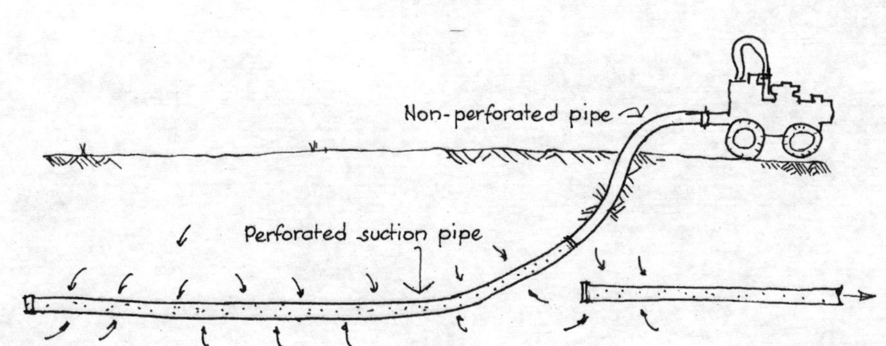

FIGURE 3.18
Horizontal system of ground water control

Pumping from wellpoints

The wellpoint system of dewatering is perhaps one of the most well known on civil engineering projects since it is used very frequently for ground water control in non-cohesive soils (Figure 3.20). The system consists of a number of small diameter vertical wells connected to a header pipe which is under vacuum from a pump. The ground water is forced out of the soil by atmospheric pressure into the header pipe, via the wellpoints, and discharged by the pump. The wellpoint itself is only a small part of the equipment, consisting of a perforated or slotted tube covered with a strainer or fine mesh, approximately 1300 mm long (Figure 3.21). The wellpoint is connected to 38 mm internal diameter mild steel riser pipe which in turn is connected by swing connection to the header pipe.

FIGURE 3.19
Trenching machine laying perforated pipe (Stuart Well Services Limited)

FIGURE 3.20
Typical conditions in which the wellpoint system should be used (Stuart Wells Services Limited)

Note: Partial collapse caused by sump pumping

A recent development is the fully disposable wellpoint (Figure 3.22), which consists of a 65 mm diameter perforated plastic inner strainer with a nylon filter sleeve: the strainer is capped top and bottom, the former to receive a 40 mm diameter flexible riser pipe. The riser pipe is also made of plastic and is disposable: since the riser pipe is flexible it eliminates the swing connection normally used in conventional systems.

Jetting the wellpoints prior to connecting them to the header pipe is achieved by a powerful pump (see Section 2.8.3). The wellpoint is screwed on to the riser pipe and connected to the pump by means of a high-pressure flexible hose (Figure 3.23): the water pressure from the jetting pump washes out the soil below the wellpoint, which is lowered into the ground as the process continues. When disposable wellpoints are employed, a jetting tube is used to form the hole (Figure 3.24), and the wellpoint is lowered into position before the tube is withdrawn. The holes that result from jetting may be 150 mm to 200 mm in diameter and these are backfilled with coarse-grained sand to form a supplementary filter. This process is known as 'sanding-in' the wellpoints.

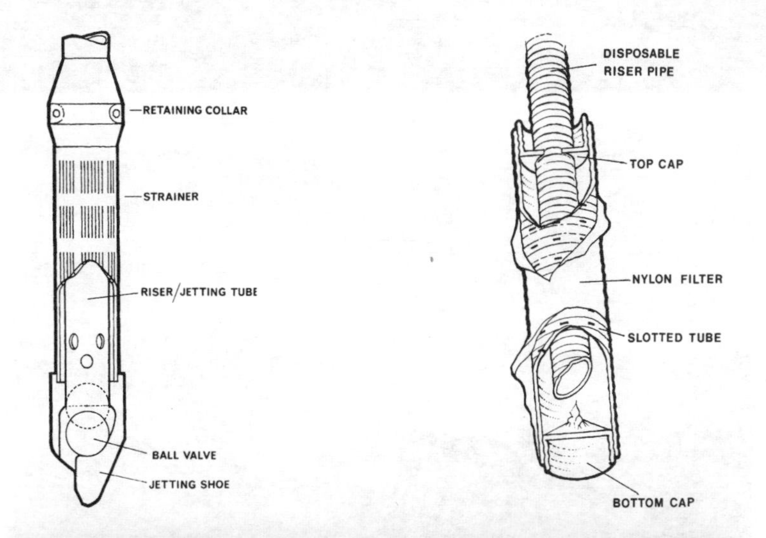

FIGURE 3.21 *(Left)* *Standard wellpoint*

FIGURE 3.22 *(Right)* *Disposable wellpoint*

FIGURE 3.23 *Jetting a non-disposable wellpoint* *(Millars International)*

FIGURE 3.24
Jetting the wellpoints, using a jetting tube
(Stuart Wells Services Limited)

FIGURE 3.25
Ring main wellpoint installation
(Stuart Well Services Limited)

Wellpoint installation

There are two types of wellpoint installation, the 'Ring-main' or 'Progressive Line'. With the former method the header pipe is placed around the excavation to be dewatered (Figure 3.25) and connected to the two pumps: two pumps are connected to the header pipe, one for the pumping operation and one as a standby should the first pump break down, since a delay in connecting a second pump could result in collapse of the excavation.

In the case of the Progressive Line method, sometimes called the rolling method (Figure 3.26), the header pipe is laid along the side of a continuous excavation, eg a trench; pumping is confined to a length of header pipe which allows work to progress without hazard. Further wellpoints are jetted ahead of the excavation to receive the progressive header pipe and when backfill has been completed the rear section of wellpoints is withdrawn. For narrow excavations it is sufficient to have a header pipe on one side of the excavation only; wide trenches or trenches in soils containing impervious materials will require header pipes on each side of the trench.

The wellpoint system is limited in its suction lift to a practical height of 6 m maximum. Any attempt to lift water by wellpoints above this height results in loss of pumping efficiency, owing to air being drawn into the system through the joints in the pipes. Where dewatering is necessary in depths over 5 m, it is common practice to introduce multi-stage wellpoints (Figure 3.27). These allow dewatering to any depth, providing the pump can work against the head involved; but practical aspects of excavation limit the number of stages employed. The use of multi-stage wellpoints involves the formation around the excavation of platforms on which the header pipe is situated; it may also involve battering the sides of the excavation to a safe angle of repose.

While the safe angle of repose for dewatered sands is much steeper than that for normal soil, it still produces a large area of excavation for the first stage. The number of stages to be employed can be reduced if the excavation is taken down to the

FIGURE 3.26
Progressive line or rolling wellpoint installation
(Ground Water Services Limited)

original ground water level before installing the first header pipe: this operation will eliminate one stage of wellpoints (Figure 3.28).

Regardless of the number of stages involved in dewatering a deep excavation, the depth of the inclined layer of soil being dewatered is limited to approximately 5 m and is therefore subject to seepage pressure from the mass of soil surrounding the excavation, which could cause instability. This pressure can be reduced by the use of deep wells positioned at the edge of the upper slope (Figure 3.29).

The wellpoint system of drainage is suitable only for dewatering non-cohesive soils with a minimum grain size of 0.1 mm: below this grain size the normal wellpoint system fails to produce the desired results. Where the grain size is below 0.5 mm, another method of wellpoint pumping may be employed, ie the 'vacuum' filter method.

FIGURE 3.27
Multi-stage wellpoint installation
(Millars International)

FIGURE 3.28
Excavation of basement to natural water level before positioning wellpoint

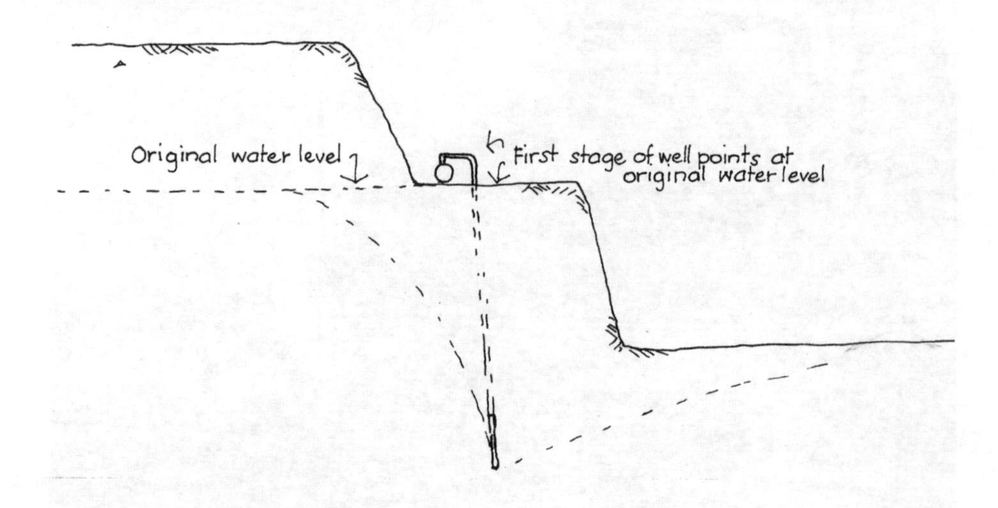

The vacuum filter method is similar to the normal method in that wellpoints are jetted into the ground and sanded in, but it differs in that the top metre of the hole is sealed with clay. The clay seals the wellpoint hole and when pumping commences creates a vacuum in the sand filter around the wellpoint. The ground water, which is being held in the very fine-grained material by capillarity, moves towards the vacuum as the force of atmospheric pressure attempts to equalise the pressure in the soil surrounding the wellpoint.

On large contracts the type of pumping and the number of wellpoints required can be determined from a pumping test, used to assess the permeability and storage co-efficient of the soil. The test consists of sinking observation wells in the area to be drained and dewatering the area by a small diameter ring main system. The level of water in the observation wells is recorded at given times.

Sand drains or sandwicks may also be used to lower the level of natural ground water and can be used to prevent hydrostatic pressure causing heave or 'boiling' in open excavations. They are also used to accelerate the settlement of clays and silts which may be subjected to applied loads, and they have particular value in the design and formation of embankments and roads over waterlogged ground.

The principle involved is the reduction of the drainage path in the layers of clay or silt, thereby allowing faster movement of water as the load is applied: this results in reducing the settlement time of the soils under load. The work is carried out by specialists who bore holes, 150 mm to 300 mm in diameter, at 2 to 5 m centres depending on ground conditions, and fill them with sand: in some cases a long nylon sleeve (sandwick) is filled with sand and lowered into the holes. Where fill material is placed directly above the drains, a horizontal blanket of granular fill must be provided over the sand drains to allow movement of the rising water.

3.3.2 Electro-osmosis

This technique of dewatering is used in soils which are cohesive in nature, eg silts and clays, and where the vacuum method of pumping does not produce satisfactory results. The unsatisfactory movement of water in cohesive soils is caused by the natural balance of electric charges in the particles of soil and molecules of water. Every particle of soil carries a negative charge of varying intensity, which attracts the positive (hydrogen) ends of the water molecules and thereby creates a balanced state. The fundamental principles in the complicated balance between soil and water are covered fully by Terzahgi and Peck.

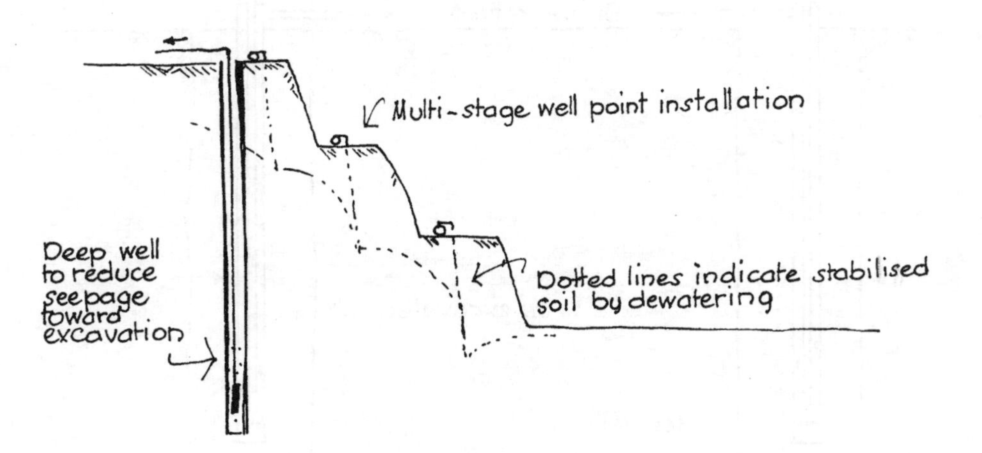

FIGURE 3.29
Combination of deep wells and wellpoint system

The balance described above has to be disturbed to cause the water to flow: this is done by inserting two electrodes into the saturated soil and passing a direct electrical charge between them. The positive electrode (anode) can be sheet piling or steel rods driven into the ground; the negative electrode (cathode) is normally a wellpoint. When the current passes through the ground between the electrodes (Figure 3.30) it disturbs the balance of the water in the capillaries causing the outer film of water (which is positive), together with any free water, to flow towards the negative point: the water is then pumped off from the wellpoint.

This method of dewatering was developed during the Second World War for the stabilisation of soils in deep cuttings for railway works and in the construction of U-boat pens at Trondhjem. The cost of using such a system of dewatering in the UK has proved prohibitive when compared with other methods. The power requirements are in the region of 0.5 to 1.3 kw per m^3 of soil dewatered in large excavations, and up to 12 kw per m^3 of soil on small excavations.

3.3.3 Freezing methods

Freezing as a method of stabilising waterlogged sands and gravel has been known for over a hundred years: it was first used in 1862 in Wales to prevent the ingress of waterlogged soils into mine shafts. The basic principle of the process is that it changes the waterlogged soil into a solid wall of ice which is completely impervious, thus allowing personnel and machines to work inside the wall without danger. To produce the low

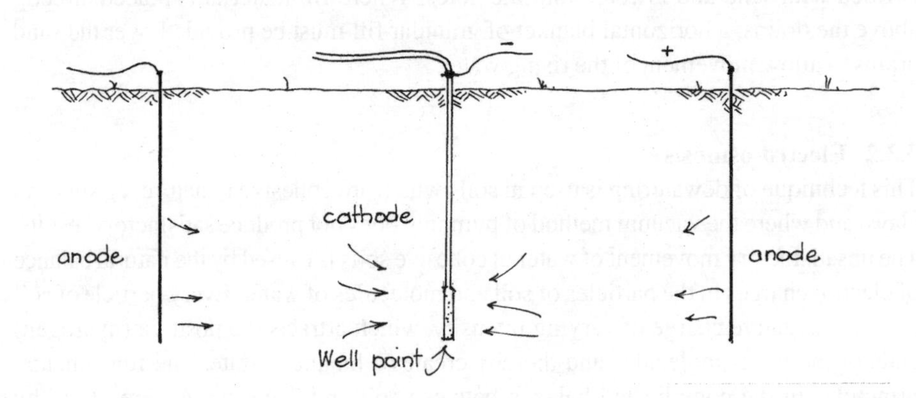

FIGURE 3.30
Electrodes used in dewatering

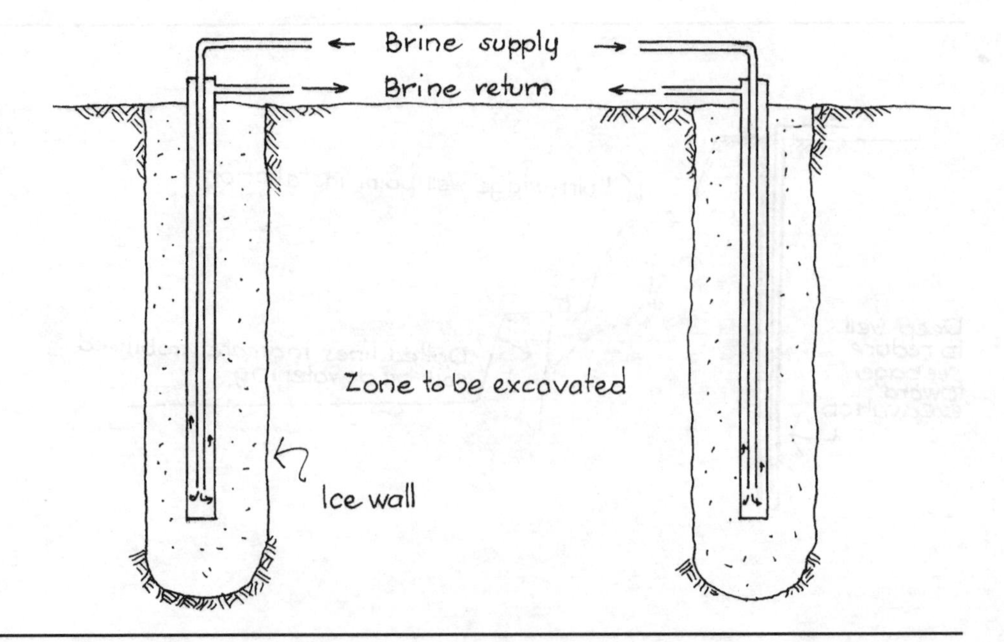

FIGURE 3.31
Freeze pipes in position – ice wall formed, ground ready for excavation

temperatures required to freeze the water content of the soil, steel freeze pipes are installed at approximately 1 metre centres around the site to be excavated: these pipes consist of an inner and outer tube. The outer tube, 100 to 150 mm in diameter, is sealed at the bottom and connected by a gland to a return pipe at the top. The inner tube, 38 mm to 75 mm in diameter, is open at the bottom and connected to the flow pipe at the top (Figure 3.31).

The flow and return pipes around the site carry chilled brine which is pumped down the inner tubes and back up through the annular space between the tubes to the return pipe. The temperature of the brine circulating in the freeze pipes ranges from –15 to –25^0 C, and therefore all pipes above ground level are insulated with polyurethane. The freezing medium must have a freezing point well below this temperature range and to achieve this a solution of calcium chloride or magnesium chloride is normally used. The liquid is cooled by a refrigeration plant, shown diagrammatically in Figure 3.32, and constantly re-circulated through the pipes. This causes the surrounding soil to freeze.

Trailer-mounted refrigeration equipment that requires little installation work makes this technique an economical proposition for many a problematic situation. In other cases where rapid freezing is required, the freezing medium used is liquid nitrogen.

The system is competitive with other forms of ground support and dewatering systems for excavations over 7 m deep. Unlike other systems, it is cheaper as the depth increases. The system does not require a high moisture content in the soil before it can be employed: a moisture content of 8 percent of the voids is all that is required. Another misconception is that the length of time taken to freeze the ground makes the system prohibitive. This is totally misconceived, since the use of liquid nitrogen as a circulating medium allows work to commence within days. Even using a brine medium, the time to obtain a wall of ice is reasonably short and depends on the spacing of the freeze pipes, the quantity of refrigeration used, and the type of soil being frozen.

At a spacing of 1 metre, a frozen wall 1 metre thick in sand and gravel can be achieved in 10 to 12 days; the same wall in clay would take 15 to 17 days. A completely continuous frozen wall is essential before excavation commences: this can be determined by sinking an observation borehole in the centre of the treated area. The borehole is

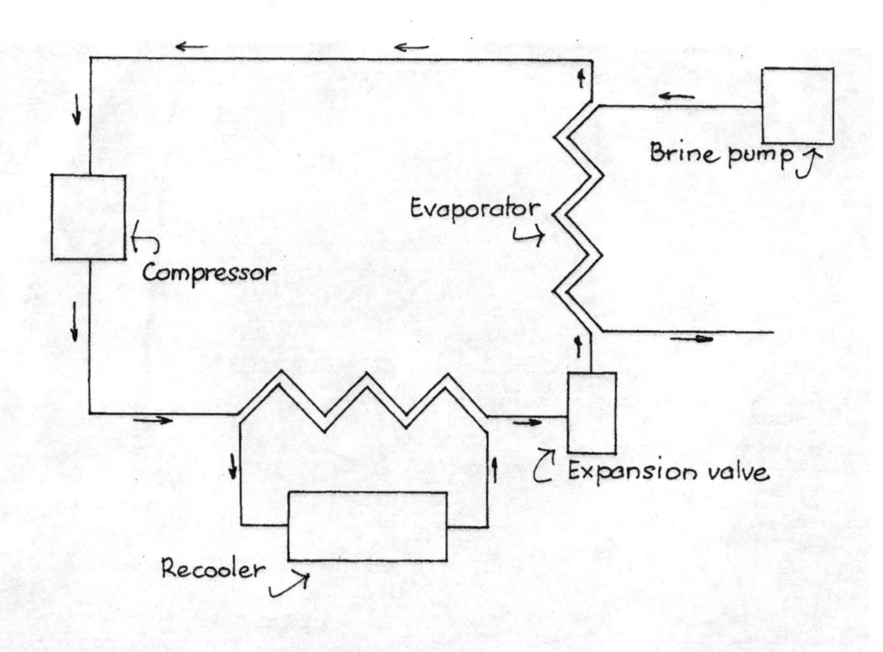

FIGURE 3.32
Diagrammatic presentation of the freezing cycle

lined with a perforated lining which allows the ingress of water. As the wall freezes, it forces the water inside the enclosure into the borehole: the water rises up the borehole and sometimes overflows on the ground. The rise in the level of water will indicate the closure of the wall.

This method of ground support is particularly suitable for the sinking of deep shafts (Figure 3.33). Shafts have been sunk to a depth of 600 m using this method, driving tunnels (Figure 3.34) and freezing large excavations (Figure 3.35 and Figure 3.36). The problem of volumetric increases in the soil due to freezing occurs only in

FIGURE 3.33
Freezing tubes in position for shaft sinking
(British Drilling and Freezing Company Limited)

FIGURE 3.34
Freezing method employed by tunnel driving
(Delmann-Haniel)

clays and silts: other materials such as sands and gravel do not suffer measurable volumetric increase when frozen artificially. Silts can exhibit a volumetric increase of up to 2 percent and precautions have to be taken when adjacent structures could be otherwise affected by soil movement.

The need for insulation of exposed ice-walls will be determined by the conditions. The completed walls may be covered with white reflective polyethylene film to prevent thawing out by radiant heat (Figure 3.35) or insulated with fibreglass blankets sandwiched between twin polythene sheets.

FIGURE 3.35
Aerial view of large excavation (17m deep) employing freezing method – note reflective polyethylene sheeting
(Delmann-Haniel)

FIGURE 3.36
Freezing installation for large excavation inside existing factory at Blackpool
(British Drilling and Freezing Company Limited)

3.3.4 Compressed air

Compressed air systems of ground water control are used in conjunction with caisson sinking (see Section 6.2) and tunnel driving (see Section 5.1) in waterlogged ground. The safe working in such conditions is of paramount importance. In the UK the air supply to caissons is governed by the Factory Acts and the Work in Compressed Air Special Regulations, which requires a delivery of 0.3 m^3 of fresh air per minute per person in the working chamber. The plant normally used for supplying the compressed air is a reciprocating compressor twin cylinder with single-stage piston. The capacity of air supply must be at least 50 percent above the normal requirements to allow for emergencies, and standby compressors with an alternative supply of power should be included in the plant set-up.

The air pressure required to keep water out of the excavation must equal the value of the hydrostatic pressure in the pore water at the level of the cutting edge of the caisson. For all practical purposes, the air pressure in which personnel can work is 340 kN/m^2 (3.4 bar). This will normally allow working at depths of up to 35 m below the water table.

3.3.5 Grouting methods

Grouting methods can be used in situations where the permeability of the soil would create a heavy demand on pumping or where the ground conditions would make the boring of wells and sinking of wellpoints very costly. The basic method is to inject the soil or rock with fluids which, on setting, seal or reduce the permeability of the material. The grout used will depend upon the particle sizes of the soil or the size of fissures in the rock formation. Since this process is costly, it must be carefully controlled to prevent wastage of material. This is achieved by additives in the grout which control the gelling properties, thus limiting their spread in the ground. Attention must be given to the position and vulnerability of existing underground structures such as sewers, basements etc, because pressure grouting can penetrate the fine cracks that may be present and so create problems.

The choice of grouting materials used in this method include:

- Cement grout
- Bentonite grout
- Chemical grout
- Resin grout
- Bituminous grout.

Cement grouting

Cement grouting is suitable for injecting into coarse materials which have a high permeability. A 'grout curtain' is formed by boring holes into the ground around the excavation area and injecting cement grout of varying consistencies. It is usual to commence grouting with a batch of thin grout and then to increase the viscosity of the grout as the process continues, by reducing the water-cement ratio. Secondary holes are bored between the lines of the primary boreholes to ensure complete grouting of the curtain.

The grout used may be composed of neat cement and water or a mixture of sand and cement in the ratio of up to 4 parts sand to 1 part cement, the latter giving better economies in the consumption of materials. A system of grout curtains constructed at the nuclear power station beside the Severn estuary at Oldbury is a good example of controlled water flow: the ingress of water was reduced from 4500 litres per minute to 16 litres per minute.

PFA can be used in conjunction with cement grouting, or it may be used in lieu of sand or as a partial cement replacement. The spherical particle shape of this material improves the flow quality of the grout. On one contract, gritstone, which was adjacent to an earth dam, was grouted satisfactorily with a solution of 1 part PFA: 1 part cement: 2 parts water, by weight. The grout was pumped through a 38 mm diameter pipe for a distance of 400 m without blockage or appreciable drop in pressure. Cementitious grouts are often referred to as 'filler grouts'.

Bentonite grouting

Bentonite or clay grouting is used in ground conditions where the particles of the soil are too small for cement grouting. While clay adds little if any strength to the soil, it has a high resistance to water flow and therefore produces an excellent barrier. Bentonite is produced from montmorillionite clay, which has thixotropic properties: when it coagulates it forms a gel which is highly resistant to water, and if mixed with certain additives, such as Portland cement or soluble silicates, the barrier formed will be permanent. It is particularly useful for grouting alluvial soils beneath dam foundations to prevent the seepage of water under the finished structure.

Chemical grouting

The chemical grouting or consolidation process is used in sandy soils of medium to coarse grading. The materials are liquid when mixed prior to injection and form into gels or solids by chemical reaction which takes place between the base substance and the hardener.

The time taken for solidification depends on the particular system being used. There are two main processes, 'two-shot' and 'one-shot'. In the 'two-shot' process (known widely as the Joosten and Guttman process, after the engineers who developed the process), pipes are driven into the ground at 600 mm centres, and the first chemical, normally sodium silicate, is injected. This is followed immediately by the injection of the second chemical, calcium chloride. The reaction between the two chemicals is immediate, resulting in a tough, insoluble 'silica-gel'. The process gives considerable strength to the soil and greatly reduces its permeability.

The 'two-shot' process has been largely superseded by the 'one-shot' process, which consists of mixing together prior to injection two chemicals whose gel time can be sufficiently delayed to allow full penetration of the soil before gel occurs. The extent of the delay can be accurately controlled by varying the proportions of the two chemicals. The extra time available for placing this grout allows wider spacing of the boreholes.

A typical 'one-shot' chemical grout consists of a liquid base which is further diluted with water before use and a liquid catalyst added at a rate to give a pre-determined gel time. The mixed solution has a viscosity that will penetrate fine sands. The resultant gel is permanent and is not washed out by running water; it gives a relatively high strength in sand, having a crushing strength ranging from 1 to 4 N/mm^2.

Chemical grouting has several advantages over other methods of grouting, some of which include:

- Stricter control of gel time, which can range from a few seconds to many hours.
- Economies in the boring of grout holes, since fewer holes are required.
- Greater penetration of the grout.
- Greater flexibility in grouting time.

Resin grouting
The term 'resin grout' is used for grouts which are formed by interaction of soluble materials. They are of low viscosity and are formed by adding a catalyst to a base solution. The difference between resin grouts and chemical grouts is one of viscosity: the former have a very low viscosity capable of penetrating fine sands in which silicate solutions are of little value.

Resin grouts include tannin-based grouts, phenol-formaldehyde and resorcinolformaldehyde. The type to be used depends to a certain extent on the chemical content of the ground-water. The chemical content of the ground-water may affect the setting of the particular grout.

Bituminous grouting
Bituminous solutions such as cut-back bitumen emulsion can be used as a suitable grouting medium: they can be injected into fine sands to form an impermeable barrier to water. Such solutions are suitable for forming cut-off walls beneath dams and similar structures but add no strength to the soil: for this reason they have no value in underpinning work.

Methods of injection
Almost all grouting work is executed by driving pipes or boring holes in the ground and pumping the grout solution through tubes at high pressure. The spacing of the holes will vary according to the type of grout used and the ground conditions encountered, but as a general guide the spacing ranges from 600 mm centres for the 'two-shot' process in sand to 10 m centres for cement grouting in rock.

The area to be grouted is first investigated to determine the required extent of grouting and this is followed by the calculation of a drilling pattern, which takes into account the size, spacing and depth of holes required. Holes or pipes are then sunk to the required depth by means of pneumatic tools, diamond drills or wash-boring, depending on site conditions; holes formed in alluvial soils are cased to prevent collapse.

The pressure at which the grout is pumped into the ground varies according to soil conditions and the reasons for grouting, but can range from 1 N/mm^2 for sands to 7 N/mm^2 for grouting fissures in rock. The most suitable pressure for grouting is difficult to determine in advance and tests may be necessary in-situ before the final choice is made.

3.3.6 Comparison of methods and costs
The choice of a method of ground water control will depend in the main on the type of soil involved and the depth to which the ground is to be excavated. For most shallow excavations where the sides of the excavation are battered or supported by timbering, normal pumping from sumps will suffice: where the soil is non-cohesive and the flow of water to a sump could create problems with soil movement, a wellpoint system can be employed. However, where the soil consists of coarse gravel and the flow of water is very heavy, some form of cut-off walling may be necessary before the wellpoint system can produce satisfactory results.

Where adjacent structures could be endangered by the excavation of soil or movement of water, special treatment will be required: this may be achieved by a permanent cut-off wall, such as a diaphragm wall, or by freezing the soil. The use of chemicals in the formation of cut-off walls will normally prove uneconomical because chemical grouting is 6 to 8 times more expensive than cement grout per cubic metre of soil treated. Since the soil particle size greatly influences the choice of method, reference should be made to a chart which identifies the various processes in conjunction with soil grading. Suitable charts have been produced by Glossop and Skempton which indicate the type of process or method of control to be used (Figure 3.37 – see also Table 3.2).

FIGURE 3.37
(a) *(Top) Ground-water lowering and compressed air*

FIGURE 3.37
(b) *(Bottom) Artificial cementing*

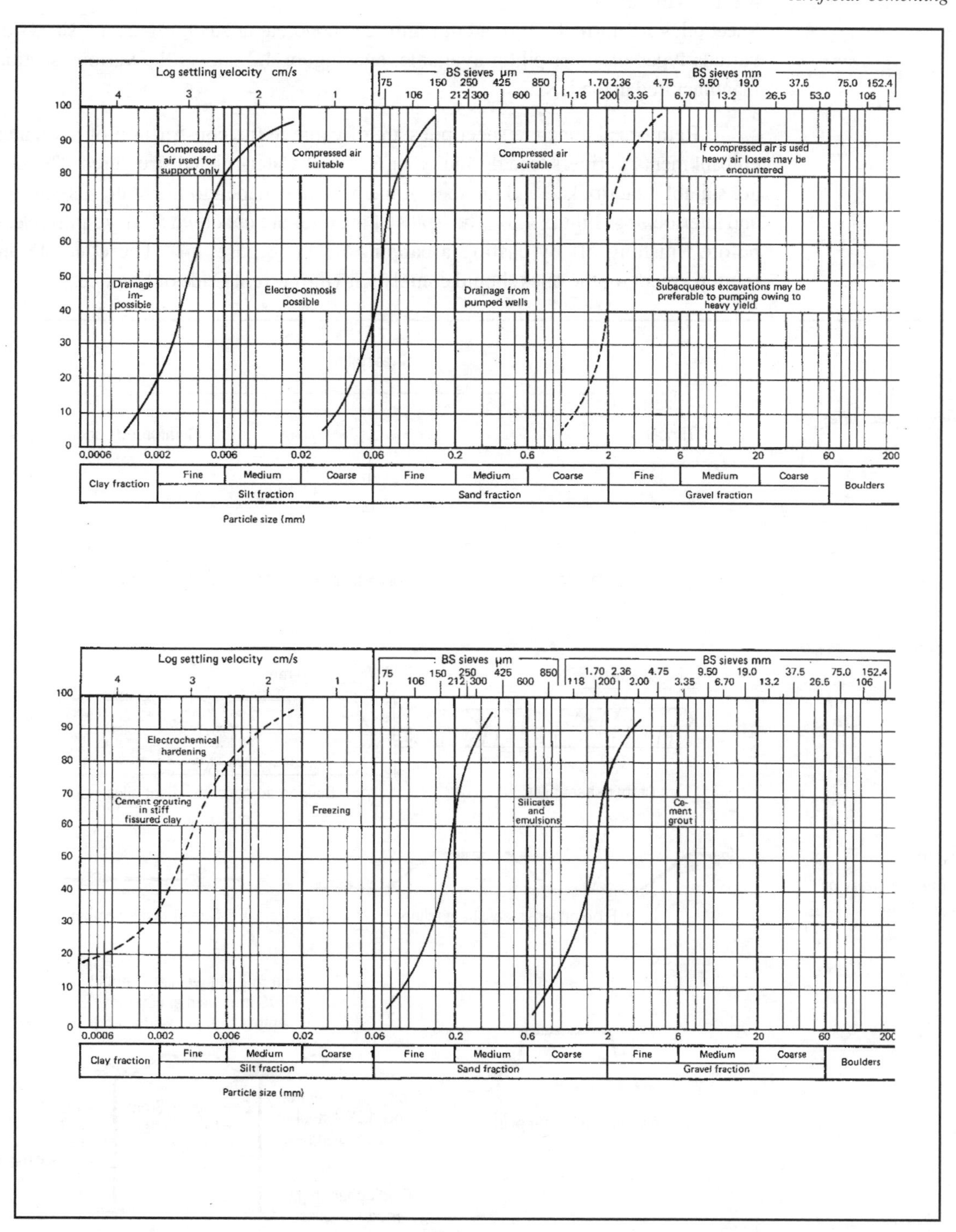

CHAPTER 4

Piling, Diaphragm and Retaining Wall Systems

4.1 SHEET PILING

4.1.1 Types of piles

Sheet piles are normally formed of reinforced concrete or steel; Figure 4.1 shows the various types of sheet piling available, and Figure 4.2 shows the various sections manufactured in the UK.

Reinforced concrete sheet piles are of value in the construction of permanent embankments to rivers, canals and other forms of water-related structures. The piles are suitably interlocked and the toes of the piles are shaped to facilitate easy driving and interlocking (Figure 4.3). The heads of the piles are cut down to the required level before finishing off by casting a capping beam (Figure 4.4). The concrete and reinforcement for the piles should comply with the current Code of Practice.

FIGURE 4.1
Types of sheet piles

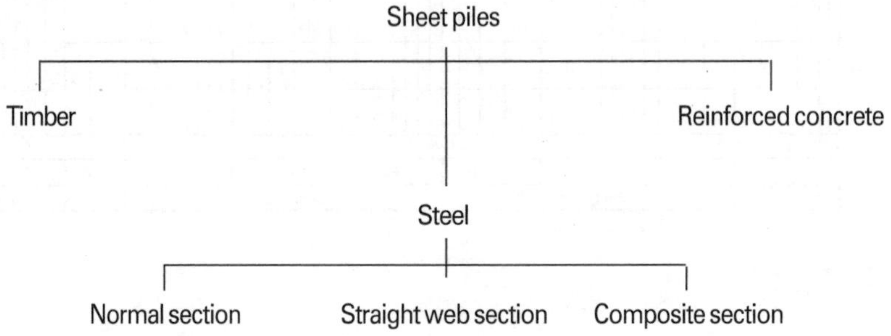

FIGURE 4.2
Various sections of sheet piling
(British Steel plc)

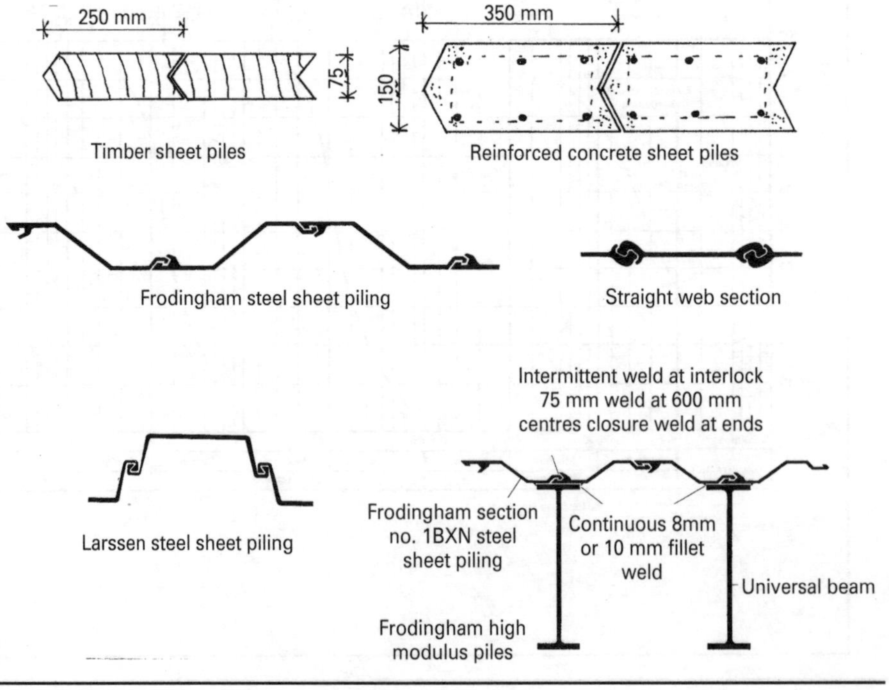

Steel sheet piling is the most common form of sheet piling and is used in both temporary and permanent works. It is used in such structures as cofferdams, retaining walls, river frontages, quays, wharves, dock and harbour works, land reclamation and sea defence works. It has an advantage over other forms of sheeting in that it has high structural strength combined with watertightness and can be easily driven into most types of ground. The sections (Figure 4.2) are interlocking and can be driven to depths which provide adequate cut-off to prevent piping (sub-surface boiling of soil due to water pressure) in waterlogged soils.

Steel sheet piles are available in four basic forms in the UK:

- Normal sections
- Straight web sections
- Box sections
- Composite sections or high modulus piles.

Normal section sheet piles include the well-known Larssen and Frodingham sheet piles. Larssen piles were named after an engineer who worked in Bremen at the turn of the century. He developed the principle of interlocking sheeting for temporary work. The sections were first rolled in one operation in the UK in 1929. The Frodingham sheet pile is the English name given to a section which was designed by Hoesch (a German company) as an alternative to the Larssen pile; it appeared in the UK in 1937. The sections are designed to provide the maximum strength at the lowest possible weight, consistent with good driving qualities.

The interlocking of steel sheet piles facilitates ease of pitching (positioning the pile ready for driving) and driving; it also results in a close-fitting joint which forms an effective water seal. In some cases, however, the joints may be sealed by brushing a sealant into the joints prior to pitching, the sealant expands to many times its original thickness, thereby forming a watertight joint.

A wide range of sheet pile sections is produced, as shown in Tables 4.1 and 4.2. These sections are available in various grades of steel, including copper-bearing steel for increased corrosion resistance. Tables 4.1 and 4.2 show the steel grades and their corresponding strengths. It will be seen from the Tables that the Larssen steel sheet pile

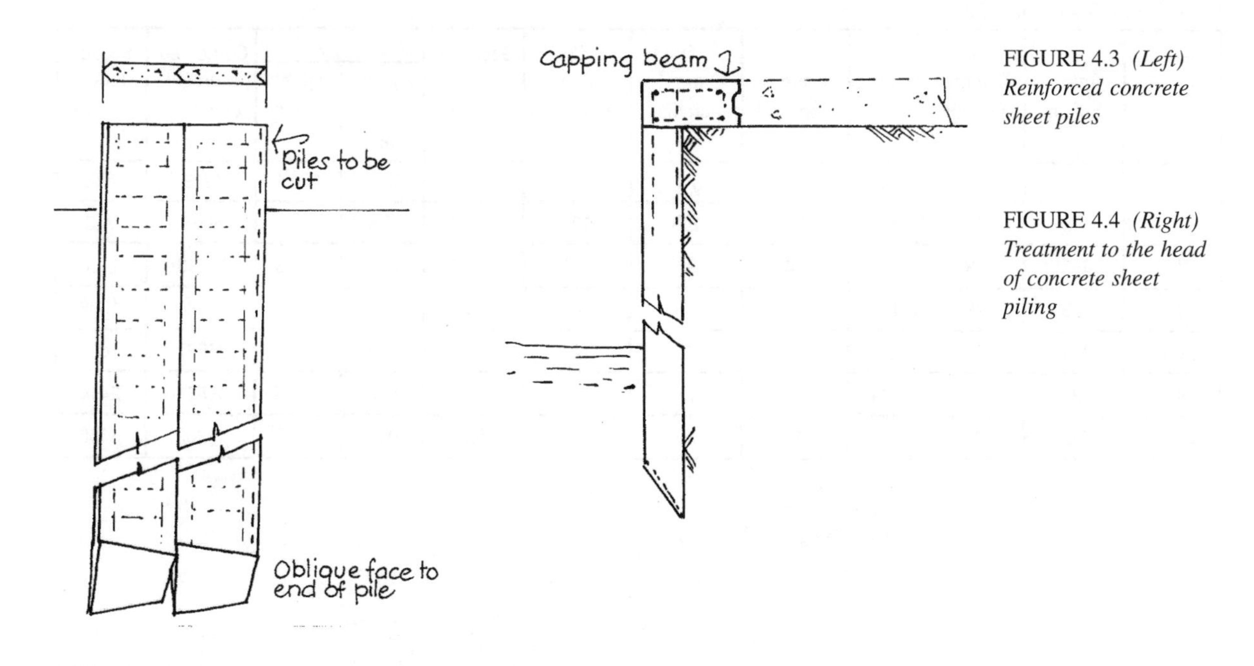

FIGURE 4.3 *(Left)* *Reinforced concrete sheet piles*

FIGURE 4.4 *(Right)* *Treatment to the head of concrete sheet piling*

TABLE 4.1 **LX and Larsson sheet piles** (British Steel plc)

Section	b mm (nominal)	h mm (nominal)	d mm	t mm (nominal)	f Flat of Pan mm	Sectional Area c m^2/m	Mass		Combined Moment of Inertia	Section Modulus cm^3/m
							kg per linear	kg/m^2 of wall		
LX8	600	310	8.2	8.0	250	116	54.6	91.0	12861	830
LX12	600	310	9.7	8.2	386	136	63.9	106.4	18723	1208
LX16	600	380	10.5	9.0	365	157	74.1	123.5	31175	1641
LX20	600	430	12.5	9.0	330	177	83.2	138.6	43478	2022
LX25	600	450	15.6	9.2	330	200	94.0	156.7	56824	2525
LX32	600	450	21.5	9.8	328	242	113.9	189.8	72028	3201
GSP2	400	200	10.5	8.6	265	153	48.0	120.0	8740	874
GSP3	400	250	13.0	8.6	271	191	60.0	150.0	16759	1340
GSP4	400	340	15.5	9.7	259	242	76.0	190.0	38737	2270
6 (122.0kg)	420	440	22.0	14.0	248	370	122.0	290.5	92452	4200
6 (131.0kg)	420	440	25.4	14.0	251	397	131.0	311.8	102861	4675
6 (138.7 kg)	420	440	28.6	14.0	251	421	138.7	330.2	111450	5066

TABLE 4.2 **FX and Frodingham sheet piles** (British Steel plc)

FX Steel Sheet Piling

Section	b mm (nominal)	h mm (nominal)	d mm	t mm (nominal)	f1 mm (nominal)	f2 mm nominal	Sectional Area c m^2/m of wall	Mass		Combined Moment of Inertia cm^4/m	Section Modulus cm^3/m
								kg per linear metre	kg/m^2 of wall		
FX13	675	300	9.5	9.5	127.7	154.5	137.9	73.1	108.3	19693	1313
FX18	675	380	9.5	9.5	135.4	160.3	147.6	78.2	115.9	34201	1800
FX26	675	430	13.2	12.2	127.1	146.6	194.9	103.3	153.0	55983	2603
FX36	675	460	18.0	14.0	149.0	158.3	244.8	129.7	192.2	82915	3605

Frodingham Steel Sheet Piling

Section	b mm (nominal)	h mm (nominal)	d mm	t mm (nominal)	f1 mm (nominal)	f2 mm nominal	Sectional Area c m^2/m of wall	Mass		Combined Moment of Inertia cm^4/m	Section Modulus cm^3/m
								kg per linear metre	kg/m^2 of wall		
1BXN	476	143	12.7	12.7	78	123	166.5	62.1	130.4	4919	688
1N	483	170	9.0	9.0	105	137	126.0	47.8	99.1	6048	713
2N	483	235	9.7	8.4	97	149	143.0	54.2	112.3	13513	1150
3N	483	283	11.7	8.9	8.9	145	175.0	66.2	137.1	23885	1688
3NA	483	305	9.7	9.5	96	146	165.0	62.6	129.8	25687	1690
4N	483	330	14.0	10.4	77	127	218.0	82.4	170.8	39831	2414
5	425	311	17.0	11.9	89	118	302.0	100.8	236.9	49262	3168

is the strongest available section, having a section modulus of 5066 cm^3/m for Section No 6, compared with a section modulus of 3168 cm^3/m for the Frodingham Section No 5. A series of corners, junction and closure piles is also available to facilitate piling to various plan shapes. Some engineers prefer the Larssen pile from a driving point of view, due to the uniform shape of the section.

Slinging holes are provided in both types of sheet pile; the Frodingham piles have a 32 mm diameter hole located 75 mm from the top of the pile, whereas the Larssen pile has the lifting hole located 150 mm from the top of the pile. Frodingham sheet piles are normally supplied interlocked in pairs, which saves time in handling and pitching; Larssen piles are normally supplied as single piles.

Straight web piling is used to construct cellular cofferdams (see Figure 4.5 and also Chapter 6, Section 6.1). Such piles are interlocked and driven to form cells which are then filled with gravel or broken rock. The outward pressure of the fill material develops high circumferential tensile forces in the piling. The Frodingham straight web pile is designed to resist these forces by virtue of the shape of the interlock. This 'crane-hook' shape gives a high tensile strength in the plane of the piling while at the same time permitting angular deviation between one pile and the next. The normal trough-shaped sections are not suitable for cofferdam work, where tensile forces have to be resisted, because the interlocks are not designed for tensile strength and therefore would deform and open out when subjected to stress. Junction piles are provided to facilitate the jointing of the cells (Figure 4.6).

FIGURE 4.5
Straight web cell jetty construction
(British Steel plc)

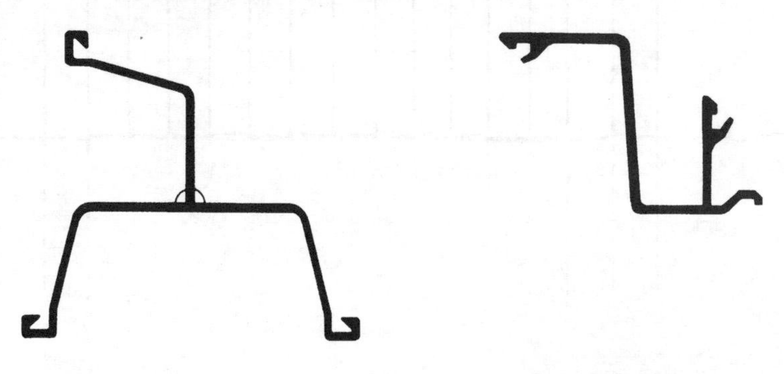

FIGURE 4.6
Typical junction piles
(British Steel plc)

Box piles are formed from two or more sheet pile sections welded together. They are used where local heavy loads are anticipated and can be positioned in a normal section pile wall so that the appearance is unaffected. Box piles can be used as individual units for open jetties and dolphins where bending moments are high. They can also be used as bearing piles for foundation work.

Composite sheet piling or high modulus section piling has been developed to support bending moments which are in excess of the capacity of normal sheet pile sections. This is of particular importance in waterfront protection where larger ships need increased wharf height beyond the limit provided by Larssen and Frodingham sections. The ability of high modulus section piles to support large bending moments and heavy axial loads simultaneously makes them suitable for quays carrying heavy cranes and for permanent load-bearing abutment walls. A typical composite or high modulus section pile is the Frodingham high modulus section (Figure 4.2), which consists of a double Frodingham section welded to one flange of a Universal beam.

Any of the larger Universal beams can be used together with any length of sheet pile to meet driving requirements and designed loading. The system produces a wall of identical units which can be produced in a range of sizes, giving optimum economy.

Lightweight trench sheeting is used for supporting the sides of trenches and excavations, cofferdams in shallow depths of water and small retaining walls. The sheeting is available in lap-jointed and interlocking sections (see Chapter 3, Figure 3.2) and is obtainable in standard lengths from 2 to 8 metres, in 0.5 increments.

4.1.2 Methods of driving

When piles are being driven they have a tendency to lean in the direction of driving; this tendency must therefore be restricted by some form of guide control. There are various ways in which piles may be guided during driving but the two principal methods in popular use are:

- Driving in panels
- Use of trestles and walings.

FIGURE 4.7
Driving sheet piles in panels

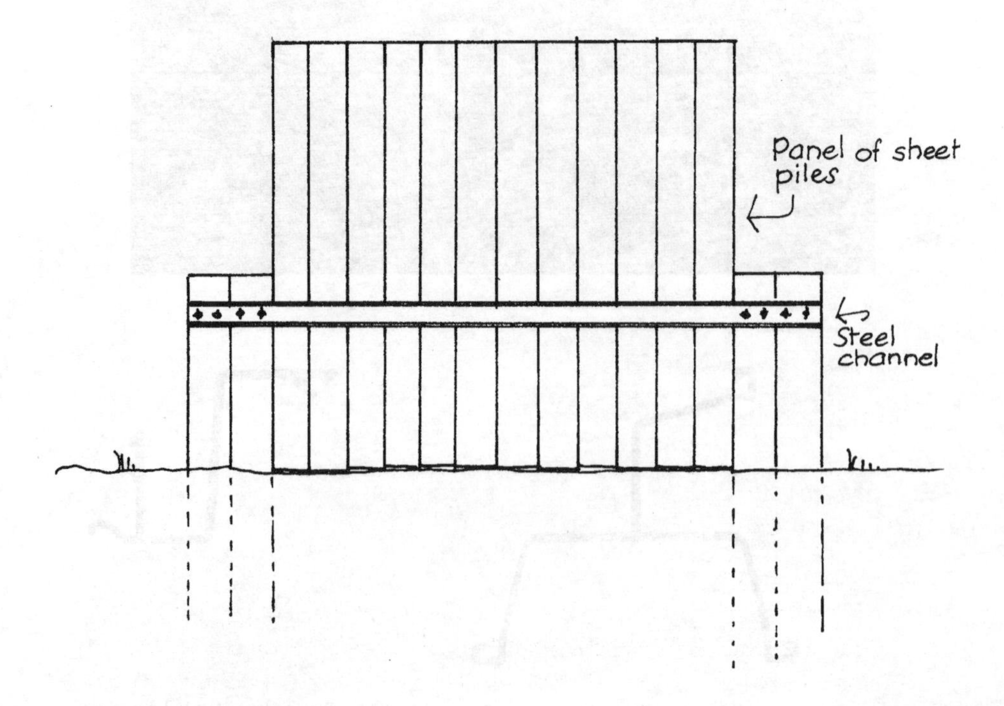

Driving in panels is a very satisfactory method of positioning sheet piles and ensuring that the piles do not creep out of verticality. Work is commenced by pitching and driving a pair of piles to part-penetration, care being taken to maintain correct position and verticality. A panel of piles, from six to twelve pairs, are then pitched and interlocked in position. The last pair of piles in the panel are driven and then guide walings are bolted between the first and last pair of piles (Figure 4.7) to support the panel during driving.

The remaining pairs of piles are then driven to their final position. The last pair of piles are left in a partly driven state to form the support of the next panel of piles. When driving long piles it is preferable to use a light hammer for the first stage, following up with a heavy hammer for final driving. This is best achieved by using two cranes, one handling the smaller hammer, the other following behind with the heavy hammer.

Trestles and walings are the common alternative method of supporting sheet piles during driving. The method involves the use of very heavy trestles which have to be moved and positioned by cranage, which in turn support long heavy walings. Trestles are normally constructed in steel (Figure 4.8). Where steel piles are to be driven through water, the guide walings may be supported on temporary timber piles (Figure 4.9) which form a heavy duty or light duty guide. With this second method of support the piles are often driven in pairs directly after pitching and there is a greater possibility of vertical creep than in panel driving. A slope of more than 1 in 300 may be difficult to close, although special tapered piles may be used. This method of support is therefore suitable for soft or loose ground conditions where control of verticality is satisfactory.

In addition to providing support for the driving of sheet piles, further support will be required in the form of spacer blocks (Figure 4.10) to obtain a good line of piling and to control the width of each pair of piles.

FIGURE 4.8
Trestle guide in steel (British Steel plc)

Driving in restricted headroom

Steel sheet piling can still be installed even when overhead clearance is limited; the procedure, however, is slow and costly, and consideration should be given to the possibility of removing the obstacle. However, two methods should be considered: the first being to drive the piles in two or more lengths, and the second to jack the piles into the ground. In the first case the piles may be plated or welded together as the sections are driven. The second method employs the reaction, say a bridge, as a resistance for hydraulically jacking the piles into the ground.

When sufficient headroom has been achieved by jacking, a normal hammer may then be employed. The pitching of piles in a restricted area can be facilitated by using travelling chain blocks suspended from steel beams.

Pile driving equipment

Pile driving equipment is also covered in Chapter 2, Section 2.4.3. For sheet piling this equipment falls into three basic categories:

- Percussion drivers – which includes diesel and air hammers
- Hydraulic drivers
- Vibratory drivers.

Percussion drivers can be used in any suitable ground including soft rock, which would require high frequency blows. These include hammers driven by steam, compressed air or diesel, the last being the most popular.

Hydraulic drivers work on the principle of pushing sheet piles into the ground by means of a hydraulic ram acting against a firm reaction. The reaction is provided by the skin friction of piles already partially driven. A fine example of this plant is the 'Still Worker' (Figure 4.11). Designed by Tosa, in Japan, the 'Still Worker' is a hydraulic

FIGURE 4.9
Guides for pile driving in water

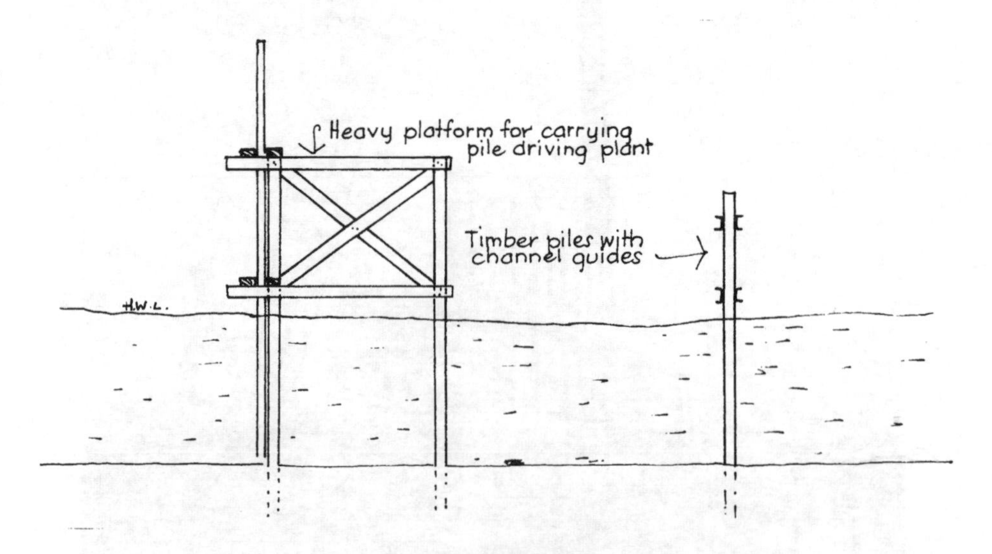

FIGURE 4.10
Guide blocks to prevent pile spread

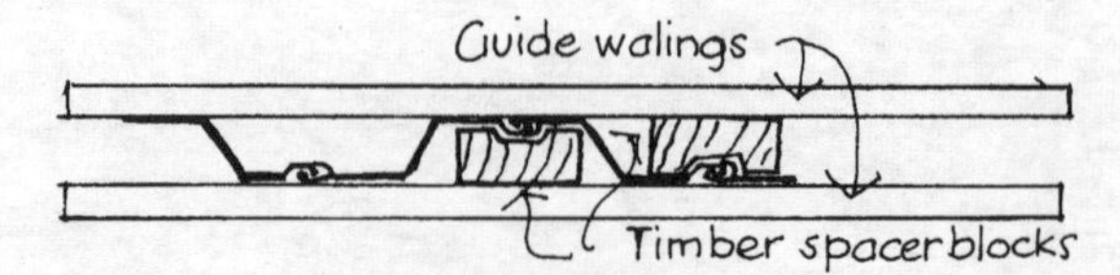

piling machine which installs and extracts sheet piles and H-piles. The hydraulic action virtually eliminates the noise and vibration commonly associate with the installation of piling. The machine is very compact and weighs between 3.7 and 10 tonnes, depending on the model. The machine can operate at ground level: it is lifted on to a 'reaction stand', a standard accessory which allows the machine to drive the first 3 or 4 sheet piles. After the installation of the first three piles the 'Still Worker' supports itself on the last driven pile and then moves its travel carriage off the 'reaction stand' to travel independently on the installed piles – see sketch in Figure 4.12 for sequence of operations.

Alternatively, the first four piles can be driven by some other method and the Still Worker lifted on to the four piles to start the hydraulic sequence of installation. The sheets are fed into the chuck of the Still Worker and the chuck grips the sides of the sheet and pushes it down approximately 1 metre into the soil. The chuck then releases its grip, slides up the sheet and then grips the sheet again, pushing it down until the process is complete. The TGM-130 model is the standard model for driving 500–600 mm wide sheet piles. This model has a pushing/pull out force of 130 tonnes. It can also be used for driving H-piles.

Vibratory drivers (Figure 4.13), are very suitable for the initial driving of sheet piles, H-piles and tubular piles but must be avoided in heavy clays since the clay tends to dampen the vibrations. The unit vibrates the piles, which reduces the resistance of the ground around them, allowing the piles to pass into the ground. The vibrations can be either low or high frequency depending on the nature of the soil – high frequency can reach 2300 vibrations per minute. The vibrators are driven by diesel-hydraulic power units, which are specially designed for the purpose. The vibrating units weigh between 1400 kg and 10 000 kg. The units are environmentally friendly in that the vibrating technique is not noisy. Furthermore, there is less likelihood of damage to adjacent property compared with other driving methods, especially when high frequency units are employed.

FIGURE 4.11
'Still Worker'
(Watson and Hillhouse (Plant Hire) Limited)

Extraction of piles

Piles that have to be extracted should have greased joints during driving. The speed and method of extraction will depend upon many related factors, such as:

- The section and length of pile.
- The length of time the piles have been in the ground.
- Soil and water conditions.
- The method of driving and weight of hammer used.

The types of extractor available are:

- Inverted double-acting hammer.
- Heavy duty extractor.
- Vibrators.

Further reference should be made to Section 2.4.4. where extracting plant is discussed. Figure 4.14 shows a hydraulic vibrator being used to extract H-piles.

4.1.3 Corrosion and protection

Any exposed steel structure may deteriorate as a result of the formation of rust, and sheet steel is no exception. Corrosion is of particular importance in the case of permanent steel sheeting because some parts are embedded in the ground and become therefore inaccessible for painting and maintenance. Research over a period of years has indicated that the useful life of a steel-pile retaining wall can be based on an average reduction in thickness of about 0.09 mm per year in marine conditions, ie one side soil and the other side salt water. Where the piling is subject to marine water on both sides, the corrosion may rise to 0.15 mm per year. Where the back face of the piling is in contact with the ground and the other side open to the atmosphere, the corrosion will be around 0.05 mm per year.

In some situations such corrosion rates will be acceptable, but in other situations they will be seen as significant. Where such rates of corrosion are seen as significant, the following methods of protection should be considered:

FIGURE 4.12
'Still Worker' – sequence of operations (Watson and Hillhouse (Plant Hire) Limited)

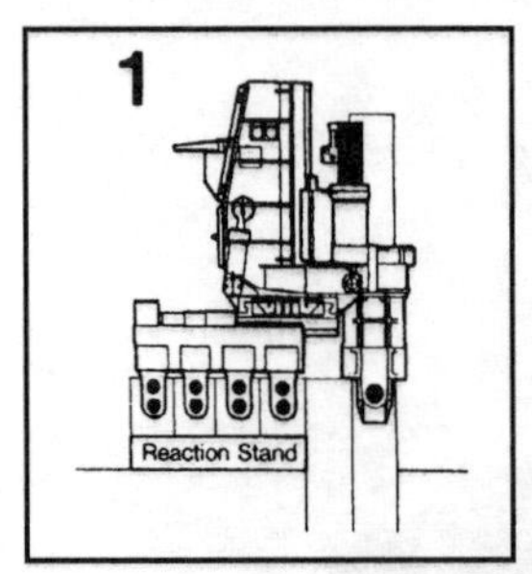

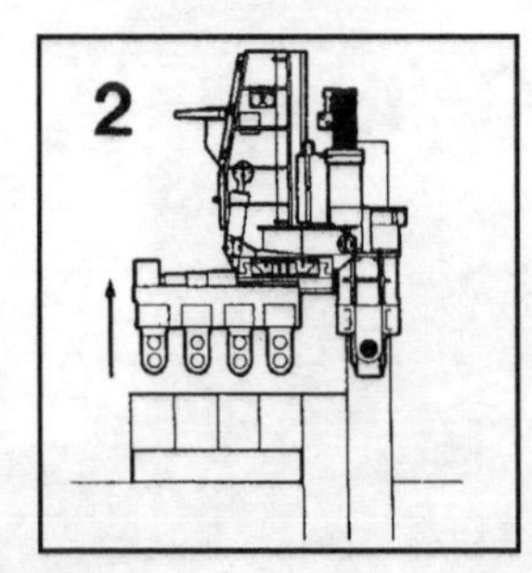

1 The Still Worker is set on the reaction stand for the installation of the first two sheet piles.

2 The Still Worker moves by elevating its travel carriage while supporting itself on the last installed pile.

3 The travel carriage then slides forward.

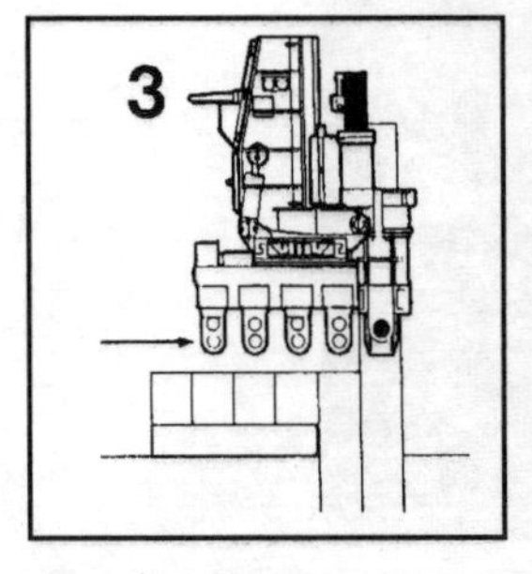

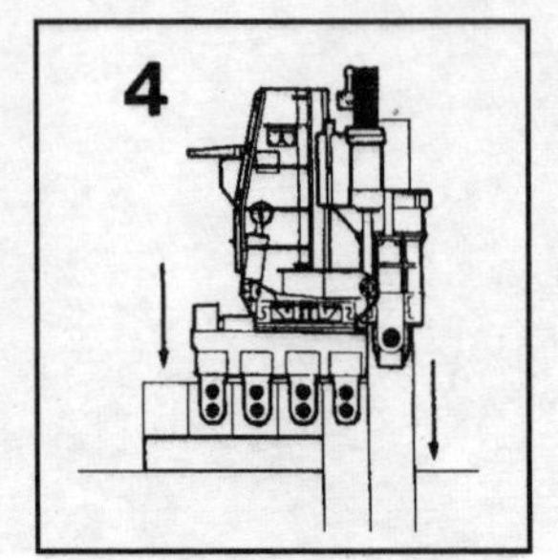

4 The travel carriage lowers itself and drops onto the installed sheet piles and continues its hydraulic installation process. After the third (or fourth) pile is driven, the Still Worker moves off the reaction stands to travel independently on the piles.

- Use of heavier sections of piling.
- Use of high yield steel.
- Application of organic coatings.
- Cathodic protection.
- Concrete encasement.
- Cathodic protection.

The effective life of steel sheet piling can be increased simply by using heavier sections of piles, thereby allowing for reduced strength by corrosion. An alternative is to use high-yield steel. Whilst high-yield steel corrodes at the same rate as mild steel, it can allow a 30% loss of thickness to occur without detriment to the strength factor.

Organic coatings include paint and painting systems. Steel piles are normally coated under shop conditions. The paints are applied to the cleaned surfaces by airless spraying and then left to dry or cure, which is a rapid process, allowing the thickness of the protection to be applied in a few coats. Sheet piles may be primed and finished with

FIGURE 4.13
Hydraulic Vibrator
(Watson and Hillhouse (Plant Hire) Limited)

FIGURE 4.14
Hydraulic Vibrator being used to extract H-piles
(Watson and Hillhouse (Plant Hire) Limited)

coal-tar pitch solutions, to BS 1070; but these coating are thin, up to 50 microns, soft and easily damaged. More substantial coatings have been developed by British Steel:

- Tar Vinyl – an aromatic pitch modified with suitable vinyl resins, which can be applied up to 150 microns in a single coat;
- High-build, Isocyanate Cured Epoxy Pitch – coal tar pitch modified with epoxy resin, which can be applied up to 400 microns in a single coat;
- Coloured Vinyl Ether Finish, applied up to 150 microns in a single coat. This material is available in a range of colours for piling which is installed as a permanent feature.

The disadvantage of surface treatments is that they are subject to damage by stone scouring during driving. The advantage is that they can be readily renewed in marine works where the water level can be lowered for maintenance purposes.

Concrete encasement will depend on the extent and nature of the work. It is a useful protection technique in marine environments where the concrete capping beam can be extended down below the splash zone on the face of the sheet piles. It is important to ensure that the concrete covers the whole height of the splash zone, otherwise there is a danger of increased corrosion at the concrete-steel interface.

Cathodic protection (BS 7361) is based on the principle that all metals have electro-chemical potential and that material in the electro-chemical series can be protected by materials that are higher in the series (ie towards the anode end of the scale). The material used for protecting steel sheeting, known as the anode, must be high in the electro-chemical series; magnesium is often used. The material is connected electrically to the piling, which acts as a cathode, and the current escapes via the anode into the soil. This makes the whole structure cathodic and steel is preserved while the magnesium is sacrificially corroded away. The anode must be replaced from time to time.

Alternatively, a power-supplied system may be used, in which the anode takes the form of lumps of carbon or scrap iron. A DC current is passed through the anode to the cathode by means of a generator, or AC transformer-rectifier, causing the sacrificial wastage of the anode. The cathodic system is suitable for protecting steel sheet piling below water level but it does not prevent atmospheric corrosion. Therefore piling above low water level must be protected by some other means.

4.1.4 Selection and use of sheet piling

This will depend upon factors such as:

- Type of work – permanent or temporary.
- Site conditions (eg headroom, type of soil).
- The depth to which piles must be driven.
- The bending moments involved.
- Nature of the structure (eg circular cofferdams or straight piling).
- The type of protection required, where necessary.

Most of these factors can be solved by simple reference to standard tables which include the properties of sheet piling, but in the case of reinforced concrete sheet piling the size of the pile will have to be calculated from first principles. Where steel sheet piling is used in permanent structures, such as pump houses, the interlocks may require sealing. This can be achieved by a propriety treatment applied to the joints before driving.

4.2 BEARING PILES

4.2.1 General considerations

Piles which wholly support vertical loads are termed 'bearing' piles, while those that restrain loads may be termed 'anchor' piles. The term 'bearing' pile is used for driven piles as well as bored piles and serves to distinguish all such piles from sheet piling. This section deals only with bearing piles since the principles of piling used for other functions can be easily determined from the data given. Piling is used in many different circumstances, some of which are:

- The support of a structure which would otherwise over-stress the allowable bearing capacity of the soil at normal foundation depths.
- The unpredictable settlement of underlying strata, where rock is to be found at a depth which can be reached economically.
- Differential settlement likely to be caused by strata changes across a site or by the close proximity of other structures.
- The seasonal shrinkage and swelling of the upper layers of soil.
- Building over water or waterlogged ground.
- The resistance of forces, such as uplift or overturning, which may be created by wind pressure or cantilever structure.
- The resistance of lateral forces, such as strata slip or other soil movement, which may cause instability of the structure.
- The underpinning of existing structures.

Piles must be driven, bored or screwed to the desired depth without damage to the pile shaft; this applies particularly to driven piles where the driving stresses may exceed the permissible working stress. The stresses during pitching and handling should not exceed the safe bending stress.

Loads should be applied concentrically with the axis of the pile on the centre of gravity of a pile group. Some difficulties may arise from the inaccuracy of pile positions, particularly in the case of isolated piles, and allowances must be made in the design since such conditions may result in eccentric loading of the pile. The usual tolerance, on plan, for all types of piles is 75 mm. Where the eccentric loading is of a significant nature, the pile cap should be restrained from lateral or rotational movement; the restraint should also be sufficient to resist the eccentric loadings.

Where vertical piles are subjected to substantial horizontal forces, the top stratum of ground should be able to resist the stress without permitting excessive lateral movement. Where the top stratum is incapable of providing such resistance, the piles must be connected by horizontal beams; if this measure is insufficient, raking piles should be used.

Preliminary work

Investigation of the substrata should be carried out as outlined in Chapter 1. The borings, supplemented by insitu tests where appropriate, should reach depths which allow the exploration or testing of soil both around and beneath the toe of the pile. Samples of the soil at these lower levels should be tested for strength, compressibility and other characteristics which will assist in determining the length and spacing of the piles.

Certain soil conditions do not permit adequate point bearing at an economic depth; this will involve the determination of skin friction at various levels in the ground before an economical design can be achieved. With the pressure of ground water it will be necessary to establish the source and the water table gradients between boreholes to determine its effect, if any, on soil stability.

Where other structures are in the proximity of the proposed pile foundations it will be necessary to survey the structures in question, since the choice of pile may be influenced by the effects installation may have on the adjacent structure. The influence of noise and ground vibrations would have to be given due consideration from an environmental point of view.

Preliminary piles

Preliminary piles for determining the ultimate bearing capacity should be installed near the bore holes; the piling data can then be studied and compared with the site investigation data. The preliminary piles should be of the same materials and dimensions as the working piles, to ensure comparable behaviour under load. The testing for load-bearing capacity and other characteristics are covered in Section 4.2.5.

Where driven piles are to be used, a special preliminary pile, designed to withstand hard driving, is employed. A record is kept, showing the number of blows per unit of linear measurement of penetration into the ground. This record will show the variations of soil resistance at various depths. Driving will continue until an acceptable 'set' has been reached (the 'set' is the desired final penetration of the pile, normally not more than 5 mm per blow of the driving hammer). The soil is then allowed to recover from the dissipation of excess pore water pressure. The time for recovery will depend on the nature of the soil and will vary from a few hours in non-cohesive soils to two days for clays. Re-driving is then started and continued until the resistance is similar to that previously achieved. This information, together with the specialised knowledge of ground conditions, will provide the engineer with data to complete the design and specification of the piling installation.

4.2.2 Types of bearing piles

The main classification of bearing piles is related to their effect on the soil. There are two main types: **displacement** piles and **replacement** piles (sometimes referred to as 'non-displacement' piles – BS 8004).

A **displacement pile** is either driven, jacked, vibrated or screwed into the ground; this action displaces the soil outwards and downwards but material is not actually removed. There are two types of displacement pile: large displacement piles, which include all solid driven piles, and small displacement piles, in which very little soil is displaced. This would include the screwed piles and H-piles.

In the case of the screwed pile it may be argued that very little soil is displaced. This is true, but nevertheless some soil is displaced and in principle it should be found in this category.

The **replacement pile** consists of forming a hole in the ground, by any of the various methods, and replacing the spoil with concrete. Classification of pile types is shown in Figure 4.15 in the form of a family tree.

Displacement piles

Displacement piles (Figure 4.15) may be sub-divided into two groups:

- Driven cast-in-place.
- Preformed.

Driven cast-in-place piles are of two types, the first having a permanent concrete or steel casing and the second without any form of permanent casing. In both cases a tube, closed at the bottom with a plug or shoe, is driven into the ground to the required set or depth. A cast-in-place pile is then formed inside the tube. Where the tube is permanent, it may be formed from a series of concrete shells about 1 metre long, adjusted for length by simply adding or subtracting shells – see Figure 4.16c. Alternatively, steel tubes may be used which can be extended by welding or shortened by cutting tools. Temporary casings are formed in steel tube which are withdrawn either during or after the casting of the pile, depending on the particular system employed.

Preformed piles are prepared from timber, concrete or steel to the design requirements before driving. Care must be taken during driving to prevent damage to the pile by stresses, exerted by the hammer, which may be in excess of the working stresses of the pile. A further problem with this form of pile is the calculation of length. The operation of lengthening piles is very costly, while the alternative of shortening of over-length piles may prove laborious and costly in both time and materials. The handling of long concrete piles must be carefully supervised since incorrect handling may result in exceeding the bending stresses, causing damage to the pile. BS 8110 covers the requirements of the provision of steel reinforcement used to resist stresses due to lifting.

Replacement piles may be classified as:

- Supported, or
- Unsupported.

In both cases a hole is formed in the ground by some form of cutting or boring tool and then filled with reinforced concrete. The unsupported hole will normally require a short tube at the top to prevent debris from falling into the concrete during placing. Support to holes may be provided by means of a medium or heavy sectional casing, screwed together as boring proceeds, or by a head of drilling mud (usually bentonite suspension). The concrete is placed in the hole by means of a tremie pipe to prevent segregation of material and pollution or weakening due to mixing with the drilling mud. In all circumstances the lower end of the tremie pipe should be placed well into the freshly placed concrete.

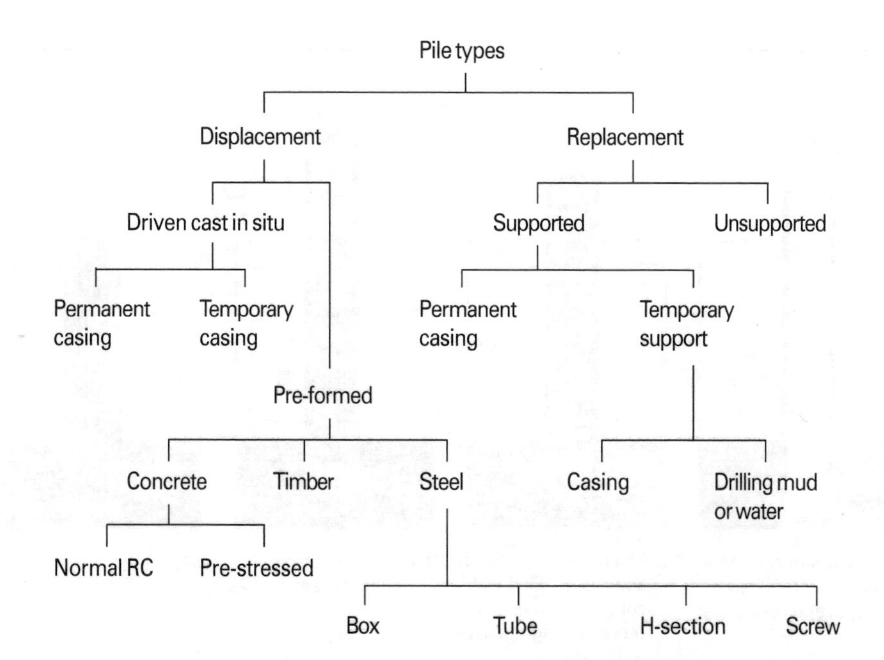

FIGURE 4.15
Various types of bearing pile

Piles 600 mm or more in diameter are commonly known as 'large diameter piles', because their construction always requires the use of large heavy plant. Large diameter piles may have their bearing capacity increased substantially by under-reaming the shaft at the base. This is achieved by an expanding cutting tool which expands and cuts a conical shaped base up to three times the diameter of the main shaft.

The various pile types shown in Figure 4.15 are now discussed in detail and various proprietary types are shown by means of sketches or photographs.

Driven cast-in-place piles

This type of pile is formed by driving a casing into the ground and filling the casing with concrete. Figure 4.16(a), (b) and (c) shows the principle employed in achieving both forms, ie permanent casing and temporary casing. In the first diagram a heavy steel casing is fitted with an expendable steel shoe and driven into the ground until the required depth or resistance is achieved.

On reaching the required depth, the casing – a cage of reinforcement – is placed and filled with concrete. The casing is vibrated as it is withdrawn, ensuring that the concrete is consolidated into the space occupied by the casing. In some systems the casing is partially withdrawn and then re-driven; this consolidates the concrete and forms a keyed surface at the soil face. Most cast-in-place piles are constructed with high slump self-compacting concrete. The permanent casing is usually formed with reinforced concrete shells, which are often reinforced with glass fibre rather than steel.

In the shell piling system the shells are threaded over a steel mandrel and bear on a precast concrete driving shoe. Alignment of shells is maintained by steel bands at each joint, the bands being coated internally with sealing compound to provide waterproof joints. The complete assembly is driven by a heavy drop hammer weighing up to 8 tonnes, which transmits the stress through the driving head to the mandrel, which in turn drives the shoe. The stress on the actual shells is just sufficient to push them into the hole made by the shoe. The mandrel can be extended and further concrete shells added until the desired depth has been reached. On completion of driving, the mandrel and spare shells are removed and the reinforcement cage and concrete are then placed. The plant used consists of an excavator fitted with a special crane head and special leaders (see Chapter 2.4 and Figure 4.16(c)). The main advantages of this type of pile are:

FIGURE 4.16(a)
Driven cast-in-place pile – temporary casing
(Keller Foundations)

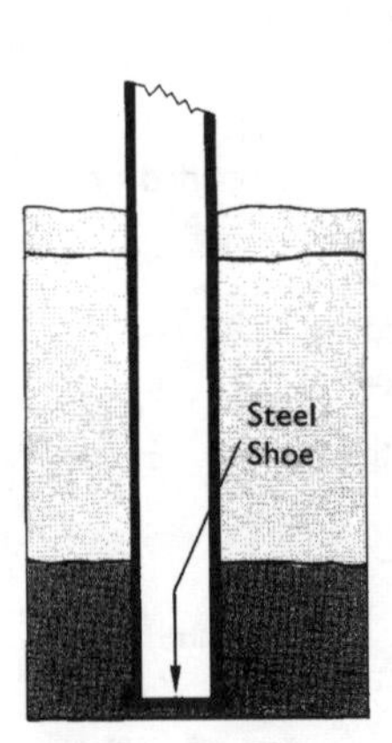

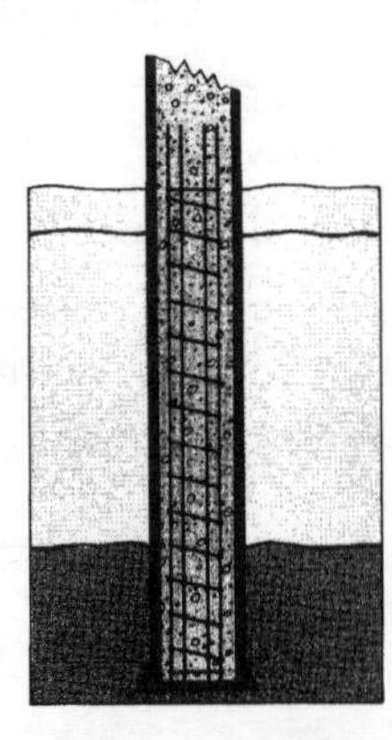

2 Reinforcement placed and casing filled with concrete to the required height above ground level.

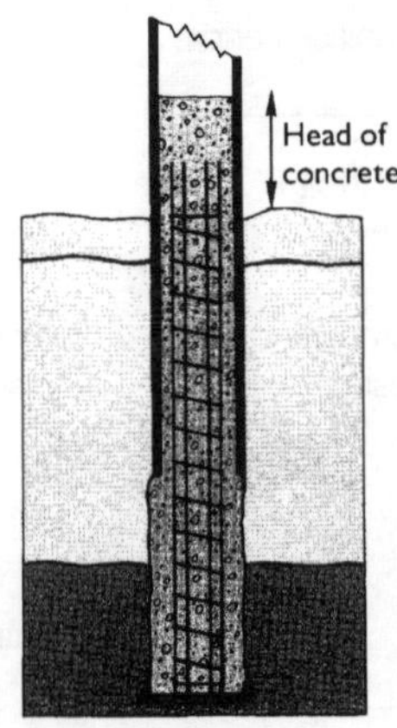

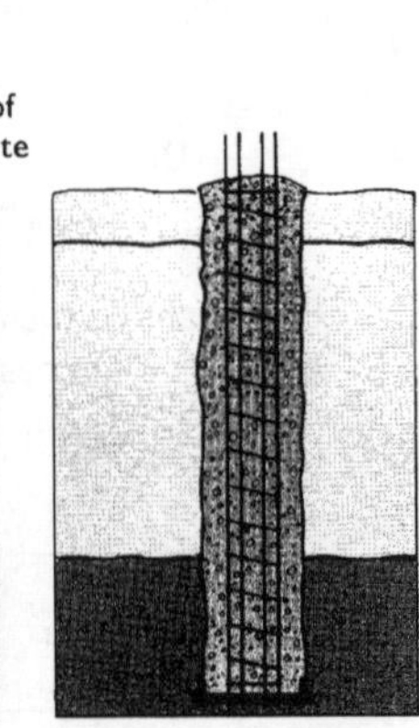

3 Casing vibrated during extraction

- Easily modified in length.
- Continuous cross-section can be maintained.
- Cast-in-place core is not subject to driving stresses.

Disadvantages include:

- Vibration and noise during driving.
- Heavy equipment and plant for driving makes it unsuitable for small sites.
- High skin friction on concrete shells can result in crushed shells, owing to increase in driving stress.

The cast-in-place core may be reinforced over the whole length of the pile, over part of the length, or simply provided with short splice bars at the top for bonding into beams or pile caps. The amount of reinforcement will depend on whether the pile is used to resist tensile or bending forces, the possibility of ground movement, and the type and form of foundation.

When the core is cast in a tube that is to be withdrawn, care must be taken not to damage freshly poured piles by driving new piles in the vicinity.

Pre-formed piles
These may be constructed using:

- Timber
- Concrete or
- Steel.

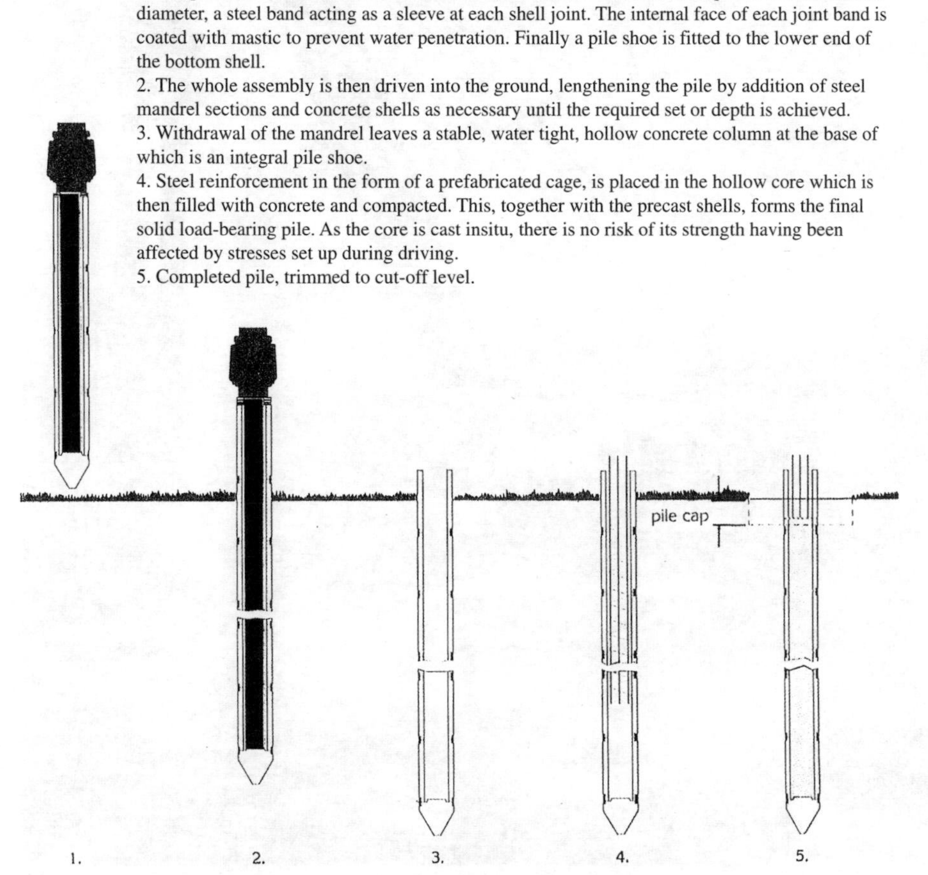

1. The pile shells are threaded on to a steel mandrel whose diameter corresponds to their internal diameter, a steel band acting as a sleeve at each shell joint. The internal face of each joint band is coated with mastic to prevent water penetration. Finally a pile shoe is fitted to the lower end of the bottom shell.
2. The whole assembly is then driven into the ground, lengthening the pile by addition of steel mandrel sections and concrete shells as necessary until the required set or depth is achieved.
3. Withdrawal of the mandrel leaves a stable, water tight, hollow concrete column at the base of which is an integral pile shoe.
4. Steel reinforcement in the form of a prefabricated cage, is placed in the hollow core which is then filled with concrete and compacted. This, together with the precast shells, forms the final solid load-bearing pile. As the core is cast insitu, there is no risk of its strength having been affected by stresses set up during driving.
5. Completed pile, trimmed to cut-off level.
pile cap
1.
2.
3.
4.
5.

FIGURE 4.16(b)
Driven cast-in-place pile – permanent casing
(Westpile Limited)

- **Timber pre-formed piles**

Timber piles are not used for permanent structural support in the UK but they are frequently used for supporting temporary platforms, such as gantries for river and maritime works, and for the support of falsework for in-situ bridges over rivers. The most common timber used is Douglas Fir, which is available in sections up to 400 mm square and 15 metres long. Pitch pine is also used; this is available in sections up to 500 mm square and also 15 metres long. Where timber piles are required for permanent works, the timber must be highly resistant to rot, and in the case of marine works, resistant to marine borers.

Before driving, a timber pile should be fitted with a steel or iron ring at the head (Figure 4.17a) to prevent 'brooming' (ie crushing and spreading of the timber) and splitting. The toe of the pile should also be protected by an iron shoe (Figure 4.17b) unless the driving is wholly in soft ground. The pile is guided by a pile frame and driven by standard equipment such as a drop hammer, which should be equal in weight to the pile being driven, for hard conditions, or half the pile weight in soft conditions. The splicing of timber piles can be achieved by using steel channels, plates or purpose-made box sleeve sections, the butt ends of the piles being accurately squared off.

FIGURE 4.16(c)
Driving equipment for West shell piles
(Westpile Limited)

- **Concrete pre-formed piles**

These are further sub-divided into:

  - Normal reinforced concrete and
  - Pre-stressed concrete.

In both cases the normal method of driving is achieved by heavy hammer blows while the pile is guided by a leader or pile frame. The piles may be driven by any type of hammer, provided that penetration to the prescribed depth is achieved without damage to the pile. Failure due to excessive compression occurs mainly at the head of the pile and this can be eliminated by the correct head cushion and weight of hammer used.

The weight or power of the hammer should be sufficient to ensure the required 'set', which will be a final penetration of between 5 mm and 25 mm per blow, depending on the soil. The size of the hammer will depend on whether the pile is to be driven to a given resistance or depth, but it should not be less than half the weight of the pile. Protection of the head of the pile is essential: this is normally achieved by means of a resilient packing, sand, or a plastic disc, which is held in position by a steel helmet.

Jetting or pre-drilling for pre-formed piles. Jetting may be employed in the placing of pre-formed piles when the frictional resistance at the sides of the pile or at the toe creates very hard driving conditions. Central jet pipes have proved to be more efficient than the exterior fixed pipes. However, jetting is effective only in non-cohesive soils, where the water can readily displace the material and dissipate without creating problems to adjacent structures or piling plant; provision for leading away any water that rises to the surface may be necessary. The pile is suspended from the pile leaders and lowered gently into the ground by means of the pile winch, the crane leaders controlling vertical accuracy. Jetting should be stopped when the pile toe is within one metre of the estimated final position, it is then driven to the required depth or set.

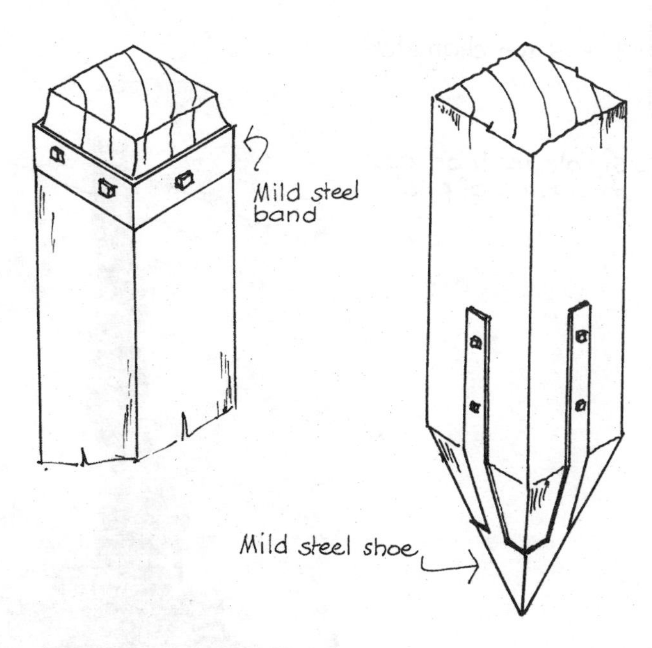

FIGURE 4.17
Treatment at the head and toe of timber piles

Pre-drilling. For very dense soil conditions where conventional driving would set up unacceptable noise or vibration, the ground can be pre-drilled. This is achieved by using a continuous flight auger attached to a hydraulic crane. The soil can either be loosened and left in position, which is generally the case when driving sheet piles, or extracted in the case of solid piles – Figure 4.18 indicates the jetting and pre-drilling principles.

All pre-formed piles should be driven to an accuracy of 1 in 75 when vertical, or 1 in 25 for a specified batter. Where this is not achieved the pile cap may have to be redesigned to cope with eccentricity: at the discretion of the engineer some piles may have to be replaced or supplemented by additional piles.

Stripping and lengthening concrete piles. When pre-formed piles have been driven to the required depth, the concrete at the head of the piles is stripped to a level that allows a 50 mm to 75 mm projection into the pile cap. The reinforcement is bent down into position within the pile cap and bonded with the cap reinforcement. Before stripping commences, a check should be made on the level of the pile to ensure that it has not risen; piles are subject to rising as a result of ground heave or the driving of adjacent piles. Risen piles must be re-driven to the original depth or resistance. Piles may be lengthened by welding extra reinforcing bars on to the newly stripped reinforcement; the pile head should be stripped to expose the original reinforcement for a length of 200 mm to prevent spalling of the concrete during welding operations.

If the extension of the reinforcement cannot be achieved by butt welding because of site conditions, it may be extended by overlapping the steel. For this the reinforcement

FIGURE 4.18
Methods of dealing with hard ground

(a) Jetting method

(b) Pre-drilling method (Dew Group Limited)

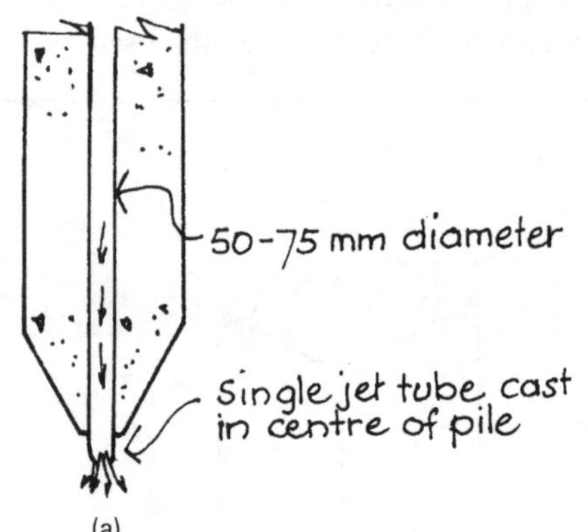

at the head of the pile should be exposed and lapped for a distance of at least 40 times the bar diameter to make a satisfactory joint. Alternatively, special couplers are available to join the reinforcement.

Other methods of lengthening piles may be achieved by one of the following:

- Using a mild steel splicing sleeve which fits over the pile and receives another precast pile for extension; the sleeve should be of a length equal to four times the pile width.
- Drilling the head of the pile and connecting the extension pile by means of dowels and epoxy resin.
- Using a modular piling system.

Modular concrete piles. The problem of lengthening pre-formed concrete piles has led to the development of precast modular piling. This system of piling consists of standard interchangeable precast piles which have a unique jointing system. The piles, which vary in length from 2.5 m to 10 m, incorporate steel connections at each end of the pile for simple assembly. One such system is produced by Westpile Limited, known as the 'Hardrive' precast modular pile (Figure 4.19), which employs a central aligning sleeve and four locking pins providing a flexural strength equal to that of the pile (Figure 4.20). Any two sections can be quickly locked together to allow driving to continue with a minimum of delay (Figure 4.21).

FIGURE 4.19
'Hardrive' piles being driven
(Westpile Limited)

FIGURE 4.20
The Hardrive standard steel joint. Four H-section pins are used in the joint to lock the pile sections together (Westpile Limited)

FIGURE 4.21
Two sections of 'Hardrive' piles being locked together (Westpile Limited)

- **Steel pre-formed piles**

Steel pre-formed piles can be formed in a wide variety of sections and can be adapted to suit almost any ground condition. The most common types of steel bearing pile are:

- Box piles
- Tubular piles
- H-section piles.

Box piles are formed by welding sheet piles together or by producing purpose-made sections. Two basic types are available in the UK, namely:

- Larssen box piles
- Frodingham box piles.

Larssen box piles are formed by welding together two sheet pile sections (Figure 4.22a) with either continuous or intermittent welds.

Frodingham box piles are available in various different sections; standard box piles, plated box piles (Figure 4.22b), and double box piles. The standard box piles are symmetrical in section and of constant wall thickness.

Box piles in general are driven open-ended. Soil displacement is small in the upper strata but the open end usually gets plugged with soil, resulting in high displacement. Shoes or plates can be provided for all box piles; and may be used to advantage when maximum resistance from a soft stratum is required or where ground heave is acceptable. The head of the pile is protected by a helmet, without a cushion packing between pile and helmet but including a dolly or packing between helmet and hammer. Any type of hammer may be used for driving.

Tubular piles are very similar to box piles in principle. They are either straight seam welded or formed by the Driam process. The Driam process consists of welding a plate which has been formed into a continuous helix. Straight seam welding is achieved in two passes, one inside the tube, the other outside, which ensures full penetration of the weld through the thickness of the plate. Since the process is continuous, the length of the tube is limited only by transportation. Outside diameters vary from 508 mm to 2134 mm and require driving equipment of appropriate diameters and weights.

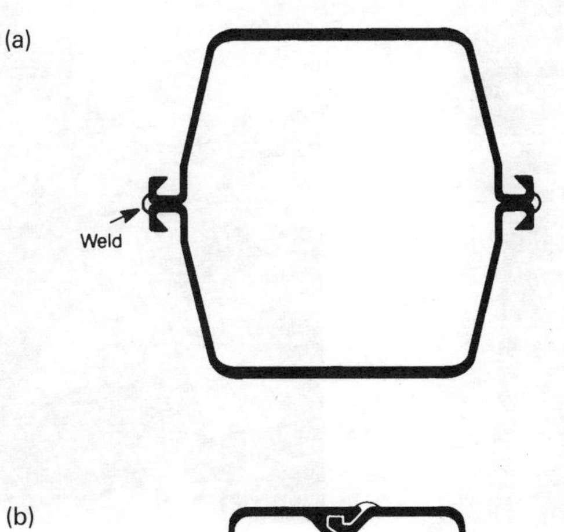

FIGURE 4.22
Types of box piles

(a) Larssen box pile

(b) Frodingham box pile

(British Steel plc)

Tubular piles are generally driven open-ended unless obstructions, such as cobbles, are likely to be encountered. A plug of soil enters the tube in the early stages of driving, which cause the piles to act as a displacement pile but without creating soil heave. Alternatively the tubes can be fitted with a welded flat plate shoe, but this may create problems with soil heave. Tubular steel piles are not normally filled with concrete since there is no real advantage in terms of load-bearing capacity.

Tubular piles and steel casings may be vibrated into the ground. Figure 4.23 shows both driving and temporary support for steel tube piles. Main advantages are:

- Very high loads are possible.
- Long piles may be installed by jointing.
- Installation is easy.
- They can be driven hard without damage.
- They have high buckling strength and energy absorbing capacity.

H-piles or Universal steel beams are being increasingly used as bearing piles. They offer several important advantages such as:

- Guaranteed integrity of the pile after driving.
- Ease of stacking and handling on site.
- Ideally suited to the support of heavy axial loads and bending moments.
- Suitable for very hard driving, and have small displacement.
- Driving required no site preparation and no head protection for the pile.

FIGURE 4.23(a)
Vibrator driving casing
(Watson and Hillhouse (Plant Hire) Limited)

FIGURE 4.23(b)
Support for steel tube pile
(BSP International Limited)

The universal section used has approximately equal depth and width in addition to having flanges and web of equal thicknesses; it can be delivered to site in lengths of up to 26 metres (Figure 4.24). H-piles should be embedded in the pile cap, to a depth of 150 mm, and need not have capping plates or dowels, or any other means of load transfer, provided that the cap is reinforced in the normal manner. However, capping plates or dowels may be used where this is included in the design. (See Figure 4.32.)

FIGURE 4.24(a)
H-pile being driven by vibrator
(Dew Group Limited)

FIGURE 4.24(b)
Line of H-piles in position for driving
(Dew Group Limited)

These piles derive their support mainly from end-bearing conditions and should be spaced to suit bearing resistance. The recommended minimum spacing for piles driven in groups is 1070 mm or three times the diagonal measurement of the pile, whichever is the greater. The piles may be driven by any type of hammer or vibrator, the only limitation being that the type of hammer must suit the angle of driving when driven on the rake.

Screw piles

A less common type of pile is the screw pile, which is a form of displacement pile in which the shaft or cylinder is fitted at its lower end with a large diameter helical blade. These piles are screwed into the ground by applying torque at the upper end of the pile shaft, which may be hollow or solid: hollow shaft piles are fitted with a square section head. Torque is applied to the pile head or mandrel by means of a cable and a powerful winch; alternatively, electric screwing capstans, which operate in pairs within a specially designed head frame, may be employed.

Screw piles have an advantage over other piles in their uplift resistance in very soft ground. This type of pile is suitable for marine work and for other forms of construction where the sub-strata are very soft. Difficulty in sinking is likely to be encountered if they are used in dense sands, owing to the resistance of the blades, which vary in diameter from 600 mm to 3 metres to suit the bearing capacity required.

Jetting may be used to assist penetration of dense sand layers and screwing can be stopped when the pile reaches a suitably dense stratum. Screw piles with wide diameter blades are suitable for foundations on very soft clays and silts. The blades or helices may be cast iron, welded mild steel or cast steel; alternatively, screw piles may be constructed entirely of reinforced concrete.

Replacement piles

These are particularly valuable on sites where vibrators and ground heave may be a nuisance or cause damage to surrounding structures or services. The ground is drilled or bored by one of several methods, and, depending on the system employed, the hole is filled with concrete either before or after reinforcement has been inserted. In the Augercast system (Figure 4.25) a continuous flight auger is employed, which has an expendable cap fitted to the tip of the auger. When the required depth of drilling had been achieved, concrete is pumped down through the hollow stem of the auger, blowing the expendable cap off the end of the auger.

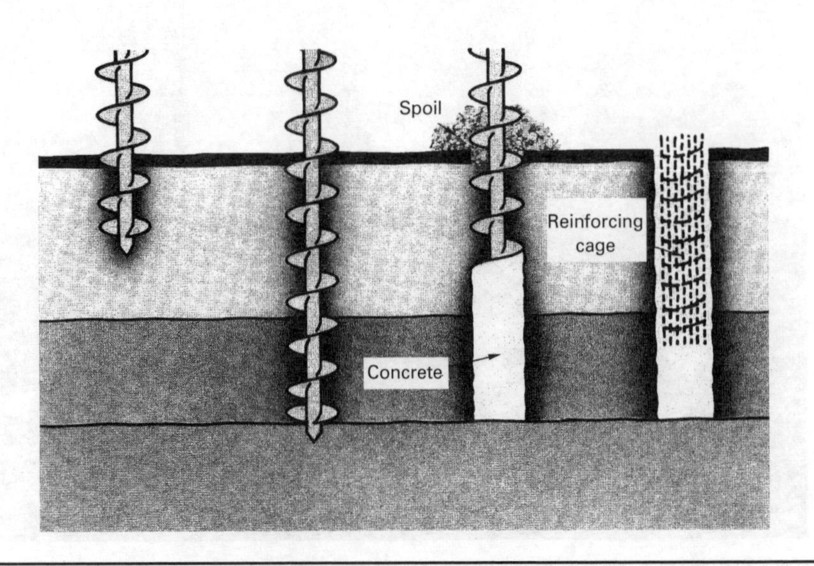

FIGURE 4.25
Sequence of operations for the augercast system
(Keller Foundations)

The auger is then withdrawn at a controlled rate whilst the concrete is placed; the reinforcement is then hung over the wet concrete and vibrated into position.

The method of drilling or boring the hole for a pile has in the past given rise to an inaccurate classification of replacement piles. They are often referred to as 'percussion bored piles' (Figure 4.26), or 'rotary bored piles' (Figure 4.27), when in fact the percussion and rotary element of the title refers only to the formation of the hole and not to the type of pile. A more recent and perhaps better description would be 'supported and unsupported' piles (see chart in Figure 4.15), since the form of support to the pile or borehole affects the formation of the pile. Variations on the formation of the pile toe, such as bulbing and under-reaming, can be carried out on both types of pile.

Unsupported piles

In soils which are stable it will often be possible to bore an unlined hole with a mechanical auger or percussion tool and to place the concrete without lining the hole. The only precaution to be taken is to line the first metre of the hole to prevent surface spoil falling into the hole. Alternatively, a special hopper with lead-in tube may be used.

FIGURE 4.26
Percussion bored piles – using tripod
(Westpile Limited)

Supported piles
These can be divided into two categories:

- Those having a permanent casing or lining.
- Those having a temporary lining or some other form of temporary support, such as drilling mud.

Where a casing is used, the hole is formed by means of percussion or by rotary drilling, depending on the accessibility and head room.

The hole is bored either by percussion methods or rotary boring. In the first case a heavy cutting tool on a small tripod (Figure 4.26) is dropped from its raised position so as to cut out a cylinder of earth. The heavy cutter is raised by a diesel-operated winch and the operation is repeated until the hole has been sunk to the required depth. As the cutting proceeds a thin sectional lining is introduced into the hole to prevent its collapse. The lining is formed with screwed joints to facilitate alignment and to prevent ingress of water. In the second case the rotary drill operates within the casing or lining, as shown in Figure 4.27.

Temporarily supported piles are very popular and offer a wide variety of choice in the finished pile. In many cases, the hole is supported by a screw-jointed steel lining which is retrieved either when the concrete has been placed or during the placing of the concrete: the tube may be winched or jacked out of the ground (Figure 4.28).

Where large diameter piles are sunk through unstable ground to a suitable bearing stratum, the ground may be supported by the use of drilling mud. This mud consists of bentonite suspension with thixotropic properties which restrains the particles of soil and forms a membrane over the sides of the borehole. The membrane is kept in place by the hydrostatic pressure created by filling the hole with the liquid. The first stage of the

FIGURE 4.27
Rotary bored piles (Keller Foundations)

borehole is bored by rotary drill and lined with a temporary steel casing. This short length of casing prevents the collapse of loose surface soil, the accidental kicking of materials into the hole, and the loss of the drilling mud in made-up ground which may overlay a site. On completion of the first stage boring, the hole is filled with bentonite suspension from storage tanks.

FIGURE 4.28
Sequence of operations in percussion drilling

Commencement of boring and sinking tube

Use of tube lifter for placing or removal of sections of tube

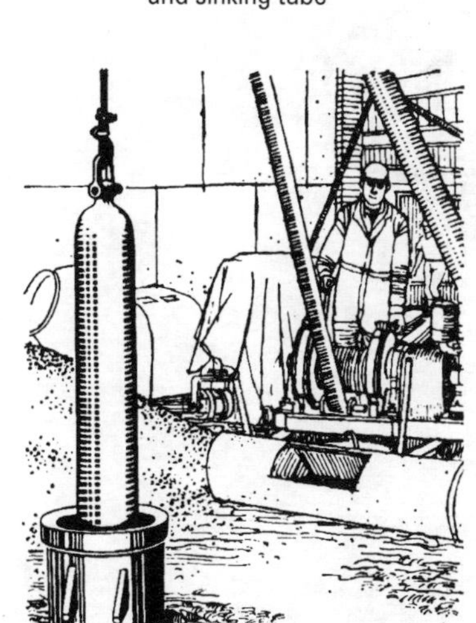

Hammer compacting concrete base

Placing cage of reinforcement after completion of base

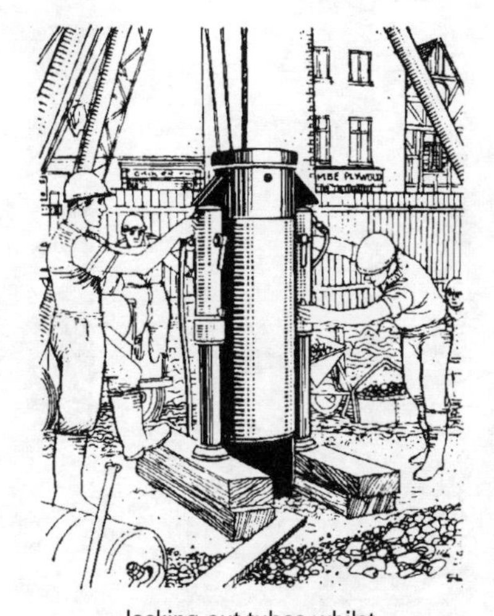

Jacking out tubes whilst concreting pile

The boring continues through the bentonite, which is continuously fed into the hole as boring proceeds. On reaching the required depth, reinforcement is lowered through the bentonite suspension and concrete is placed through a tremie tube. The concrete displaces the mud, which is pumped back into the storage vessels as it rises up the borehole; it is then strained to remove soil particles before re-use. The short temporary casing is withdrawn as the concrete reaches the upper level of the hole.

Under-reaming

Replacement piles can be enlarged at the base to carry an increased load. This method of enlarging the base is known as 'under-reaming' or 'belling' and is achieved by a belling bucket rotated by the drilling rod (Figure 4.29). The bucket can have arms hinged at the top or bottom to allow the sides of the bucket to be jacked out to the desired position. The bucket with top-hinged arms is the most popular tool, since it cuts a conical shape which is suitable for maintaining stability in fissured soils. Most buckets can under-ream a base diameter of up to 3.5 metres, but it is possible to obtain buckets which will form a 'bell' of up to 5.4 metres in diameter. The inspection of larger diameter boreholes and other aspects related to their construction should strictly comply with BS5573, which covers all safety aspects for this type of work.

FIGURE 4.29
Under-reaming tool (Cementation Piling and Foundations Limited)

4.2.3 Methods of driving

Driving hammers

A wide range of hammers and vibrators are available and these have been discussed in Chapter 2. The tendency is to move towards a greater use of hydraulic hammers and vibrators because they are completely self-contained and require no ancillary equipment such as boilers or compressors. However, air hammers are still used and are available as single-acting and double-acting hammers. Single- and double-acting diesel hammers are also widely used, varying size and weight from 3600 kg to 12 000 kg. The hammers can be used in conjunction with piling frames, hanging leaders and short rope-suspended leaders. In the latter case the hammer is suspended from a crane jib on a single rope and a trestle frame is provided for guiding the pile. With hanging leaders, hydraulic rams are sometimes incorporated for control of the slewing and raking angles; this gives a greater control in movement. Figure 4.30 shows a hydraulic hammer driving steel tubular piles and Figure 4.31 shows a diesel hammer driving sheet piles.

FIGURE 4.30
Hydraulic hammer driving steel tubular piles
(BSP International Foundations Limited)

Bored piles

The drilling of bored piles can be divided into two clear categories: rotary and non-rotary drilling. Large diameter piles, better known as caissons, can be sunk by rotary drilling. Modern drilling rigs are capable of sinking shafts up to 60 metres deep by means of telescopic kellies, and diameters of up to 2 metres are common. Depending on the required depth and diameter, this type of drill is either crane-mounted or truck-mounted. Where soil conditions pose problems in drilling – for example, collapse or water seepage – some form of support must be employed. This could include support from bentonite suspension, or some form of temporary casing which would require sinking by vibrator. Non-rotary drilling includes the normal tripod-mounted percussion rigs and special grabs which work within a semi-rotary moving casing. These machines are best suited to soils which are difficult to bore with rotary machines, such as soils containing coarse gravel and cobbles, or boulder clays. The casing is given a continuous semi-rotary motion to keep it sinking as the grab is advanced in depth. Holes up to 1.5 metres in diameter and 30 metres deep can normally be achieved using this form of plant.

4.2.4 Caps and capping beams

Pile caps are usually constructed of concrete to such a depth as will ensure full transfer of load to the piles and, at the same time, resist punching shear. Since it is almost an impossibility to bore or drive piles exactly vertical or to an exact rake, the pile cap should be large enough in plan to accommodate any deviation in the final position of the pile heads. The piles should project into the pile cap and, in the case of concrete piles, have the pile reinforcement bonded to the cap reinforcement. As discussed earlier, steel H-piles may be fitted with cleats to ensure full transmission of load from cap to pile; alternatively, the pile may be bedded in the pile cap (Figure 4.32).

FIGURE 4.31
Diesel hammer driving sheet piling
(Watson & Hillhouse (Plant Hire) Limited)

Capping beams should be used to connect a series of caps together, when the number of piles being capped is less than a group of three. Three is the least number that will ensure stability against lateral forces, with the exception that caisson piles provide their own stability and very rarely require even a pile cap. Capping beams are also suitable for distributing the weight of a load-bearing wall, or of close-centred columns to a line of piles. The piles in this case may be staggered (Figure 4.33) to allow for any eccentricities that may occur in loaded conditions. Where eccentricity is likely to be only slight, owing to light loading, the piles can be driven in a line beneath the centre of the capping beam. If the purpose of using piles is to overcome the problem of swelling and shrinkage of the subsoil, the capping beam must be kept clear of the ground. This can be achieved by casting the capping beam on 50 mm of polystyrene or other compressible material: this will allow an upward movement of the ground without consequent damage being caused to the beam.

The size of the pile cap will be determined by the spacing of the piles which form the pile group. This will depend a great deal upon structural considerations. The general rule for spacing piles in clay is a minimum spacing of three times the diagonal measurement of the pile. The pile cap should overhang the outer piles by a distance of 150 mm. Economies can be achieved in the construction of caps and beams by constructing concrete block walls to the exact sizes of the cap and filling the space with concrete, having first backfilled the wall with soil (Figure 4.34). This method achieves a saving over methods which require formwork and extra excavation for positioning the formwork.

4.2.5 Testing for load-bearing capacity

The bearing capacity of a pile will depend upon several factors, such as the size, shape and type of pile, and the particular properties of the soil in which the pile is embedded. The ultimate bearing capacity is that which equals the resistance of the soil; further loading than this will cause the pile to penetrate still further into the ground. BS 8004 states:

> 'For practical purposes, the ultimate bearing capacity may be taken to be that load, applied to the head of the pile, which causes the head of the pile to settle 10% of the pile diameter, unless the value of the ultimate bearing capacity is otherwise defined by some clearly recognisable feature of the load/settlement curve.'

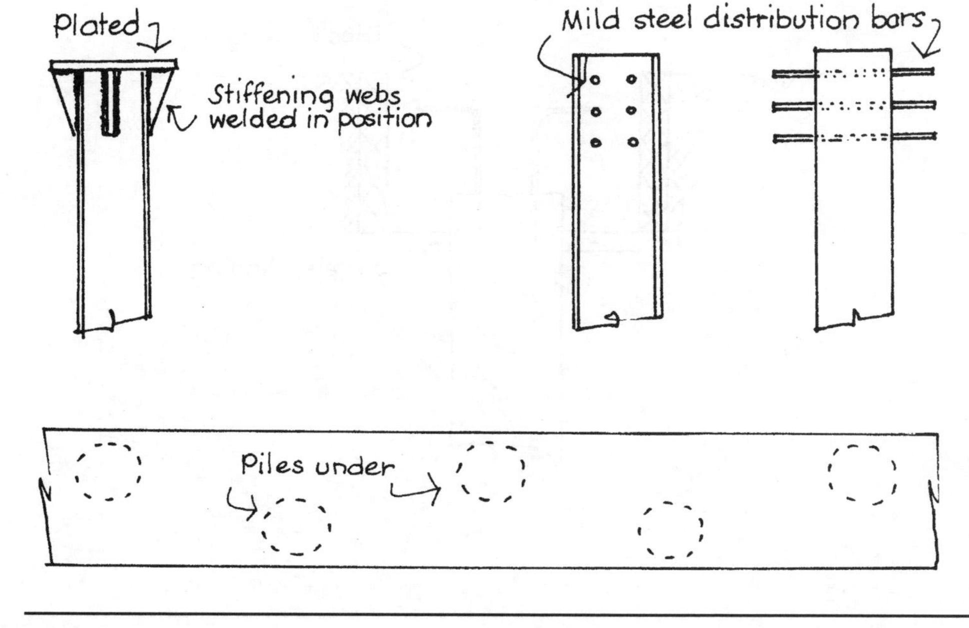

FIGURE 4.32
Capping to steel H-piles

FIGURE 4.33
Plan of capping beam showing staggered piles

This statement has been qualified by experts as being a settlement of 10% of the diameter for end-bearing piles in clay, but as little as 1% of the pile diameter for a friction pile.

The method of calculating ultimate bearing capacity of a pile will depend upon the magnitude of the work involved, the type of soil and the specification laid down by the design engineer. The following alternative methods of calculation may be used:

- Dynamic pile formulae.
- Static formula.
- Test loading.

Dynamic formulae. These are used for calculating the approximate ultimate bearing capacity of piles mainly in non-cohesive soils and are based on certain assumptions, namely that the resistance to driving is determined from the energy delivered by the driving hammer together with the movement of the pile under a blow; and that the resistance to driving is equal to the ultimate bearing capacity for static loads.

There are several ways in which these formulae can be derived, but the basis of each is the same: namely, that the energy delivered by the hammer on impact is equated with the work done in overcoming the resistance of the ground to penetration. The most popular formula for driven piles is the Hiley formula, which can be found in most comprehensive text books on piling. This formula is intended only for the calculation of the bearing capacity in non-cohesive soils, and though it may be used in hard clay it must not be used for calculations in other cohesive soils unless adequate experience is available for the site from previous pile loading tests. On achieving the ultimate bearing capacity a factor of safety must be applied before calculating the safe working load. The safety factor can vary according to rate of settlement of the pile permitted at working load, this being subject to variation according to the size of pile and the compressibility of the soil. It has been suggested by some engineers that a factor of 2 will be adequate for most circumstances. Since dynamic formulae are based on the mechanics of a falling weight, it is unsuitable for certain other types of driving equipment such as vibrators and diesel hammers.

Static formulae. In the UK the large number of pile load tests have increased the reliability of static formula for the design of piles in various soils. The advantage of the soil mechanics method of calculation is that allowable loads can be assessed from

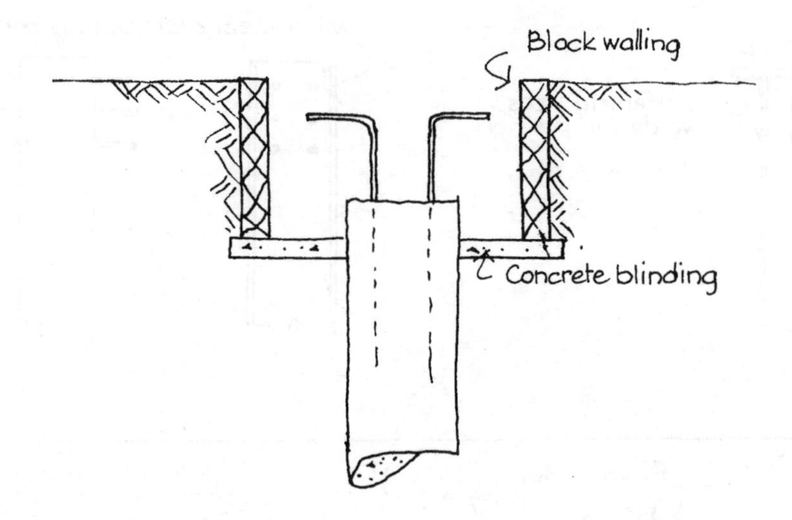

FIGURE 4.34
Method of forming pile cap using blockwork as permanent formwork

the properties of the soil before piling work commences. The normal tests for non-cohesive soil include the standard penetration test and the Dutch or cone penetration test. In the latter test a cone 36 mm in diameter is fixed on the end of a rod enclosed in a tube having the same diameter as the base of the cone; the assembly is pushed into the soil and the forces required to advance the cone and the tube independently are measured. The ultimate bearing capacity is taken as being equal to the resistance of the cone. If the soil is cohesive, laboratory tests are more applicable for estimating the shear strength values of the soil. With soft clays and silts, the friction or adhesion to the pile may increase over a period of time, but the amount of increase will vary with the type of soil and pile material; a loading test should be applied to verify the specified working load.

Test loading. This serves a two-fold function: firstly to check the ultimate bearing capacity of the pile, and second to check the workmanship involved in forming the pile. The test is carried out on a trial pile in close proximity to the borehole used for the other tests, so as to obtain an accurate correlation of the various tests. The pile head should be cut off or built up to the necessary elevation and suitably capped to produce a horizontal bearing surface. The pile is then tested by one of two tests:

- The maintained load test, or
- The constant rate of penetration (CRP) test.

In the maintained load test the load may be applied either by means of a jack, which obtains its reaction from kentledge heavier than the test load (Figure 4.35a), or by means of a jack which obtains its reaction from anchor piles (Figure 4.35b). When using the anchor method, the anchor pile should be at least 3 test-pile diameters away from the test pile (centre to centre). The load is applied in increments of about 25% of the working load, up to the full normal working load. Smaller increments are then added until the specified limit has been reached.

Settlement is measured until it has ceased, or is as little as 0.08 mm in 20 minutes. When this state has been reached a further increment should be applied and the final settlement noted. At each stage of loading the settlement and time scale is plotted. Loading may be stopped at working load, at one-and-a half times working load, or at ultimate bearing capacity.

FIGURE 4.35(a) *Test loading using kentledge (Keller Foundations)*

In the constant rate of penetration test the pile is made to penetrate the soil at a constant speed by continually increasing the load. The pile movement creates stresses in the soil until it fails in shear, thus having reached the ultimate bearing capacity of the pile. Since the purpose of this test is the determination of the ultimate bearing capacity, it is difficult to establish the exact settlement under any given load. The equipment used in this test can be the same as that used in the maintained load test. The reaction achieved by kentledge or anchor method should be greater than the estimated ultimate bearing capacity of the pile.

The force from zero to ultimate bearing capacity is applied through a hydraulic jack which has sufficient travel to accommodate the total movement of the pile. The jack is operated to give a uniform rate of penetration, ranging from 0.6 mm/min in clay to 1.2 mm/min in sand or gravel. By plotting load against penetration the ultimate bearing value can be established.

4.2.6 Economics and selection

The economic selection of any particular piling system is not based solely on the cost per unit length of pile. Other factors which have to be considered are:

- The possibility of having to sink one or more piles to a greater depth than is anticipated.
- The contractor's experience of very difficult sites.
- The contractor's ability to complete the piling work within a phased construction programme.
- The cost of extensive test loading if required.

These and other factors, which may vary from site to site, will influence the final economic choice of the piling system. However, in addition to the economic factors, there are certain physical aspects that affect the choice of the piling system:

- Type of soil.
- Surface gradients.

FIGURE 4.35(b)
Test loading using reaction piles
(Westpile Limited)

- Site difficulties – water-bearing ground, obstructions, adjacent structures.
- Nature of superstructure.
- Location of site.

The soil type will have some influence on the choice of pile; rock sub-strata may, if sloping, require a pile that can be bored or driven into the upper layer to anchor the pile toe. Soft rocks can easily be bored for replacement piles. Cohesive soils are suitable for all forms of displacement pile, but where boulder clays and clays with shale occur it may be necessary to use replacement piles, in some cases using the grab method of excavation. Non-cohesive soils, such as sand and gravel, are suitable for displacement piles since the displacement of soil will be negligible in most cases. Some care will be necessary when driving in very dense sands if in close proximity to other structures.

The surface gradient of the site to be piled may have a great influence on the selection of system. Unless the site is at a gradient less than 1 in 20, it may have to be levelled before piling work can commence. The exception to this is the use of small diameter percussion piles which can be installed on gradients up to 1 in 5 without difficulty. Rotary bored piles can also be installed on gently sloping sites without too much difficulty in setting up the plant.

Water-bearing ground does not affect the progress of piling, but it does affect the cost. Where bored piles are used, the borehole must be lined as the work proceeds, increasing in cost with the depth of the bore. In some cases the lining will need to be of a permanent nature to prevent damage by water flow to the newly-formed pile. In such cases displacement or partially pre-formed piles may be used. If the soil is waterlogged silt or sand, its shear strength may be greatly reduced by vibration, inducing failure of existing services.

Obstructions create the biggest problem in piling work. In some cases the obstruction will be reduced headroom, which may require a system such as small-diameter percussion boring to be used. Where the headroom extends to 10 metres, it will be possible to employ rotary boring equipment, providing that the diameter is not too large. Underground services and existing foundations are the main forms of obstruction and accurate location is necessary before piling commences. Where the service or structure is subject to damage through vibration or ground heave, it will be necessary to use some form of bored pile.

The type of superstructure has less influence on the type of piling system than might be imagined. This is because larger diameter piles can be used, either singly or linked together by caps or beams, to carry any form of structure and load involved. Light structures can be founded on short-bored piles, whilst heavy high-rise structures may require large diameter piles or groups of piles. Where loads are high and adequate bearing capacity can be found only at great depth, it may be economic to form piles by boring or by driving sectional piles; this will overcome the problem of transport and handling of very long precast piles. If the number of piles required on a site is less than fifty, some form of mobile boring rig or percussion boring method may be more economic than other proprietary methods of piling, which require heavy plant and result in correspondingly higher transport costs. The location of the site may influence the choice of method of piling. For example, noise and vibration can cause nuisance, particularly in urban areas, and thus give rise to claims and complaints.

Concrete piles are suitable for all types of soil and loading conditions, but if precast they require additional reinforcement against the handling and driving stresses. Their greatest disadvantage is that they are costly to extend or cut back to the required length. Steel piles are suitable for all types of work and can be extended and cut back easily.

4.3 VIBRO-REPLACEMENT AND VIBRO-COMPACTION

4.3.1 Introduction

Where structures cannot be safely founded on loose soils or fill material, piling may be considered as a means of transferring the loads to suitable levels. However, piled foundations are not the only means of achieving satisfactory foundations in such situations – the engineer may wish to consider geotechnical processes or vibratory processes. Geotechnical processes are discussed in section 3.3.5 as a means of ground water control, but they can also be used to increase the bearing capacity of the ground.

Of the processes discussed, the most economical is likely to be cement injection. However, vibratory processes can also be used to consolidate and strengthen ground conditions at a comparatively low cost; this is achieved by stabilising the soil so that greater loads can be carried without risk of settlement. It also allows simple, shallow foundations to be used on otherwise poor sites. Vibratory processes can also be used on recently filled sites which contain brick rubble, soil, concrete or other miscellaneous material. This means that sites which were considered totally unsuitable for construction operations can now be considered.

4.3.2 Methods and materials

Two principal methods are employed for the compaction of the ground, namely vibro-replacement and vibro-compaction.

Vibro-replacement

In the vibro-replacement method stone columns are constructed through weak soils to improve their load-bearing and settlement capacities. There are three processes, the dry process, the bottom feed process, and the wet process. In the dry process a heavy vibratory unit is allowed to penetrate the weak soil to the designed depth and the cavity is filled with stone, the stone is compacted in stages. The vibrator is up to 4 metres long and weighs about 2 tonnes. It is either suspended from a crane or held in crane leaders.

In the bottom feed process the vibrator is mounted on a specially designed track-mounted machine, which employs leaders to guide the vibrator. A heavy duty tube on the outside of the vibrator carries stone to the tip of the vibrator. Stone is fed into the tube via a stone reservoir, which in turn is fed by a skip travelling up and down the leaders. The vibrator is powered by a diesel generator mounted on the machine. The vibrator remains in the ground during the construction of the stone columns – the stone being supplied to the tip of the vibrator. Stone columns up to 15 metres deep can be formed in this way (Figure 4.36).

In the wet process the vibrator is suspended from a crane and the weak soils are removed by water jetting. Stone backfill then replaces the weak soil and is compacted into the surrounding ground. This process requires a water supply of 10 000 to 12 000 litres per rig hour. The stone columns formed by these processes can carry loads of 10 to 40 tonnes, but a safe bearing factor of 3.0 is applied to the calculated ultimate loading.

Vibro-compaction

This process is based on the fact that non-cohesive soils, such as sand, can be compacted into a denser state by vibration. The action of the vibrator, which is often accompanied by water jetting, reduces the forces between the granular material allowing it to consolidate to the optimum density. This form of compaction is permanent and can be

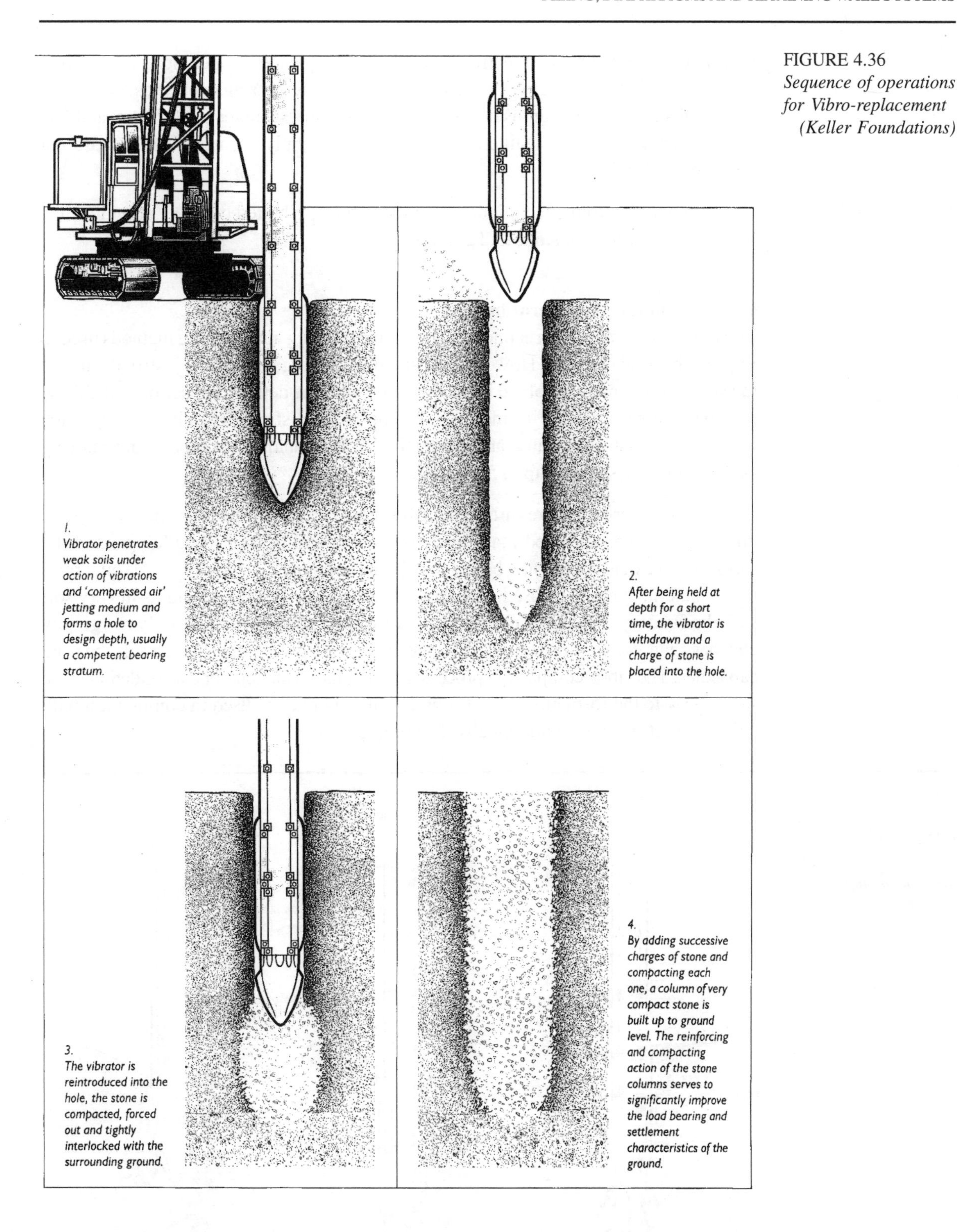

1.
Vibrator penetrates weak soils under action of vibrations and 'compressed air' jetting medium and forms a hole to design depth, usually a competent bearing stratum.
2.
After being held at depth for a short time, the vibrator is withdrawn and a charge of stone is placed into the hole.
3.
The vibrator is reintroduced into the hole, the stone is compacted, forced out and tightly interlocked with the surrounding ground.
4.
By adding successive charges of stone and compacting each one, a column of very compact stone is built up to ground level. The reinforcing and compacting action of the stone columns serves to significantly improve the load bearing and settlement characteristics of the ground.

FIGURE 4.36
Sequence of operations for Vibro-replacement (Keller Foundations)

used in loose soils up to 29 metres in depth. The sequence of operations is shown in Figure 4.37. The high relative density leads to high safe bearing capacities and permit the use of shallow foundations designed to bear pressures between 250 kN/m^2 and 500 kN/m^2. The process is not suited to cohesive soils simply because such soils do not respond to vibration.

In addition to the techniques described above, vibrated concrete columns may be employed instead of stone and gravels.

4.3.3 Economic considerations

These processes may be an economic alternative to piling and grouting methods used to improve bearing capacity. However, the site must be large enough to justify the use of the special equipment involved in the process. Since the depth compaction using these methods is approximately 12 metres, it can be used satisfactorily only on sites which will provide suitable resistance at these depths. Vibro techniques are capable of achieving safe bearing pressures of up to 500 kN/m^2.

While these pressures are suitable for most spread foundations, they may prove unsuitable for concentrated loads such as those found in framed buildings, and the formation of extensive capping beams may be uneconomic. In both cases large quantities of fill material have to be used. Where stone is used for forming columns in very soft soil, the quantity of material used may produce cost figures which are only marginally cheaper than conventional piles, and the latter will give much higher safe bearing capacities. The vibro-compaction process is particularly valuable for consolidating loose sands prior to the formation of raft foundations and may be used in conjunction with raft construction more economically than piling.

FIGURE 4.37
Sequence of operations for vibro-compaction (Keller Foundations)

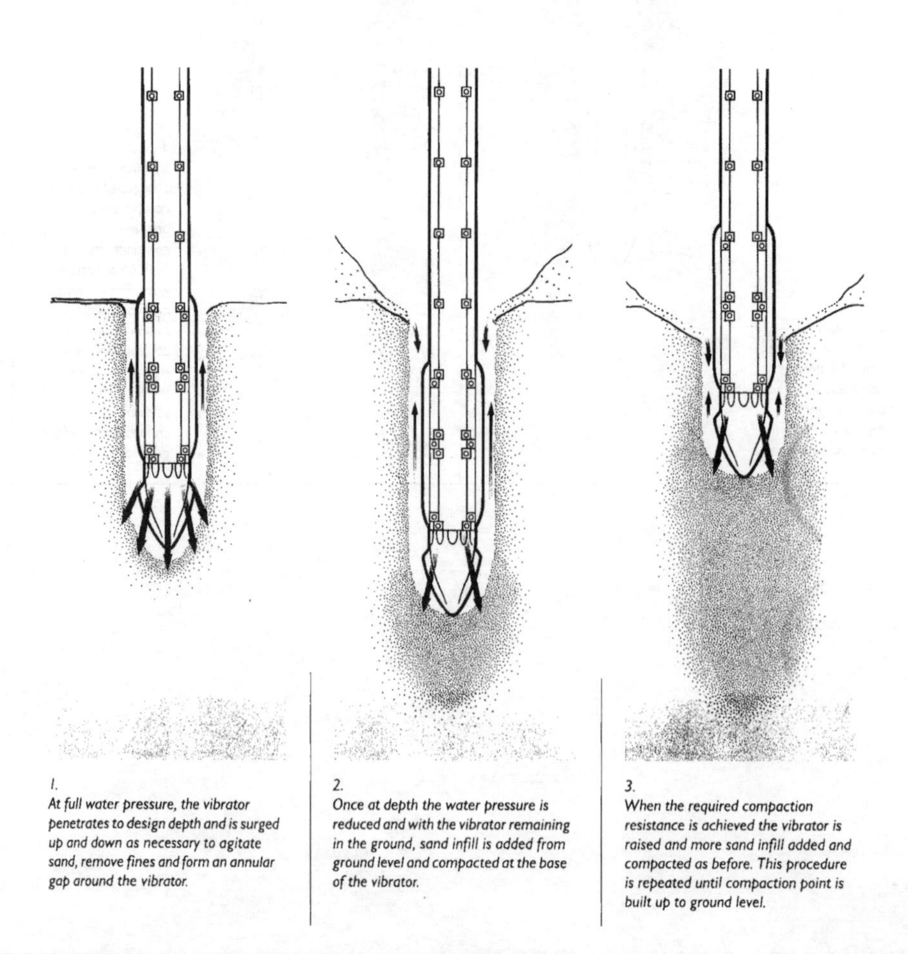

1.
At full water pressure, the vibrator penetrates to design depth and is surged up and down as necessary to agitate sand, remove fines and form an annular gap around the vibrator.
2.
Once at depth the water pressure is reduced and with the vibrator remaining in the ground, sand infill is added from ground level and compacted at the base of the vibrator.
3.
When the required compaction resistance is achieved the vibrator is raised and more sand infill added and compacted as before. This procedure is repeated until compaction point is built up to ground level.

4.4 DIAPHRAGM WALLING

4.4.1 Introduction

'Diaphragm walling' is a term that was first used for the construction of continuous in-situ concrete walling. The process, sometimes called the slurry trench method, involves the excavation of a narrow trench to the required depth and filling the trench with reinforced concrete. The most interesting factor is the method of trench support during excavation, which is achieved by means of a thixotropic mud called bentonite. Bentonite, or fullers earth, is composed of the clay mineral montmorillonite and consists principally of silica, alumina and traces of sodium. The bentonite powder is mixed with clean water to form a creamy slurry. The slurry has a specific gravity of about 1.2 and is therefore suitable for maintaining a head of pressure which will prevent ingress of water and soil. A shallow perimeter trench provides storage for the slurry while at the same time providing a guide for the grab or drilling equipment to excavate the trench. The trench is excavated in sections and the slurry follows the trenching equipment as it digs, thus supplying a constant head of pressure.

The process was originally used for sinking oil wells in desert areas and was first used as a means of wall construction in the UK in 1961. It has a tremendous advantage in the construction of basement walls in built-up areas, since it can be carried out without first underpinning adjacent property. Walls can be excavated alongside existing buildings and anchored back as the excavation takes place to provide support for existing structures and clear working space for the construction of the new basement. Where this slurry method of support is used, precautions have to be taken against loss of mud, which could lead to a collapse of the trench. Such a precaution may include a supply of bentonite, stored in a separate storage tank, mixed with some lightweight material such as polystyrene, which would act as a sealing material.

This form of construction may also be used as retaining and cut-off walls in civil engineering work. Diaphragm walling, as a term, has been developed in recent years to

FIGURE 4.38 *(Left)* *Cast in place guide wall* *(Soletanche Limited)*

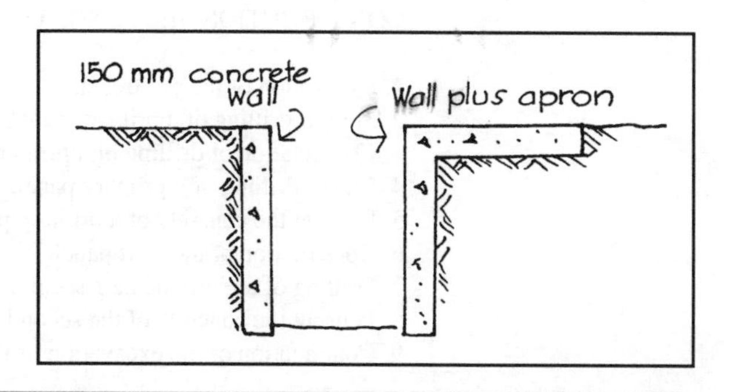

FIGURE 4.39 *(Below)* *Construction of perimeter trench with apron*

include other forms of construction which perform similar functions to those mentioned. These forms include mix-in-place walls, precast diaphragms and flexible diaphragms.

4.4.2 Methods of construction

Thick cast in-situ diaphragms

The use of bentonite or slurry wall diaphragms is suitable for the formation of deep basement walls or retaining walls where there is water in the soil and heavy loads in the vicinity of the excavation. The first step is to construct a perimeter slurry trench, between 1 metre and 1.5 metres deep, and the width of the proposed wall 450 mm to 1 metre. When the perimeter trench has been excavated formwork is fixed and the trench lined both sides with 150 mm concrete (Figure 4.38). In some cases a concrete apron may be provided (Figure 4.39).

The trench is then filled with bentonite slurry from large storage tanks, and excavation is carried out by suitable equipment in predetermined sections. Each section is approximately 2.4 metres wide and the excavation is taken to the full required depth, commonly 35 to 50 metres deep, but depths of 100 to 150 metres are possible. The sequence is shown in Figure 4.40.

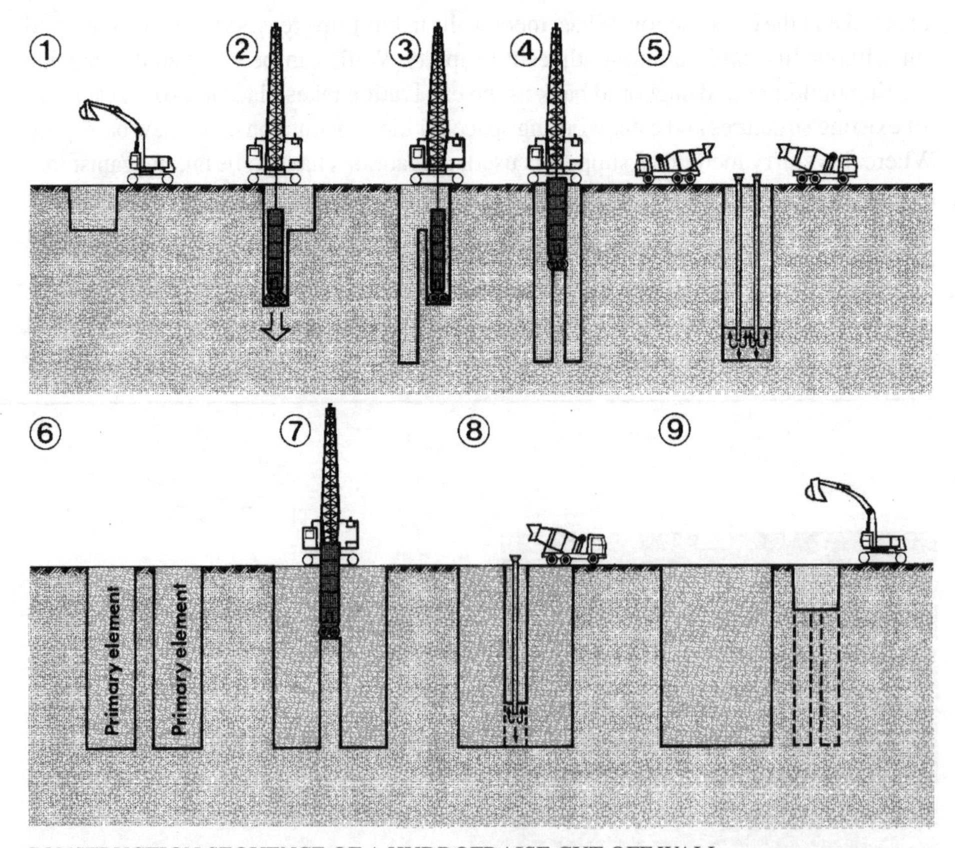

FIGURE 4.40
Sequence of operations in constructing diaphragm wall
(Soletanche Limited)

CONSTRUCTION SEQUENCE OF A HYDROFRAISE CUT OFF WALL

1. Excavation of the pre-trench
2. Start of drilling of a primary panel, 2nd element
3. Continuation of drilling of a primary panel, 2nd element
4. End of drilling of a primary panel, 3rd element (wedge)
5. Pouring the concrete of a primary panel
6. Construction of the next panel
7. Drilling of an intermediate secondary panel
8. Pouring the concrete of the secondary panel
9. Continuation of the excavation of the pre-trench

As the excavation proceeds the slurry flows from the perimeter trench into the deep section under construction, thus giving the necessary support. Reinforcement, made up into cages on site, is then lowered through the bentonite, spacers on the reinforcement cage ensuring correct positioning to allow the specified cover of concrete. When the reinforcement is in position a steel stop-end is assembled, which incorporates a water-bar. When the stop is removed, one half of the water-bar remains in the set concrete and the other half projects to receive the new concrete – see Figure 4.41.

With the developments of more advanced excavating machinery, the need for stop ends has been reduced – reverse circulating plant can cut a groove in the adjacent panel. The concreting of the trench takes place through tremie pipes which displace the bentonite up into the perimeter trench or storage vessel. Because the density of the slurry is below the density of the concrete, the slurry will be readily displaced. This process is carried out in alternating sections along the line of the perimeter trench until the wall is complete. Care must be taken to ensure that the density of the bentonite is kept below the density of the concrete. This is achieved by recycling the bentonite through a shaker and hydro-cyclones to remove fine material.

The above method of construction uses either grab type equipment or reverse circulation equipment such as the 'Hydrofraise' to remove the soil (Figure 4.42). The reverse circulation method is a principle well-known in oil and water well drilling. The 'Hydrofraise' is a drilling machine powered by three down-the-hole motors, with reverse mud circulation. A heavy metal frame serves as a guide, fitted at the base with two cutter drums, which rotate in opposite directions and break up the soil. A pump placed just above the cutting drums evacuates the loosened soil, which is carried to the surface by the drilling mud; the mud is continuously filtered and poured back into the trench. The 'Hydrofraise' is suspended from a heavy crawler crane. The crane also carries the power packs supplying the hydraulic power, which is conveyed through hoses to the three down-the-hole motors, two of them driving the cutter drums and one driving the pump.

The 'Hydrofraise' is suspended from the crane and controlled by the operating cable to give either a constant rate of advance or a constant weight on the cutter drums, the weight of the 'Hydrofraise' is 16 to 20 tonnes. A special feature of the cutting tool is its ability to cut into existing concrete panels, thereby eliminating the need for construction joints.

Grab type equipment is also used for excavating the trench. These are operated either by hydraulics or mechanical cable. The hydraulic grabs can be computer controlled

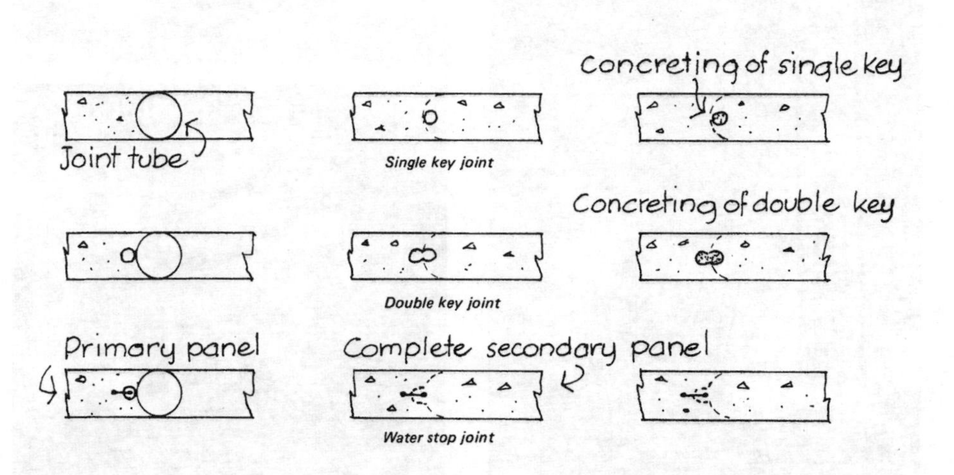

FIGURE 4.41
Methods of forming joints in diaphragm walls

so that the grab excavates and then returns exactly to the vertical digging position to continue the excavation to the required depth. (See Figure 4.42(b))

Where the normal 'Hydrofraise' cannot be used because of limited headroom, the 'Compact' or 'Low headroom' hydrofraise can be employed (Figure 4.43). This development allows diaphragm walling to take place where headroom is limited to around 5 metres.

Counterforts

Where diaphragms cannot be tied back or supported internally, counterforts are provided at the back of the wall. This allows extra reinforcement to be placed to deal with high bending moments.

Thin diaphragm walls

Some diaphragms are mainly required as a screen against water. Where space on a site permits, it is usually more economic to place the screen far enough out from the excavation to allow space for a berm (a horizontal ledge at the top or bottom of an earth bank to ensure stability), and for ground at an appropriate slope to give the necessary structural support to the screen on the inside.

Under these conditions, the diaphragm can be an unreinforced concrete slurry trench, which can be excavated using long-reach backactors. Alternatively a grouted diaphragm may be employed. The latter system is particularly well suited to work in made ground or in very permeable gravel strata where bentonite processes are not satisfactory.

FIGURE 4.42
Excavating machinery for diaphragm walls

(a) *(Left) Hydrofraise*

(b) *(Above right) Hydraulic bucket grab (Soletanche Limited)*

FIGURE 4.43
(Below, right) The Compact or Low Headroom Hydrofraise (Soletanche Limited)

There are several methods of forming a thin diaphragm. The most common method used for forming thin diaphragms employs the use boreholes and pressure grouting. A grout curtain can formed by pressure grouting through sleeved tubes. Micro-fine cements have high penetration in fine soils. An alternative method is to employ steel H-piles, which are either driven into ground or lowered into a slurry-filled trench; the steel piles are then grouted on completion.

Top-down construction

A further development with insitu diaphragms is the formation of tunnels, underpasses and other subways. Two diaphragm walls are constructed to form the outer walls of the subway and then the slab forming the roof of the construction is cast on top of the walls. The placing of the top slab provides stability to the two wall diaphragms and allows excavation to take place under the slab. This technique is very suitable for 'fast track' construction when public roads are involved and the disruption to traffic must be kept to a minimum. It is also suitable for basement excavation where the main superstructure is advancing at the same time as the basement excavation. In the case of tunnelling, the technique is sometimes referred to as the 'pre-deck' method (see Chapter 5).

Precast diaphragms

These diaphragms are formed by placing pre-cast units in a grout-filled trench and effectively sealing the joints. The wall units are reinforced concrete units of a plan section which suits the particular perimeter wall. The dimensions of the units or panels are limited only by the site cranage available; panels weighing 40 tonnes can be precast. The panel units (Figure 4.44) can either be cast on site or brought to the site from a casting yard. The joints are made watertight by the use of an 'expandable water-stop' joint, which can be pressure grouted on completion. However, where watertightness is not essential, plain edged panels can be used (Figure 4.45). Stability is achieved by anchoring the diaphragm back to the soil as the excavation proceeds. The system uses a conventional slurry-supported trench, which is excavated by methods discussed earlier.

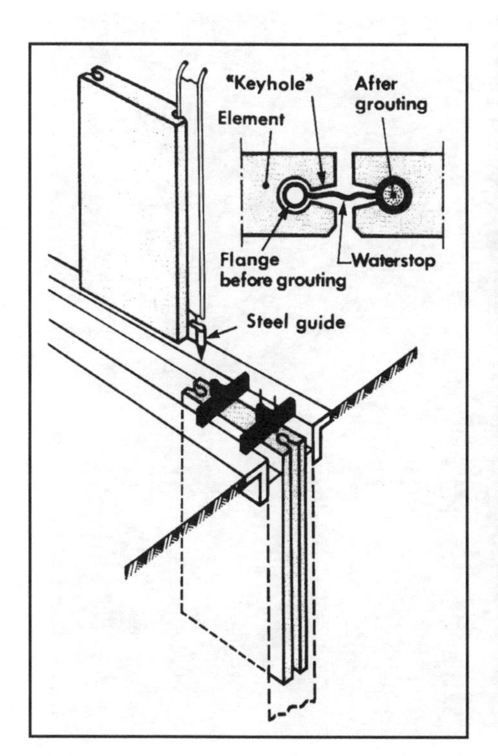

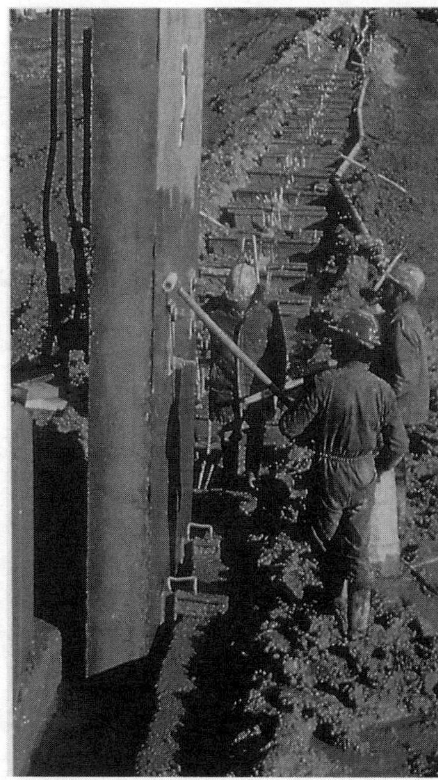

FIGURE 4.44 *(Left)* *Precast panel diaphragms with water-stop joint* *(Soletanche Limited)*

FIGURE 4.45 *(Right)* *Plain edged panels being installed* *(Soletanche Limited)*

Other pre-fabricated units are available for use in conjunction with conventional cast in-situ walls or other forms of construction. Precast column units, with continuity steel protruding are lowered into boreholes supported by cementitious mud. When the mud has hardened, the soil between the columns is excavated, along with the main excavation, to a depth of 2 metres, and an in-situ wall is cast between the columns. This is one sure method of obtaining continuity in the horizontal reinforcement.

Major diaphragm work is illustrated in Figures 4.46 and 4.47, where work to the access shaft of the Channel Tunnel is shown.

Contiguous piling

This is an alternative solution to diaphragm walling, in which the soil is removed by boring. Care must be taken to ensure that the piles touch for their full length, otherwise grouting will be necessary to make an effective seal between the piles. The thickness of these walls will depend on the resistance they have to provide when they are exposed by excavation; they can vary in thickness from 375 mm to 600 mm. In this method the boreholes may be supported by temporary linings or bentonite until the concrete has been placed. Special cutting tools or a heavy cutting shoe may be used to cut a slot in the newly cast piles to form a key for the intermediate piles. If the pile wall is to be part of the permanent structure, it can be lined with mesh reinforcement and sprayed with concrete to provide a smooth finish. This process is known as the Gunite process (see Chapter 8). Another method of finishing the pile wall is to cast a concrete facing wall and capping beam on the inside face of the piles (Figure 4.48). In all cases the piles are reinforced.

4.4.3 Plant and equipment

The plant used for excavating the trench is normally of a very special nature, as discussed below, but it is possible to use any of the powerful back-acting machines for trenches up to 15 metres deep. The specialised machines range from special grabs to sophisticated cutting tools, but they fall into three main categories:

FIGURE 4.46
Diaphragm walling to shaft for Channel Tunnel access
(Soletanche Limited)

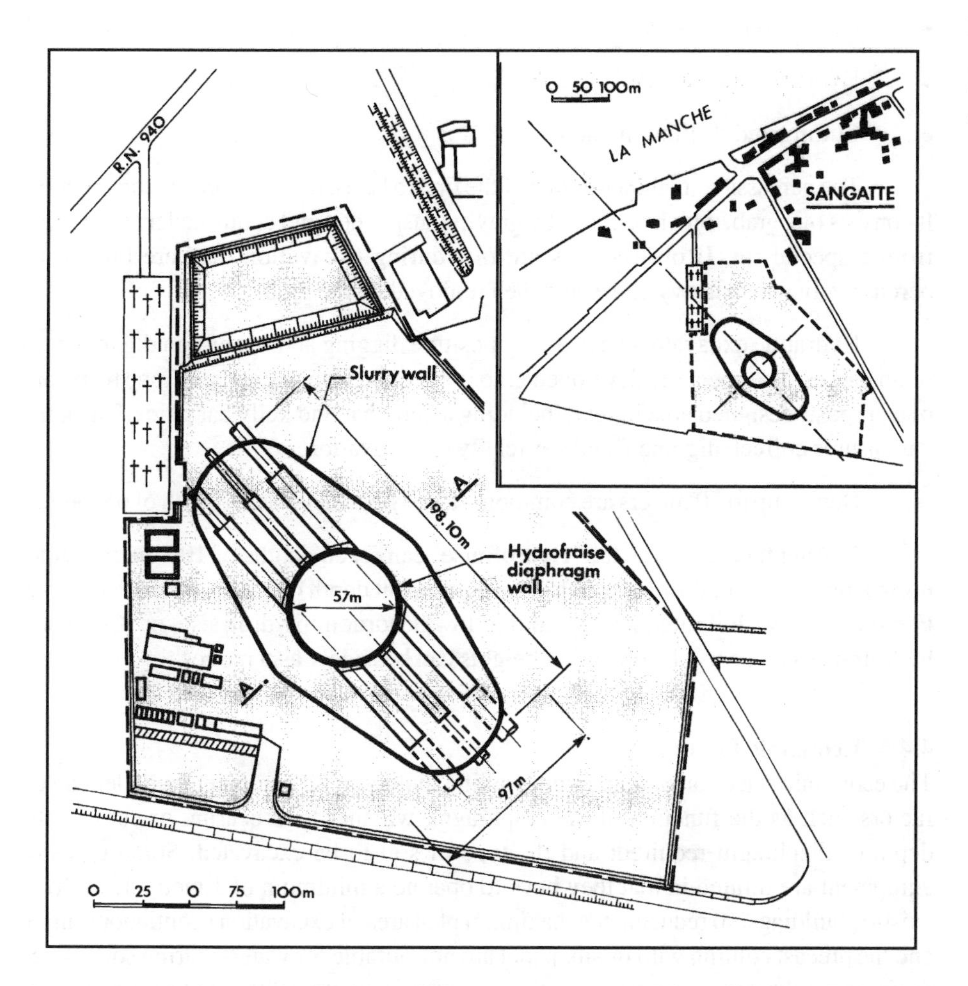

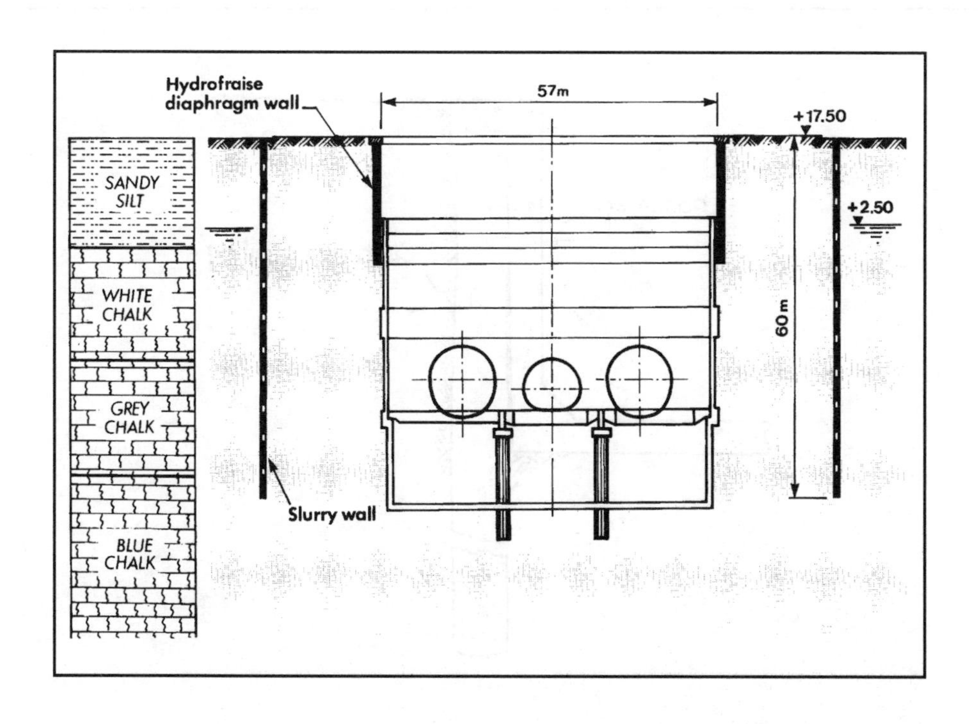

FIGURE 4.47
General view of site showing circular diaphragm for Channel Tunnel shaft (Soletanche Limited)

- Percussion methods.
- Hydraulic methods.
- Cutting and drilling methods.

The percussion method was one of the first to be used in this form of construction. It consists of a grab, which operates by gravity, suspended and controlled by steel ropes from a special rig. If obstructions are met during excavation they are broken up percussively with a heavy chisel and then removed by the grab.

Hydraulic excavators have a very positive digging action and can dig to depths of 36 metres. The specially developed grab is operated from a standard crane by means of a purpose-designed attachment, including a kelly bar and kelly bar guide. The kelly bar ensures correct alignment and verticality of excavation.

Depths up to 50 meters are common – a kelly bar can reach a depth of 65 metres.

Cutting tools, such as the 'Hydrofraise' can reach depths of 100 metres under normal operating conditions, which extra equipment depth of 150 metres are possible. For situations with limited headroom the 'low headroom' hydrofraise, or 'Compact' hydrofraise, can be used – operating heights can be as low as 5 metres.

4.4.4 Economic factors

The economic selection of diaphragm walling methods and equipment will depend on factors such as the function of the diaphragm, whether load-bearing or cut-off; the depth of diaphragm required; and the type of soil to be excavated. Some types of equipment are limited in that they have to operate a minimum of 1 metre away from existing buildings, so reducing the maximum plan area of excavation. Contiguous piling and the precast column with in-situ panel are not suitable for water-bearing soils, since grouting would prove expensive and the construction method difficult. Precast units are generally more expensive than insitu walls.

FIGURE 4.48
Facing wall and capping beam to contiguous piling

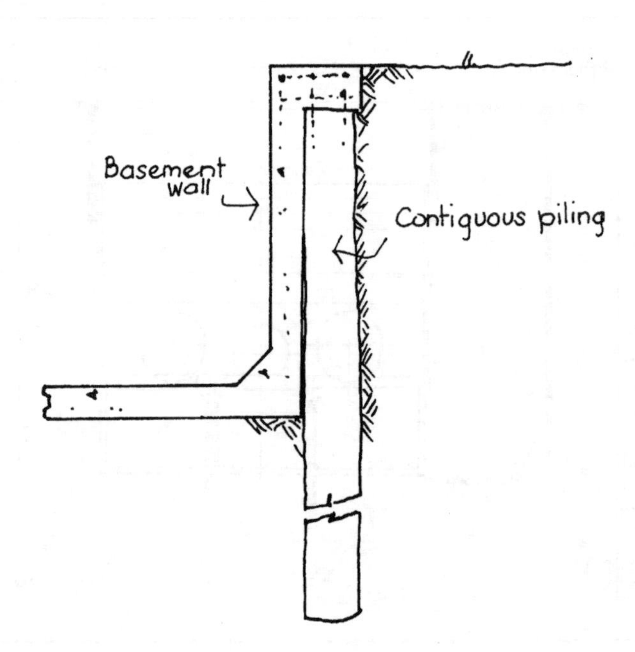

4.5 RETAINING WALLS

4.5.1 Types of walls

There are two main types of retaining wall: the gravity retaining wall and the cantilever retaining wall. The former is mainly used for the support of solids such as soil, fuel, chemical and waste materials. The latter may be used for both solids and liquids. Reservoir walls are often constructed in a cantilever form. In some cases the wall may be designed to support dead loads in addition to the normal lateral loads, but essentially these walls are designed to resist lateral movement. Both types of retaining wall must be designed to resist the following factors:

- Overturning.
- Over-stressing of the material in the wall.
- Forward sliding.
- Settlement due to over-stressing of soil under the wall.
- Circular slip.

These possible factors of failure are shown diagrammatically in Figure 4.49. Gravity retaining walls depend on their dead weight for strength and stability. They are limited in height to approximately 3 metres if inclined, or 2 metres if the wall is vertical. Over these heights the thickness of wall, to comply with the safe height/thickness ratio, would make the construction uneconomical. They are designed so that the width of wall is sufficient to distribute the resultant loads of the wall and the earth pressure to the soil below the base of the wall without undue settlement. High tensile stresses at the back of the wall can be offset by designing the width of the wall so that the resultant force is kept within the middle third of the base. A suitable width of base can be taken as between one quarter and one-half of the wall height. The width at the top of the wall can be taken as one-seventh of the wall height.

Cantilever walls are used for retaining walls up to 7 metres in height without counterforts, and if counterforted can be designed for greater heights without being excessively thick. The wall shape may vary to suit the loading and the material to be

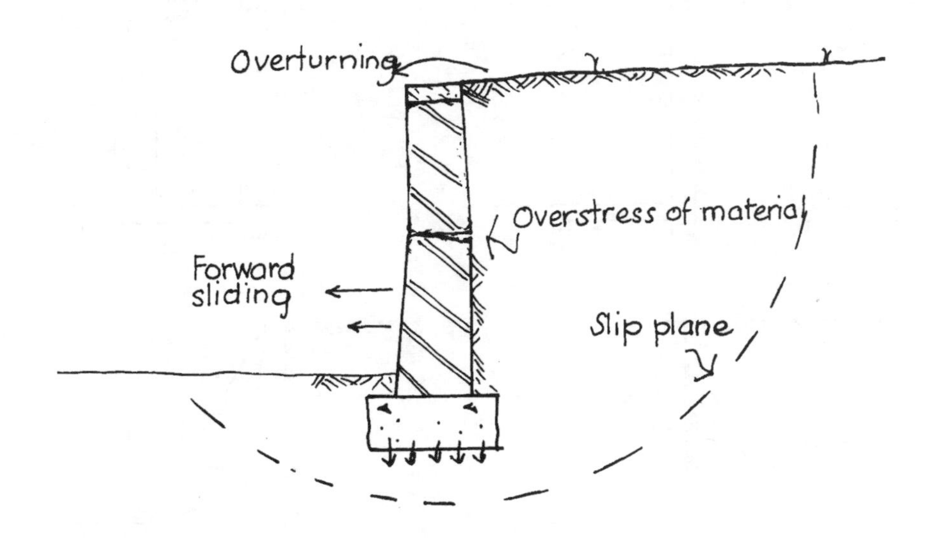

FIGURE 4.49
Possible failures in retaining walls

supported. The walling material is normally reinforced concrete, although pre-stressed concrete may be used for liquid-retaining structures. The form shown in Figure 4.50(a) is used in situations where the wall is to support soil and where it is not possible to excavate behind the wall. The form in (b) may be used where excavation behind the wall is necessary; this type makes use of the backfill for stability. When the height exceeds 7 metres, the extra thickness of the wall must be considered against the cost of constructing counterforts. The forms in Figure 4.50(c) and (d) show the use of counterforts; (c) shows a buried counterfort which reduces the cost of finished concrete (since it can be cast from rough shuttering without any finishing work), whereas (d) produces a very strong buttressed retaining wall which is not dependent on the weight of backfill material. Other forms and types of retaining walls can be used, such as:

- Diaphragm walls.
- Steel sheet piling.
- Concrete crib walls.
- Contiguous piling.
- Anchored walls.

4.5.2 Methods of construction

There are four basic factors for consideration in the construction of retaining walls:

- Retention of the soil, if excavation is necessary, while the wall is being constructed.
- The actual constructional form and materials to be used.
- Control of ground water.
- Backfilling the structure.

The retention of soil during excavation is covered fully in Chapter 3, but will depend a great deal on the proximity of other structures and the amount of excavated material that can be freely removed without danger to plant and equipment; this can be achieved by battering the sides of the excavation giving clear working space.

FIGURE 4.50
Retaining wall design

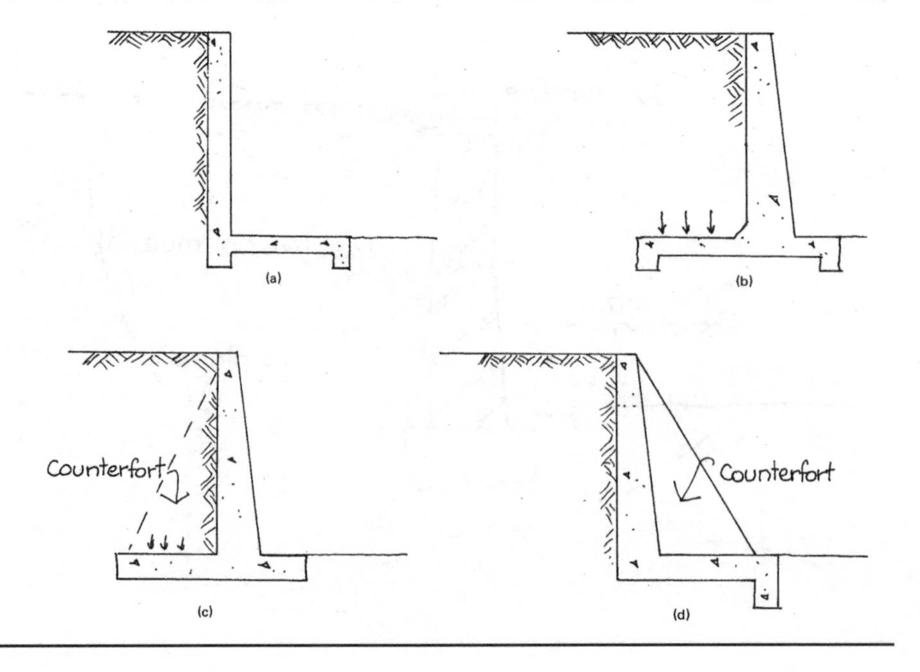

Where the excavation has to be limited but space allows steel sheet piling to be used, the wall can be constructed without disturbing the compacted soil behind the wall. Where loads and structures in close proximity prevent the driving of sheet piling and excavation as outlined above, a diaphragm walling technique should be considered.

The shape of the retaining wall should be such as to provide economic use of formwork, where in-situ concrete is used. In some cases the forms may be track-mounted for long walls, giving rapid turn-round in stripping and fixing. Some plan shapes may present difficulties in achieving the rapid turn-round of wall forms. The materials used in retaining walls may include brick and stone but in the main are constructed of reinforced concrete either in-situ or in the form of concrete blocks or units (Figure 4.51).

Pre-stressed retaining walls are of particular value in basement structures, reservoir construction and dam construction. Anchored retaining walls are very suitable for deep basement work where the retaining wall can be designed as a deep narrow strip without the toe element.

Anchored walls

Anchorage for retaining walls may take three different forms, depending on the soil conditions behind the wall. The three types of anchorage used can be classified as:

- Grouted anchors.
- Plate anchors.
- Rock anchors and rock bolts.

Grouted anchors are formed by under-ream drilling and grouting. The under-reamed cones (Figure 4.52) are cut with a special expanding cutter bit at the bottom of the borehole, and after placing the high tensile steel tendon the borehole is pressure-grouted. Injection anchors are used in non-cohesive soils. The granular material of these soils permits the grout to penetrate into large or irregular zones along the length of the drilled hole (Figure 4.53).

Plate anchors can be used in all types of soil. The plates used for anchoring are pointed so that they can be driven into the ground easily. A cable is attached to the plate

FIGURE 4.51
Precast wall units retaining the bank of a stream
(Bell and Webster Concrete Limited)

and when the cable is tensioned the plate swivels and takes the strain by gripping the soil. This type of anchor can be seen in Chapters 3 (Section 3.2.4) and Chapter 6 (Section 6.6).

Rock anchors can be constructed in the same way as clay anchors, by under-reaming the rock at the base of a borehole. Single anchors in sound rock have been tested to 200 tonnes, multiple anchors being used when excessive loads are required. Rock bolts, which are used extensively in tunnelling, provide a fixing into rock and are commonly secured with a resin fill which flows around the anchor bolt. The resin and catalyst are contained in a single capsule which is inserted into the hole and mixed by rotating the bolt. The resin has a higher compressive strength than surrounding materials and therefore bolts perform exactly as if they were cast insitu.

Figure 4.54 shows a typical system for anchoring retaining walls.

Concrete crib walls

Concrete crib retaining walls (Figure 4.55) consist of a series of concrete lintels or beams which are stacked in a grillage pattern, usually interlocking, and filled with earth or rock. They are constructed to a batter of 1 in 6 or 1 in 8 and are very suitable for retaining embankments to motorways.

The foundation for such walls is a normal insitu concrete foundation designed for site conditions and loadings. Provision should be made for any water draining through the crib walling to prevent erosion of the fill material. The fill material should be well

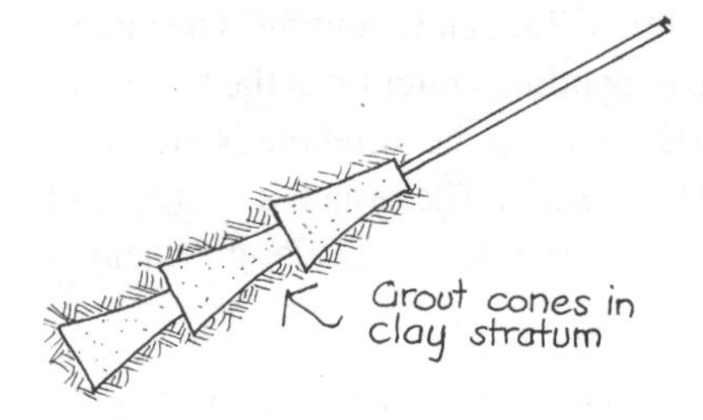

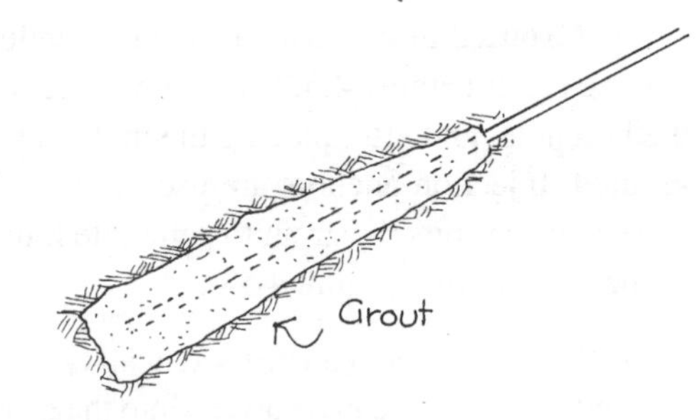

FIGURE 4.52 *(Left)*
Anchor in clay soil

FIGURE 4.53 *(Right)*
Anchor formed by injection

FIGURE 4.54(a)
Anchored retaining wall to excavation
(Dywidag Systems International Limited)

graded from 63 mm down to 10 mm and be able to produce an angle of internal friction of not less than 35°.

Ground water and backfilling

Control of ground water: if water is allowed to build up behind a retaining wall not specially designed for liquid retention, then there is a danger either of the wall being over-stressed by water pressure, or of the water creating a slip plane under the wall. This type of danger can be eliminated by placing agricultural pipes behind the wall at a low point, and connecting the drainage line to weep-holes through the wall. The area directly above the land drain should be filled with granular material and the level below the invert of the pipes should be concreted.

FIGURE 4 54(b)
Anchored retaining wall to embankment
(Dywidag Systems International Limited)

FIGURE 4.55
Crib walling
(Tarmac Precast Concrete Limited)

Backfilling retaining walls should be carried out with great care to prevent undue stresses being placed on the wall. The materials used should be of the quality which the designer has assumed in the earth pressure calculations. Inferior material, such as soft wet clay, can treble the pressure exerted against the wall compared with a dense dry fill. The materials used should be well graded granular fill which will compact easily by mechanical means.

4.5.3 Waterproofing

Retaining walls must be waterproofed to prevent unsightly efflorescence appearing on the face of the wall. This can be achieved in two ways:

- By using good quality vibrated concrete, with or without additives, depending on the quality of work.
- By applying some form of waterproof membrane to the face of the wall.

With the latter method of waterproofing, if excavation is possible the membrane will be applied to the earth-side of the wall against a protective backing of concrete blocks. If this method of construction is not possible, the membrane can be applied to the free face of the wall; in that case a supporting skin of brickwork or masonry must be built in front of the membrane to prevent the membrane being pushed off the face of the wall. Where retaining walls form part of a basement the construction may have to be vapour-proof in addition to being waterproof. This is best achieved by using an applied membrane rather than vibrated concrete. Expansion and construction joints may be formed incorporating a PVC water bar.

CHAPTER 5

Tunnelling and Underpinning

5.1 TUNNELLING

5.1.1 General considerations

Before tunnelling operations can start there are numerous factors that require detailed investigation. These factors include:

- The purpose of tunnelling
- The type of ground
- The method of construction
- Removal of debris
- Control of ground water.

Purpose of tunnelling

Tunnels are constructed for various purposes, such as road and rail transport, conveyance of water or sewage, access to mines, power houses and other underground structures, and ventilation of underground structures.

In the majority of cases the purpose will affect the size and shape of the tunnel, but small and medium sized tunnels may involve driving a larger cross-section than the ultimate size, simply to provide room to manoeuvre equipment and materials. Therefore the economic design of a tunnel for a specific purpose must involve consideration of constructional method.

The purpose of a tunnel will affect the shape in the following ways. Where the tunnel is to carry liquids the frictional surface should be at a minimum; this may lead to egg-shaped sewers or circular sewers, depending on their volume of flow. Ventilation tunnels of circular cross-section give minimum surface area and provide a form most suitable for resisting internal and external pressures. Tunnels to be used for transport may be rectangular or horseshoe in section if a reasonably level floor can be maintained; however, circular cross-sections have the greatest resistance to stress, and are therefore more usual. These will require a suspended floor for the carriageway, having the advantage of providing space for services.

Type of ground

Detailed investigation of the ground will be necessary to ascertain the type of soil or rock, whether the ground is water-bearing, and the extent, if any, of defects in the strata. Tunnelling in rock and soft ground is discussed fully in Sections 5.1.3 and 5.1.4. It should be readily appreciated that the type of ground affects the method of construction and choice of equipment. It may also affect the location of the tunnel, both its horizontal level and its lateral position – for example where the ground contains badly faulted rock or variable ground near much more reliable ground for tunnelling operations.

The final location of the tunnel will depend on the economics of dealing with ground conditions and obstructions. The problem of tunnelling in the vicinity of major obstructions, such as large sewers, underground railways and similar services, can be greatly reduced by preparing models of the proposed tunnel in relation to existing obstructions. Relevant details can be obtained from surveys and from information held by local authorities.

Method of construction

This will depend on the type of tunnel, on ground conditions, on the length of tunnel to be driven, and on the time available for construction. In the main, tunnelling can be placed in two categories:

- those in rock; which require blasting and generally need lining, weak places being temporarily supported until permanent lining is carried out;
- those in soft ground where permanent structural lining must be installed as driving proceeds and where excavation does not require the use of explosives.

Spoil handling methods depend chiefly on the size of the tunnel and the type of ground; wherever practicable mechanical means are employed.

A continuous working cycle is employed for the construction of most tunnels. In rock tunnels the cycle is: drill, fire, clear smoke, load away debris and erect support girders, and finish with the erection of the lining, which is often sprayed concrete. Soft-ground tunnels follow a different cycle, excavation and disposal being immediately followed by the erection of prefabricated linings to support the ground, the space between the excavation and the outside of the lining being filled by grouting with cement or other material after one or more cycles. In both cases several cycles may be completed in twenty-four hours.

Removal of debris

The method employed for the removal of debris (or 'mucking out') depends on the size and length of the tunnel. In rock tunnels, where blasting is employed, loading is by mechanical shovel; the type of mechanical shovel depends on the size of tunnel. Face

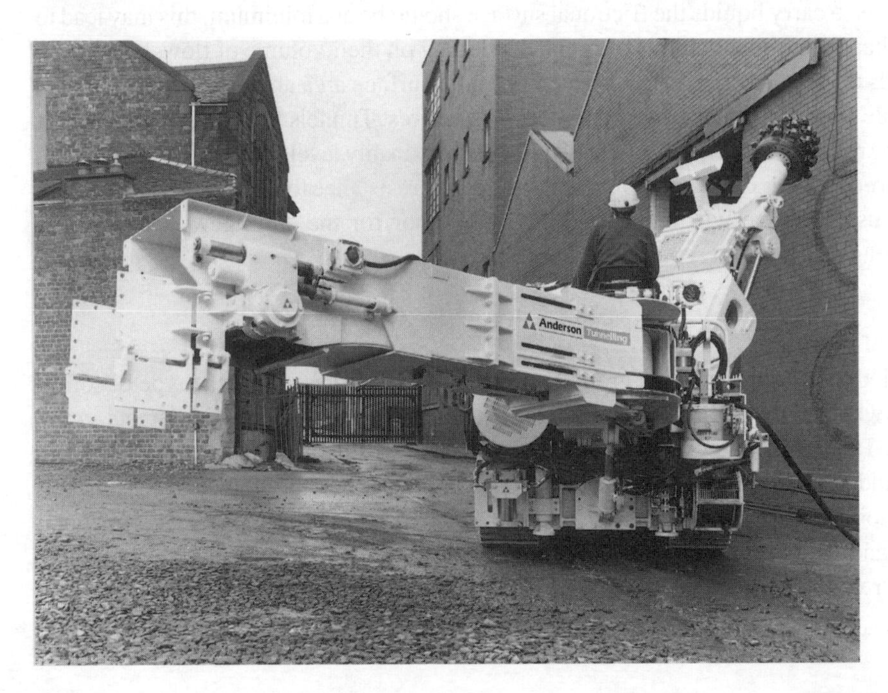

FIGURE 5.1
Anderson roadheader RH22 with slewing discharge conveyor (Anderson Group Limited)

shovels, tyred and tracked-loading shovels and specially developed rocker shovels and backacters have all been used. However, a popular method of feeding the debris to tubs or trucks is the discharge conveyor (Figure 5.1). In small, soft-ground tunnels hand-loading is still employed, but conveyors are increasingly used to raise the material high enough to load it into tubs or trucks. Mechanical excavation of soft ground is generally based on some form of hydraulic backhoe.

In both rock and soft-ground tunnels transportation of debris along the tunnel can be achieved by independent trucks (large tunnels only), by rail-mounted wagons or by long-line conveyor. It should be noted that, particularly with long tunnels, progress is frequently governed by how fast the debris can be removed.

Control of water

Tunnelling, whether in soft ground or rock, is always made more difficult if water is present. Various methods have been employed to control water; pumping is the most common. Other methods include lowering the water table, grouting, freezing and the use of compressed air.

5.1.2 Methods of tunnelling

Methods of tunnelling vary according to the purpose of the tunnel and the problems encountered in its formation. Most tunnels are formed by cutting and boring the ground and lining the tunnel as work proceeds. However, two other methods of construction are also used, which purists might argue are not tunnelling at all:

- 'open-cut' method
- 'immersed-tube' method.

The first method is suitable for shallow tunnels in both water and dry ground, whereas the immersed-tube method is suitable for tunnels crossing deep water.

The **open-cut method** (Figure 5.2), sometimes called 'cut and cover', is suitable for shallow tunnels and culverts. The process may involve the construction of a deep trench supported by diaphragm walling or, in the case of underwater construction, may

FIGURE 5.2
Open-cut method used for a large culvert (F C Precast Concrete Limited)

consist of placing fabricated sections into a trench dredged in the water-bed and backfilling the trench. A variant of the open-cut method is the pre-deck method (Figure 5.3) which is suitable for high level tunnels such as pedestrian tunnels. The tunnel walls are sunk by the diaphragm walling method and the upper surface of the ground is removed to allow the positioning of the tunnel deck. When the deck has been positioned and waterproofed, backfilling and reinstatement is then completed. Tunnel excavation can then continue from both ends of the tunnel without fear of collapse and with minimum disturbance to traffic and services.

The **immersed-tube system**, known also as a submerged-tube tunnel, consists of lowering prefabricated tunnel sections into a prepared foundation pad at sea or river-bed level. The tunnel sections may be either concrete or steel and are prefabricated, usually in a dry dock, before floating them out to their final position. Sections are then carefully lowered into place from pontoons, guided by anchor cables (Figure 5.4) and joined together prior to trench filling. When the tubes have been joined together, they can be de-watered to allow work to proceed internally.

Conventional tunnelling, in which tunnels are formed by cutting through the ground and forming the tunnel profile as the cutting proceeds, varies in method of construction depending on the nature of the ground and the length of diameter of the tunnel.

The selection of a system or method of tunnelling will involve four important components, namely, excavation equipment, muck-loading and haulage, method of lining, and ground water control. In rock or hard ground the excavating method will vary from drilling and blasting to mechanical rock cutting, whereas in soft ground a tunnel shield is generally used with excavation by machine or by hand-held pneumatic tools. These methods are discussed fully in Sections 5.1.3 and 5.1.4.

FIGURE 5.3
Pre-deck method of tunnelling

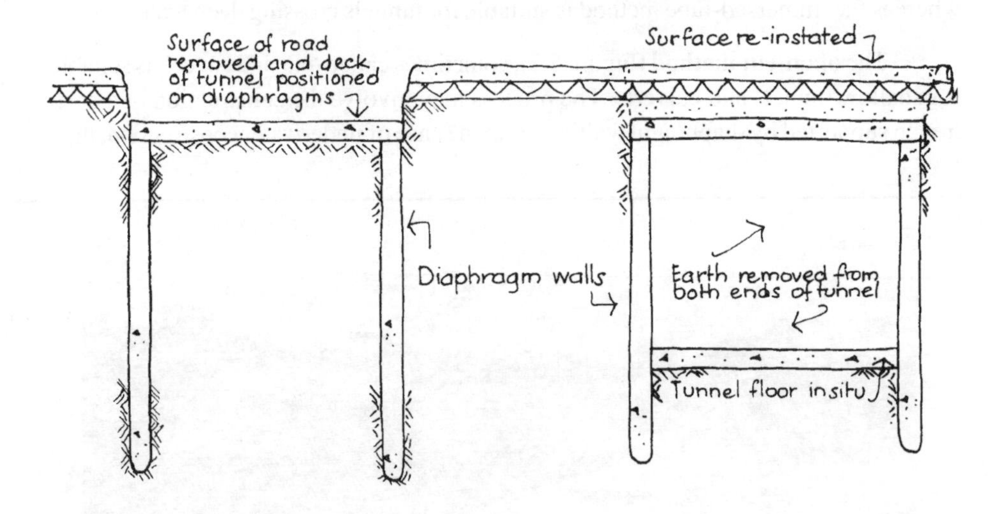

FIGURE 5.4
Submerged tunnel for shallow waters

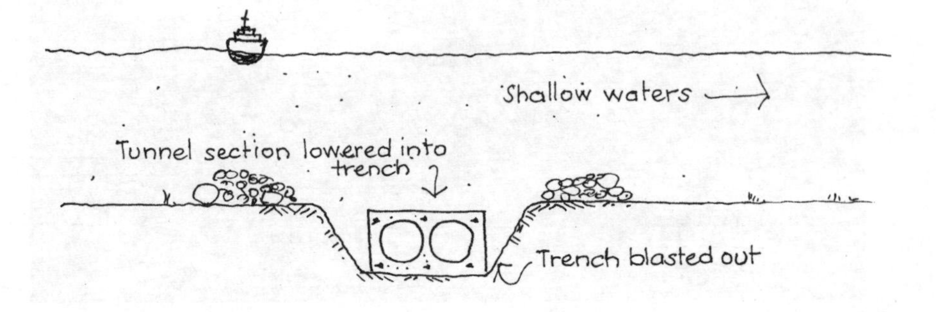

Pipe-jacking

This is a tunnelling technique that has been used in the UK since the late 1950s. The method consists in forming a pit at both ends of the proposed tunnel or pipeline and jacking sections of steel or concrete tube from one pit to the other. The equipment for jacking the tube consists of hydraulic cylinders (Figure 5.5) which thrust against a reaction wall and distribute their load to the tube via a thrust-ring. This method of tunnelling is very suitable for taking services under canals, railway embankments and roads without creating a disturbance to traffic or to the ground.

A pit of the required length to accommodate the jacking equipment and tube sections is dug and for large diameter tunnels this is followed by the construction of a thrust wall or thrust platform (Figure 5.6). The jacking equipment is installed at the desired gradient and the process commences by excavating and positioning the first section of the pipe with a drive shield. Excavation is carried out by conventional mining techniques, the drive shield controlling the working face. The excavated material is brought out through the unit, usually in special trucks which are designed to run on the invert of the tunnel. Recent developments involve the casting of the complete tunnel lining on the thrust base; this is suitable for short to medium distance jacking. The method eliminates cranage problems in handling large sections and allows greater flexibility in size because transportation of units no longer limits the designer. The slotted thrust base is employed as a reaction for the hydraulic jacks.

In the early days this method of tunnelling was limited to jacking pipe diameters of up to 2 metres, but improved jacking equipment can cope with tubes up to 4 metres diameter and box sections of a size up to 7 × 4 metres. Where the larger section unit creates problems in providing suitable jacking reaction, a modified system of jacking can be employed. This system introduces a sleeve, carrying supplementary jacks, into

FIGURE 5.5
Telescopic pipe-jacking station pushing a 2.1 m (o.d.) pipe
(J F Donelon & Company Limited)

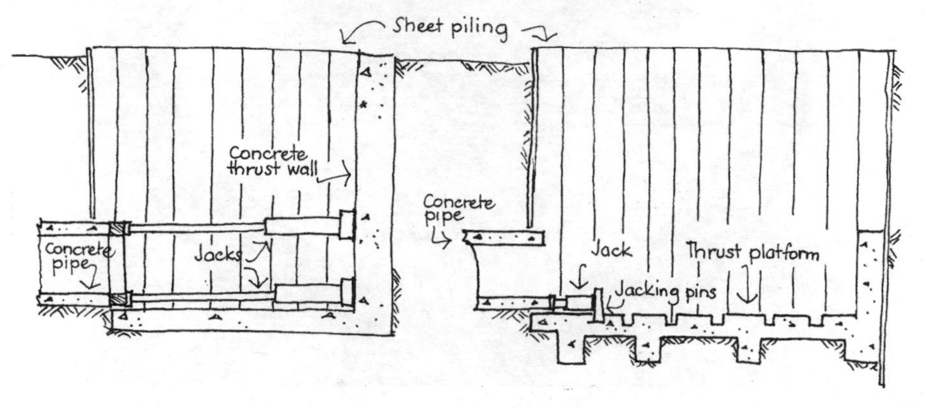

FIGURE 5.6
Thrust wall and thrust platform for jacking

the line of units being jacked. The line is then jacked forward in a caterpillar fashion (Figure 5.7), jacking the first section forward, then using the jacks in the intermediate position, followed by jacking from the pit to close the line. Alternatively, bentonite can be introduced from the rear of the driving shield to reduce friction during jacking.

Drive or thrust-shafts are generally 2 metres longer and 1 metre wider than the unit being jacked. Where the thrust-shaft is in excess of 7 metres deep, the jacking can take place from a bolted segmental shaft with an underground chamber for the jacking equipment. Reception shafts should be of sufficient size to recover the driving shield.

Trenchless technology systems

Trenchless technology systems cover steerable and non-steerable systems for constructing small diameter tunnels, such as those found in drainage and other services. The systems include:

- Auger boring
- Microtunnelling.

Auger-boring

This is method of tunnelling is for conduit installation and uses a non-steerable rotating cutting head attached to flights within a casing. The principle of forming pits and thrust walls is similar to the pipe-jacking technique, but there the similarities cease. The tube is held and located by a rail-mounted machine which augers the soil as it pushes the sleeve or tube into the ground. The auger (Figure 5.8) turns and cuts the face of the borehole and transports the excavated spoil back down the sleeve into the shaft for disposal. When the machine has inserted a length of sleeve, complete with auger, into the ground, the machine is withdrawn and another length of sleeve is welded on and the auger flight extended. This process is repeated until the desired length has been installed in the ground.

FIGURE 5.7
Jacks at an intermediate stage using the first section of the tunnel as a jacking base
(Westfalia Lunen)

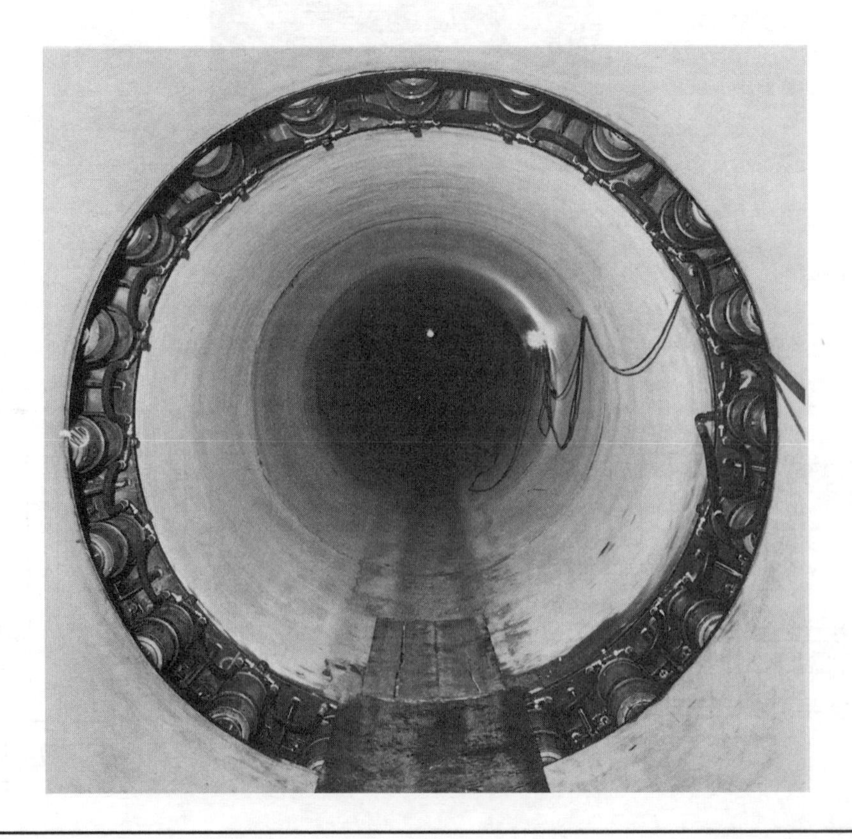

For sewer pipes the base of the pit is constructed to the desired gradient of the proposed sewer. Whereas pipe-jacking methods are capable of driving large-section units through the soil, this system is limited to a diameter of 750 mm and is therefore most suited to small sewers and pipelines up to 600 mm diameter. The finished pipeline may be threaded through the sleeve by means of roller-mounted spacing legs and the annular space between the sleeve and pipeline can be grouted with a PFA cement-bentonite mixture.

Microtunnelling

Microtunnelling is a technique used for conduit installation using steerable, remote controlled tunnel boring by pipe jacking. The excavated material is removed by mechanical auger or as a slurry. Microtunnelling machines (Figure 5.9) have been developed to work from drive shafts in almost any type of ground conditions. However, it is essential to know what the ground conditions are, so that the correct equipment can be used to cope with excavation, spoil removal and the jacking forces involved.

The only surface excavation required is for the driving and reception shafts. Spoil is normally removed from the cutting face by means of an auger running through the newly installed pipeline, to a skip at the base of the driving shaft. An alternative method of spoil disposal is to use water or bentonite to reduce the spoil to slurry at the cutting face; the slurry is then pumped to the surface where the soil and liquid is separated for disposal.

The control system for microtunnelling is normally at ground level in the form of a console and the direction of the cutting machine is monitored by means of laser equipment. Accuracy depends on the skill and experience of the operator but can be in the region of 10–25 mm in either the horizontal or vertical plane.

The system is commonly used in drainage installation where surface disruption has to be kept to a minimum. The technique provides the required degree of accuracy

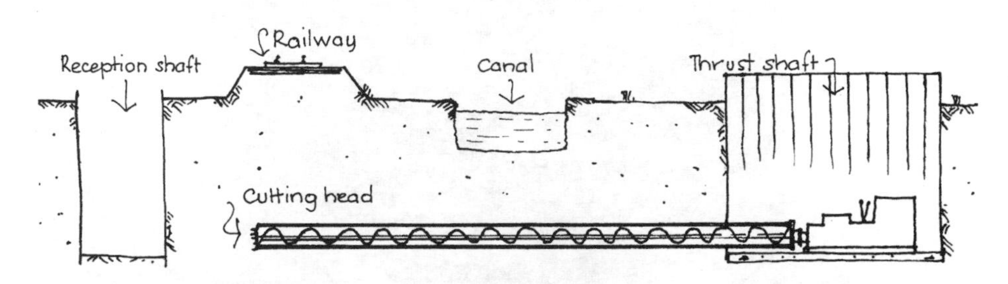

FIGURE 5.8
Auger boring operations

FIGURE 5.9
Microtunnelling machine on jacking rig (Euro Iseki Limited)

for this type of work. The drive and reception shafts are normally placed in the positions of the required inspection chambers in the finished scheme. Driving shafts (Figure 5.10) vary in size from 2.3 m diameter to 5.28 m diameter, depending on the size of pipe being driven; machine diameters range from 375 mm diameter to 1100 mm diameter; and pipe sizes range from 250 mm to 900 mm internal diameters. Reception shafts (Figure 5.11) are smaller, ranging from 2.3 m to 3.0 m diameter.

Clay and concrete pipes are used with the system and these are jacked through the bore behind the cutting tool. In some cases temporary steel tubes are jacked in and removed at the next inspection chamber position allowing the new pipeline to follow in the established bore.

The technique is concerned with boring tunnels too small for the entry of personnel and cover diameters up to 1000 mm, which constitutes 99% of all urban underground pipes and conduits.

5.1.3 Tunnelling in rock

Detailed investigation of the rock will allow the designer and contractor to classify the rock in terms of construction method. This classification would be:

- Hard rock – unsupported or lightly supported
- Hard rock – heavily supported
- Soft rock – unsupported or lightly supported
- Soft rock – heavily supported
- Soft rock containing hard intrusions.

FIGURE 5.10
Tunnelling machine positioned in driving shaft
(Euro Iseki Limited)

FIGURE 5.11
Tunnelling machine appearing at the reception shaft
(Euro Iseki Limited)

Hard rock tunnels are driven through rock with a minimum compressive strength of 100 N/mm^2. There are three methods of driving tunnels in hard rock: whole face boring, the use of roadheaders, and conventional drilling and blasting.

Whole face boring employs a machine which is specially designed to suit the particular cross-section required (Figure 5.12). The disadvantage is the high capital cost involved in the original purchase of the machine, together with the specialist maintenance required; the advantages are that a smaller workforce is employed and high rates of progress are possible. The debris from this type of machine would normally be removed from the tunnel face by conveyor.

Whole face borers are capable of cutting rock which has a compressive strength in excess of 300 N/mm^2; however, it is usually uneconomical to employ whole face boring equipment on rocks that have a strength exceeding 130 N/mm^2. The disadvantage of this method of driving is the inflexibility of the machine to corrections in alignment and changes of direction. The cross-section is always circular, 'overbreak' is minimal and hence only the theoretical amount of concrete is required when lining. Whole face machines do not perform well in badly broken ground such as fault zones.

Roadheaders are purpose designed excavating machines, with powerful cutting booms, for use in coal mining or civil engineering (Figure 5.13). They can be designed or adapted to suit contractors' shields and excavators and have the advantage over whole-face machines of being adaptable for a range of tunnel diameters. The machines

FIGURE 5.12
Whole-face boring machine
(Wirth Howden)

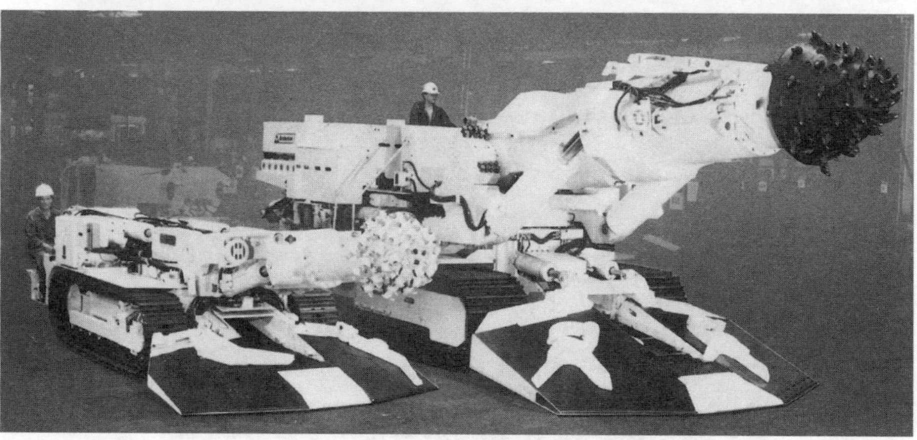

FIGURE 5.13
Roadheader for low height excavation
(Anderson Group Limited)

are track mounted and use hydraulically operated cutting booms. They employ a range of extra equipment, such as gathering arms for the debris, water jets for reducing the dust and conveyors for removing the debris from the face of the tunnel. Figure 5.14 shows a roadheader breaking through into the access shaft; after completing a 800 m tunnel excavation in strong sandstone, the tunnel diameter was 4.1 m.

Where shield protection is required during excavation the cutting boom can be fitted to a turret. These in turn are fitted to a sliding frame which allows the machine to be advanced or withdrawn depending on the activity at the tunnel face. Figure 5.15 shows an adapted machine for circular tunnel working. Alternatively, the cutting booms can be fitted to conventional hydraulic excavators providing a machine that has high reach for large excavations.

Conventional methods of driving tunnels have four stages of operation: drilling, blasting, mucking out and supporting. Drilling is achieved by percussive or rotary percussive methods. In order to reduce health hazard, the dust is suppressed by means of water or by dry extraction. The drilling equipment may be mounted on main line trolleys or gantries, depending on the tunnel diameter (Figure 5.16).

Hard rock tunnels which are heavily supported require a reduced 'round length' (length of blasted rock face) to eliminate rock fall. The amount of time involved in supporting the tunnel before another section can be drilled and blasted makes it uneconomical to employ a large heading gang, so the gang size and mucking equipment is arranged to suit the progress of tunnel support. Most hard rock tunnels are bored by an almost identical process; the rock face is drilled to a drilling pattern which will best suit the quantity of rock to be moved, and charged with the necessary explosive. The pattern, depth and direction of the drilling depends on the size and shape of the tunnel and on the type of rock encountered. The explosive used may be one of the types described

FIGURE 5.14
Roadheader breaking through into shaft
(J F Donelon & Company Limited)

FIGURE 5.15
Roadheader adapted for working in circular tunnel
(Anderson Group Limited)

in Chapter 8, which are detonated in a pre-arranged sequence, working from the centre to the circumference. When the charges have been laid and fused, the drilling equipment and personnel are moved to a safe place. Smoke, dust and fumes are rapidly cleared by an auxiliary ventilation system which ventilates the tunnel by displacement at a rate of 100 m^3 per minute per square metre of face.

The equipment for mucking will vary according to the cross-section of the tunnel. In small tunnels muck cars and appropriate loading equipment may be used, whereas in large cross-sectioned tunnels conventional track or tyre-mounted shovels are used to load dumpers or tunnel wagons. An alternative method of transporting the spoil, providing the haul distance in one direction is limited to approximately 400 metres, is by low haul dump machines; these machines are particularly useful at tunnel intersections. Conveyor belts may also be used for transporting the spoil.

5.1.4 Tunnelling in soft ground

Tunnelling through clays and non-cohesive materials present a number of problems for the engineers. The amount of water present in the ground will have a great effect on the difficulty and manner of operations. For example, some clays and dense dry gravels will stand unsupported over a short span for a limited time, whereas waterlogged silts will immediately find a way through the smallest opening in the support. Some form of tunnelling shield is required for tunnelling in soft ground. The work is carried out either by whole-face boring machines (Figure 5.17), which may employ bentonite slurry as a means of preventing the ingress of water, or by small excavators, which will depend on the diameter of the tunnel, or by hand. In most cases some form of tunnelling shield is required to protect the working operations. The essential parts of a shield are a hollow cylindrical skin supported by substantial diaphragms which provide protection near the tunnel face, a cutting edge, and a system of hydraulic jacks to drive the shield forward.

FIGURE 5.16
Hard rock tunnelling using drill and fire techniques
(J F Donelon & Company Limited)

FIGURE 5.17
Breakthrough of 3.150 m (o/d) tunnel boring machine into access shaft
(J F Donelon & Company Limited)

The diaphragms in the tunnel shield contain openings to give access to the face and house the jacks which thrust against the completed lining to force the shield forward; the cutting edge trims the ground to an accurate shape.

Excavation through the diaphragm opening can be executed by hand or by various types of mechanical excavator if the ground is able to stand unsupported for a short time. Figure 5.18 shows hand excavation during the installation of a 2.44 m (i/d) tunnel.

Loose dry sands and gravels will require temporary support at the face to control the amount of ground-flow into the tunnel. Waterlogged ground can be tunnelled by the use of compressed air techniques in which the working chamber is sealed and the internal air pressure raised to balance the pressure of the ground water; access for personnel and materials is achieved by air locks. The effect of this process is to prevent water from entering the tunnel and to stabilise the ground. Grouting such ground, in advance of the tunnel excavation, can also make the work less hazardous by increasing the strength and reducing the permeability of the ground. In addition a grout arch may be formed to reduce the over-burden pressure during tunnelling operations and stabilises the soil at the entry and exit zones in the access shafts (Figure 5.19).

FIGURE 5.18
Hand excavation in clay
(J F Donelon & Company Limited)

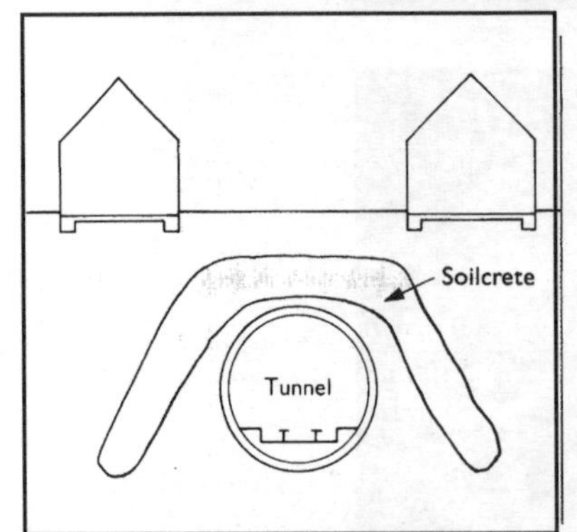

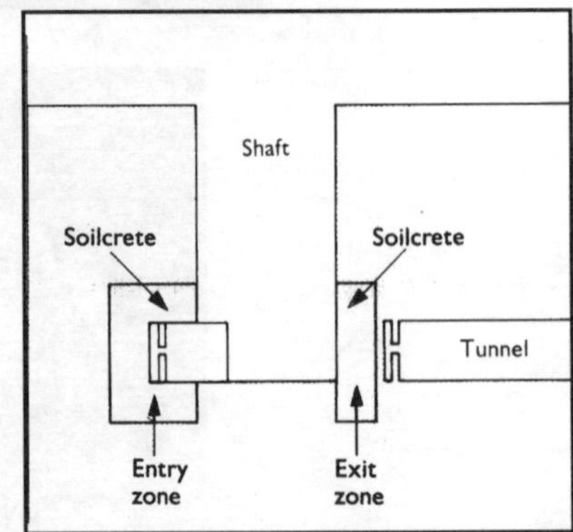

FIGURE 5.19
Soilcrete stabilisation to tunnel and access shaft
(Keller Piling Limited)

Disposal of debris is generally achieved by some form of mine car, mounted on rails in the case of smaller tunnels. Loading these cars is either done manually or mechanically; commonly some form of conveyor is used to raise the excavated material and load it into the car. Conveyors have been used as an alternative to mine cars and track. Few soft-ground tunnels are large enough to use tyred or track-laying equipment.

Tunnel linings

All tunnels have to be lined but the process depends on the type of ground. The lining can either be preformed segments or insitu concrete, the latter being either sprayed in position or cast in place.

In soft ground the linings have to support the loads which are imposed by the ground and may also have to withstand the jacking pressures from a driving shield. They may have to resist water pressure, at the same time having qualities which permit ease of handling and construction. All these factors, together with the cost and speed of construction, have led to the use of pre-formed linings as opposed to in-situ linings. However, slip-form techniques are being used to offset the time factor in tunnels which do not need temporary or early support by the linings.

A popular form of lining used in soft ground is the segmental cast-iron type; it is very durable, resistant to corrosion and strong in compression. It is used in complicated and difficult conditions or where point loads are expected. Precast concrete linings are more popular because they are less expensive (Figure 5.20), but they are more difficult than cast iron to make watertight. Cast steel, pressed steel and welded prefabricated steel linings are also used in soft ground.

Expanding tunnel linings can be used in ground that is self-supporting and has been cut to a true circular section. The linings are inserted behind the tunnelling shield, which has been pushed forward to leave an unsupported space. The lining, which consists of precast segments, is erected and expanded against the ground to a predetermined pressure. This type of lining does not require grouting.

FIGURE 5.20
Precast concrete linings on magazine rack prior to assembly (J F Donelon & Company Limited)

In-situ tunnel linings differ from other forms in that the process gives a smooth lining without undulation, at the same time giving full support to the surrounding earth without further grouting. The lining may be installed by a continuous process, such as the slip-form technique, or by traditional forms mounted on rail tracks.

In the former method of construction, the driving shield is connected to a 60 metre long box-frame which supports the forms and carries away the spoil. The long box-frame is mounted on tracks for easy movement but is supported by legs during concreting. The excavated spoil is transported by conveyor through the box-frame and out through the competed tunnel.

If the tunnel is formed in hard ground the lining may take the form of sprayed concrete. These linings may be sprayed on to mesh which has been fixed to the rock face, or sprayed between steel ribs. Further details of the spraying technique are discussed in Chapter 8. Important economies can be gained when employing this technique if an undulating lining, following the rock face as distinct from filling to a shutter line, is acceptable: the thickness can be varied to suit rock conditions, and rock bolting may be eliminated. Figure 5.21 shows a sprayed lining that follows the undulating face of the tunnel.

New Austrian Tunnelling Method (NATM)

Sprayed linings have become popular in rock tunnels with the development of the New Austrian Tunnelling Method (NATM). In this technique the sprayed concrete is finished to an acceptable line and thickness. The system was designed by Pantex Stahl AG in Switzerland, and is produced under licence by Caledonian Mining Company Limited (CMCL) in the UK. The system employs lattice girders, metal mesh and shotcrete. Shotcrete is sprayed concrete that has an aggregate size normally up to 15 mm, but can be up to 25 mm. Steelcrete, a product of CMCL, is a patented type of shotcrete, which contains some form of reinforcement – either stainless steel or polypropylene fibres (see Chapter 8 for more details).

The lattice girders, known originally as Pantex-Lattice-Girders, can have either three or four main bars, spaced by diagonals. Figure 5.22 shows the lattice girders being assembled in a rock tunnel. An alternative to the single top bar, in the three bar lattice, is a kidney-shaped bar that provides extra reinforcement. The system has been used on projects all over the world, including the Channel Tunnel and the Jubilee Line Extension for the London Underground network.

FIGURE 5.21
Sprayed tunnel lining following the undulating surface of the tunnel
(Aliva Limited)

The main advantage of the NATM system is the ease of handling the components. The lattice girders are produced in sections that can be lifted and bolted together without difficulty. The completed girders are tied to others by linking reinforcement bars, and the spaces between the girders are reinforced with mesh prior to the application of the shotcrete. All these operation are carried out with ease and speed.

Figure 5.23 shows the assembly of Pantex-Lattice-Girders and the finished shotcrete up to the last bay of the tunnel lining. Figure 5.24 shows various stages of the formation of the finished lining, the lattice girders, the reinforcing mesh and the shotcrete finish.

FIGURE 5.22
Assembly of Caledonian-Pantex lattice girders (Caledonian-Pantex)

FIGURE 5.23
Assembly of Pantex-Lattice-Girders and view of shotcrete finish to tunnel lining (Pantex Stahl AG)

FIGURE 5.24
Various stages of the formation of the NATM lining (Pantex Stahl AG)

5.1.5 Use of compressed air

The health hazards encountered in the use of compressed air are discussed in Chapter 1, Section 1.2.8, on Safety, and compressed air plant is discussed in Chapter 2. There are three basic uses for compressed air in tunnelling:

- The safety of the workman
- The safety of the property above the tunnel, and
- Adequate progress in otherwise troublesome ground.

When excavation in water-bearing ground can be undertaken in a confined space, eg in tunnelling, shafts and caissons, it is possible to exclude the water by increasing the pressure within the working space. This technique of dealing with ground water is a traditional method of undertaking underground construction work where soil conditions are difficult because of the presence of water. It is especially suitable for excluding water from fine silts or soft clays which may be difficult or expensive to treat by grouting processes. It is also suitable for work in water-bearing rock where there is too much water for pumping methods to handle.

In some cases the use of compressed air may be combined with other techniques, such as dewatering and grouting: the air pressure can then be reduced in the working chamber if the surrounding ground water level is reduced. In other cases ground treatment may be necessary to prevent excessive loss of pressure at the face; when air escapes through zones of weak soil and reaches the atmosphere or residual air at lower pressures, serious adverse conditions may arise over a short period of time.

One method of sealing the face of excavation against air loss, if the soil is very fine, is to spray the face with bentonite; this immediately seals the ground and thereby reduces air losses. A great deal of air is also lost through tunnel linings, owing partly to cracked segments and partly to the use of pervious materials. Caulking should be carried out as near to the face as possible to prevent excessive loss of air in the tunnel. Unfortunately, caulking in compressed air conditions may lead to re-caulking in free air, since the former conditions do not show leaks. It is also essential to have sufficient ground cover to balance pressures in the excavation; particular attention must be given to this problem when tunnelling below or adjacent to free water, buried channels or culverts and old pile foundations: any or all of these obstacles can cause sudden loss of air. In cases where the ground cover is insufficient to balance the air pressure, it will be necessary to load the ground surface to give suitable protection.

The limiting factors of using compressed air on its own are the pressure in which the workforce can be allowed to work and the loss of air pressure through the working face, which may prove uneconomical.

5.1.6 Safety aspects

Safety aspects in tunnel construction may be divided into two sections, namely, safety in soft ground and safety in hard rock.

Safety in soft ground

The face of every tunnel and base and crown of every shaft requires inspection by a competent person at the commencement of every shift. Materials for supporting the ground must be examined by a competent person before use and the erection and dismantling of supports must be directed by a competent person.

One of the main hazards in soft ground is the movement of soil; this may cause pipelines to settle and fracture, creating dangerous conditions above or in the tunnel. The development of high-pressure jacking equipment has brought about hazards of spilt oil and high-pressure blow-outs. The design of temporary works for holding the face of a tunnel requires special attention: all supports should fit in position quickly without the possibility of moving out of position. This reduces the time that the face is left open. Accidents can occur during the mucking operation and it is therefore essential that adequate illumination be provided at the working face and protection given against moving muck-waggons or conveyor belts. Provision should be made, if room allows, for the walkway to be separated from the haul-way. Spillage of soft material can create slippery conditions and precautions should be taken to eliminate this hazard so far as is practicable. Where laser equipment is being used for alignment of tunnels personnel must be protected from the rays emitted.

A final note on safety in compressed air working concerns the hazard of fire: this hazard is very high because of the excess oxygen in the working chamber. Whilst it is easy to prevent smoking it is not so easy to eliminate burning and welding operations. Provision should be made to deal with the outbreak of fire.

Safety in hard rock tunnels

The hazards peculiar to tunnelling in rock can be summarised as follows:

- Rock falls
- Blasting and use of explosives
- Dust and fumes
- Noise
- Movement of heavy plant.

Small rock falls may occur in large tunnels during scaling operations, where blasting has not cleared away all loose material. The decision to support the rock depends a great deal on the rock condition, and the type of support depends on the availability of plant and material, together with economic considerations.

Explosives must be used only by experienced personnel; however, even with experienced personnel dangerous practices can develop. Failure to remove sources of electric power, checking the circuits at the face instead of from the blasting position, and smoking at the face, are practices which should be avoided. When dealing with 'misfires' these must be washed out with water; alteratively relief holes may be used.

Dust and fumes are best dealt with by efficient ventilation (see Section 5.1.7). The concentration of dust and fumes should be checked regularly: there is immediate danger from fumes if the workforce return to the face too soon after blasting, whereas dust is a long-term hazard which could give rise to Pneumocosis. It is recommended that twenty minutes should elapse between blasting and a return to work, but this depends on the nature of the ventilation system; tests should be made for the presence of carbon monoxide and carbon dioxide.

Noise created by heavy plant and drilling machines produces a short-term problem of poor communications which can lead to accidents. In the long term it can have a serious effect on the hearing. These problems can be overcome by wearing ear protection. Movement of plant, especially plant concerned with muck moving, is responsible for many accidents in tunnelling. Particular problems may arise when coupling and uncoupling skips and when re-railing skips which have jumped the lines.

5.1.7 Ventilation and lighting

Tunnelling work requires efficient ventilation to clear the dust and fumes, or simply to provide fresh air at the working face. Two methods of ventilation are employed; the exhaust system and the blown air system.

The **exhaust system** draws the foul air and dust into the duct near the working face, thereby causing fresh air to flow the full length of the tunnel. With the exhaust system it is necessary to use rigid ducting to prevent its collapse under conditions of partial vacuum. Since the amount of air required for clearing blasting fumes is much greater than the amount required for normal working conditions, the ventilation system should be equipped with variable-load fans to remove quickly the dangerous fumes.

In the **blown-air system**, air is released at the face and the pressure carries the foul air back down the tunnel, which may be detrimental to workers in other parts of the tunnel. As mentioned above, a combined system may be used; this has the advantage of exhausting the noxious fumes after blasting before reverting to the blowing method. Care must be taken to avoid 'dead' areas in long tunnels.

Some installations use both systems; this is achieved by a two-way valve and duct arrangement which can be changed to suit conditions. Economy in ventilating depends on the airtightness and resistance of the duct: loss of volume or a drop in pressure lead to high running costs.

Adequate lighting in tunnels is a prerequisite to efficient working. A light intensity of 250 lux in the working area is necessary if the workforce is required to work at maximum efficiency. This service is further complicated by the fact that lamps should be withdrawn from the face prior to blasting and then returned for working. The size and spacing of lights depends upon factors such as tunnel dimensions and rock conditions. In addition to the normal lighting, an emergency reserve circuit, supplied by batteries or a generator, is desirable.

5.2 UNDERPINNING

This section on underpinning deals mainly with heavy construction and structures such as those found in civil engineering. Although some of the methods discussed apply equally to the underpinning of lighter structures, it is suggested that such work be studied as part of building construction as distinct from civil engineering.

5.2.1 General considerations

Before underpinning is resorted to, a full investigation should take place to ascertain its feasibility. The object of underpinning is to transfer foundation loads from existing levels to new safer levels at greater depths. The operation may be carried out for one or all of the following reasons:

- To allow the adjacent ground level to be lowered
- To increase the load-bearing capacity of the foundation
- To prevent or arrest settlement of a foundation.

If settlement is involved, the following questions must be considered:

- Is the damage significant?
- Is the damage progressive or static?
- If the damage is progressive, is it likely to become serious?
- Is the cause due to subsoil failure?

Investigation of the ground below the structure is essential before work commences; such an investigation will assist in the determination of:

- Conditions responsible for settlement
- The nature of ground, its characteristics and load-bearing properties
- The depth at which the new foundation can be established.

The structural stability of the building being underpinned must be checked for inherent weaknesses which may develop during the underpinning operations. Precautions should be taken to reduce the load on the structures being underpinned; this can be achieved by propping, shoring and removal of live loads within the structure. High walls should be checked for plumb before and during the underpinning work; datum levels should be established and checked frequently as the work proceeds.

Before excavation of supporting ground is commenced, the support of the excavated face must be carefully planned. This could take the form of pressure grouting with cementitious grout; alternatively, normal trench supports may be employed. The underpinning should be executed in stages which allow a gradual transfer of load without risk of settlement.

5.2.2 Methods of underpinning

The method of underpinning chosen will vary with the need for underpinning. The following examples will be considered:

- Underpinning of settlement due to shallow-seated faults
- Underpinning of settlement due to mining subsidence

- Underpinning against the effects of adjacent excavation
- Underpinning to move structures.

Settlement due to **shallow-seated faults**, such as compressible soil, is easily arrested when compared with mining subsidence. The foundation can be supported by a new concrete foundation, which is excavated in alternative strips (Figure 5.25); the new concrete foundation fills the excavation and is vibrated to pick up the old foundation.

An alternative method is the 'jacked pile' or 'mini pile' (Figure 5.26). These are often thick-walled steel tubes which can be welded if an extension is required for depth purposes. Diameters of the tubes vary from 90 mm to 250 mm, able to carry loads from 40 k/N to 250 k/N. The tubes are jacked down to a load-bearing stratum. Double angle piles may be used if access to the inside of the building is possible (Figure 5.26b).

Underpinning for **mining subsidence** differs from traditional underpinning in that it is uneconomical to sink a foundation to a stratum which is not subject to settlement. There are two broad methods of dealing with this problem:

- firstly, to form a raft or series of continuous beams under the building, thereby distributing the load over a large area and thus preventing local fracture due to ground movement;
- secondly, to construct jacking points under the building so as to permit jacking back to the vertical and horizontal positions.

The latter method has been used in the construction of large industrial complexes in which settlement due to coal mining is inevitable. The pedatifid raft system, installed by Pynford, has been successfully used on a large automatic plant near Nottingham. The structure was built on a slab approximately 49 × 18 metres. Jacking positions were arranged in four rows which divide the slab into three equal areas.

The jacking points, known as 'pedatifid pockets', are in pairs spaced at 2.4 metres centres along the rows (Figure 5.27) which are parallel to the longer axis.

The overall depth of the movable slab is 600 mm and a series of longitudinal beams, cross-beams and covering slab form a waffle plate design (Figure 5.28) around the jacks. The waffle-plate slab, which stands in a recessed sub-base, has its top surface level with the adjoining factory floor. When the floor requires re-levelling, the space beneath the raft is sealed so that air pressure can be used to reduce the weight of the slab on the jacks. After lifting the slab to the required level it is permanently supported at all the pockets.

FIGURE 5.25 *Underpinning foundations with concrete fill*

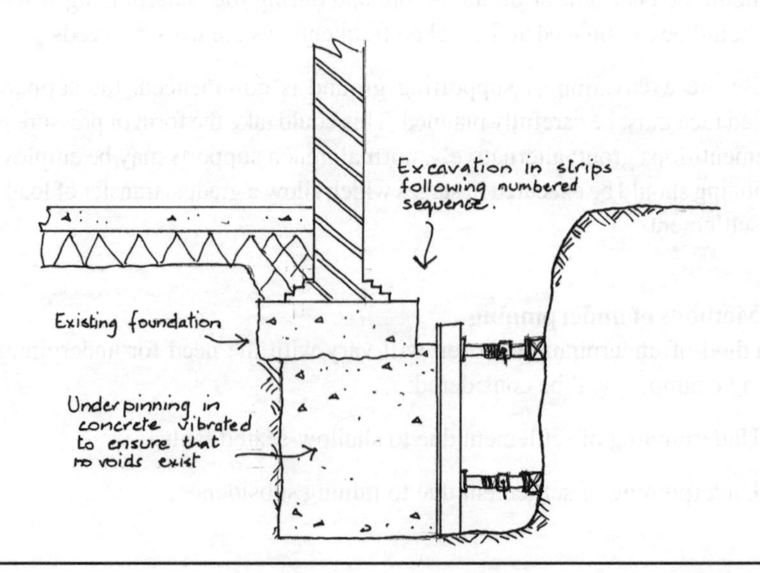

The pedatifid system is designed so that access to the jacking positions can be obtained from above the raft level by removing steel plates in the floor. These plates are bolted down to the raft with substantial bolts anchored into the slab; the jacks are arranged in pairs so that either one can carry the full load at the pocket during lifting operations. To lift the slab, jacks are mounted on top of short columns passed down through the pockets to the sub-slab; the main slab is then jacked to the desired level and supported by short columns within the pockets, which have been cut to fit between the jacking point and the sub-slab. The jacks are then removed in sequence and the cover plates are bolted down to hold the slab.

The principal advantage of this system is that it is not necessary to provide a shallow basement beneath the slab to give access to the jacks and jacking positions. Another advantage is that, due to the small size of the pockets and the fact that they can be placed anywhere to suit access above floor level, it is possible to provide many points of support thus reducing the depth of the floor beams.

Underpinning against the effects of **adjacent excavation** can be accomplished by several methods:

- Traditional underpinning executed in sections
- Pile and beam support
- Diaphragm walling
- Grouting techniques.

(a)

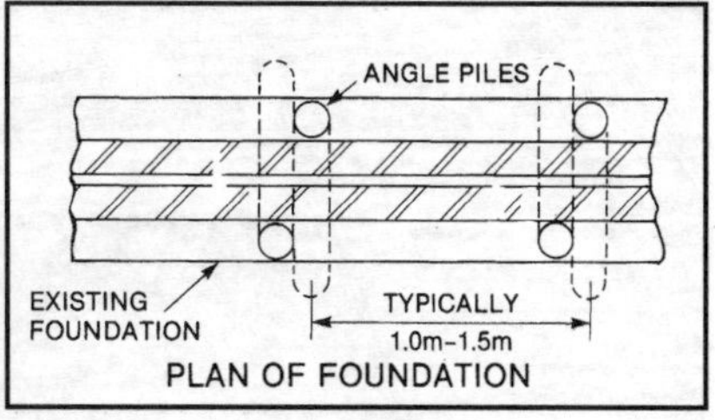

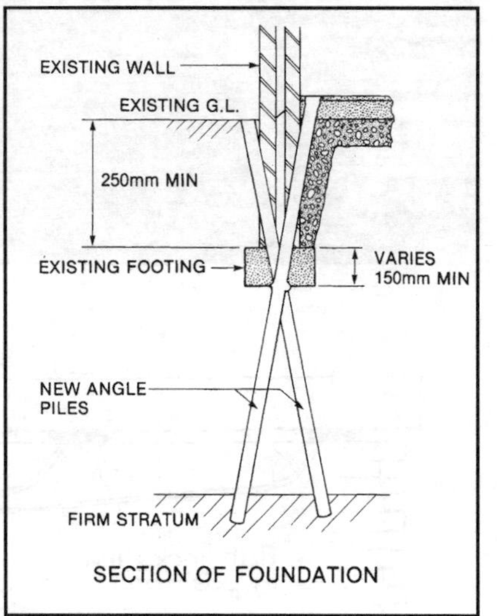

(b)

FIGURE 5.26
Underpinning with thick-walled tube

(a) Installation of external pile

(b) Layout for double angle piles
(Roger Bullivant Limited)

FIGURE 5.27
Pedatifid pockets in pairs
(Pynford)

FIGURE 5.28
Formation of waffle plate around pedatifid pockets
(Pynford)

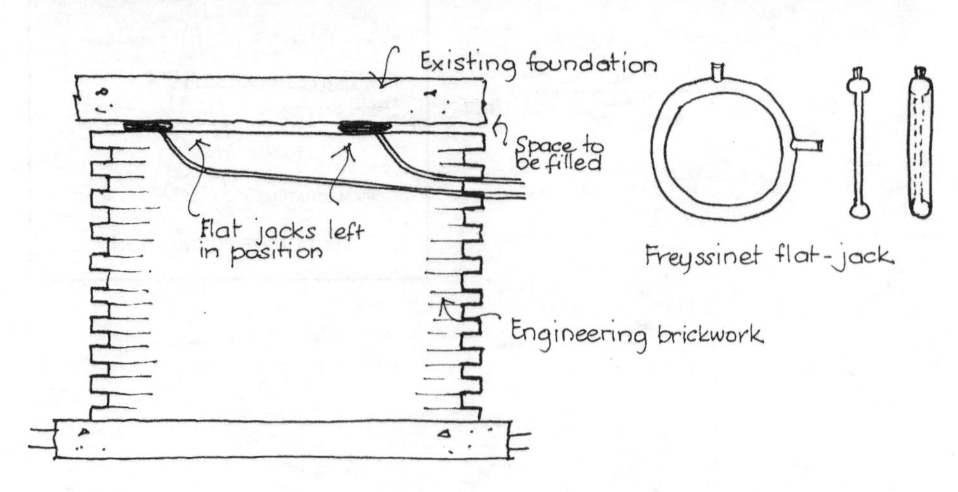

FIGURE 5.29
Freyssinet flat-jacks used in underpinning

With care it is possible to achieve load transfer from the existing structure on to the new underpinning without any settlement. This can be achieved by use of the 'Pretest' method, in which the newly formed base can be pre-loaded to prevent any initial settlement. Variations of the pre-loading mechanism are:

- Hydraulic jacking
- Freyssinet 'flat-jack'
- Pynford 'stools'.

'Pretest' methods of underpinning are used in conjunction with normal underpinning activities in which 'legs' of brickwork or concrete are brought up to a required level and after hardening are used to support the 'pretest' system. In the case of hydraulic jacks, beams support a layer of wet concrete which, when jacked up against the existing foundation, squeezes into and around all the projections. After this concrete has hardened, the space between the legs and concrete beam is filled by conventional methods. This technique ensures that the underpinning is fully supported by the new foundation.

The Freyssinet 'flat-jack' method works on the following principle: the jacks, which consist of flat, hollow, light-gauge, metal canisters approximately 250 mm in diameter, are placed between the legs of underpinning and the existing foundation. Oil is introduced under pressure and the jack expands to stress the new foundation against the existing. When the desired stress has been reached an epoxy resin is then introduced and the injector pressure pipe is sealed off. The jacks are left in place and the gap between the new wall and existing foundation is filled (Figure 5.29).

In the case of the Pynford 'stool', the new wall is built to a height which will receive the 'stools' and these are jacked up against the old foundation. Steel reinforcement is then place in position and a concrete beam is then cast. The concrete is flooded to a height above the bottom of the existing foundation to allow the concrete to pack tightly under the foundation, thus reducing any risk of settlement. Figures 5.30 to 5.32 show the Pynford stool technique being used to underpin vaulted arches and to transfer the loading via columns to a new foundation.

FIGURE 5.30
Pynford stools supporting a vaulted arch
(Pynford)

FIGURE 5.31
Reinforcement added to the Pynford stools
(Pynford)

FIGURE 5.32
Completed concrete beam supported by new concrete columns
(Pynford)

FIGURE 5.33
Grease skate
(Pynford)

Cementitious grouting is very suitable for underpinning ground of a gravelly nature. The soil is removed to the level of the top of the existing foundation and grout tubes or drill holes are inserted. The grout is injected under pressure to form a wall of solid soil which both supports the existing foundation and allows excavation without complicated shoring. This method of underpinning has been used to strengthen the foundations of cooling towers; a two-shot silicate-based process was used, sodium silicate being injected as the lances or injection tubes were driven in and calcium chloride being added during withdrawal (see Chapter 3 for further details on grouting).

Underpinning to **move structures** is employed when structures require moving without resorting to extensive demolition. The weight of the building or structure is transferred from the foundations to a series of wheeled carriages: this is achieved by a system of beams which penetrate and support the walls. The carriages are mounted on rails which are laid in the required direction of movement: changes in direction can be achieved by relieving any carriage of its load and swinging the rails into correct alignment. Such operations are carried out by specialist firms who carefully survey and brace the structure to prevent damage during the operation. New foundations are prepared to receive the structure when it has been re-positioned and the structure is carefully lowered on to the foundation by means of jacks. This method of underpinning has been used for moving framed buildings up to five storeys high for a distance of 0.4 km.

An alternative method of moving structures is by the use of 'grease skates' (Figure 5.33), which are suitable for the movement of heavy structures over short distances. Grease skates can reduce the friction during the slide to around 1% of the original load of the structure. In Norway a 2500 tonne hotel was moved, by Pynford, requiring only 250 kN pulling power to move the whole structure. Figures 5.34 and 5.35 show the technique being used to move a railway bridge.

FIGURE 5.34 *(Left)* *Preparation of the skid path to move the bridge* *(Pynford)*

FIGURE 5.35 *(Right)* *Grease skates in position under the bridge wall* *(Pynford)*

CHAPTER 6

Marine and Other Works Associated with River or Groundwater Environments

6.1 COFFERDAMS

6.1.1 Introduction

Cofferdams are usually temporary structures which may be employed to assist in the formation of foundations. Their function is to provide a working area at foundation level from which ground and water is excluded sufficiently to permit safe working. It should be noted that cofferdams do not necessarily exclude all water, since it would be uneconomical to attempt to do so.

It is common practice to use steel sheet piling for cofferdams, but although this material is ideally suited, being strong and relatively watertight, it is not the only suitable material, especially with the increased use of diaphragm walling. Material used in the construction of cofferdams should suit design requirements and these may involve earth, rock, steel or concrete.

The choice of material and type of cofferdam will depend upon the following conditions:

- Location of cofferdam, eg onshore, estuarine, offshore.
- Depth and size of excavation.
- Type of overburden and substrata.
- Volume of water, velocity of flow, tide levels.
- Availability of materials.
- Accessibility of site.

If the head of water is low, then an earth-fill cofferdam may suffice; if the water is likely to erode or undermine the dam, a face of impervious material may be used to protect the earth-fill, the fill giving weight and support to the face material. Before selecting the type of cofferdam to be used, a full site investigation should be carried out. Borings should be made and samples of soil tested, in relation to both permanent and temporary works. It is essential when constructing cofferdams to assess the insitu permeability and strength of the soil: this will assist in determining the depth to which the cofferdam sheeting can be driven to prevent 'blow' or 'boiling'. (A 'blow' or 'boil' is caused by the flow of water and soil, usually fine silt, into the bottom of an excavation, due to the water pressure outside the excavation being greater than that inside.) When cofferdams are constructed in fast-flowing rivers the site investigation should assess the problems which may occur due to current-velocities and wave action.

6.1.2 Types of cofferdams and methods of construction

The main types of cofferdams and ground support are shown in Figure 6.1, the upper part of the chart showing the usual types, the lower part showing those which utilise steel sheet piling. It will be seen from Figure 6.1 that certain types are suitable for both water and land cofferdams whereas others are suited only to one or other; for this reason the selection of the type to be used requires detailed consideration, and for this a full site investigation report is necessary.

Selection of cofferdams

When selecting a suitable cofferdam for a specific problem the following factors should be considered:

- Site investigation report
- Whether the cofferdam is required on land or in water
- The size of the working area required inside the cofferdam
- The nature of the permanent works to be built
- The amount of water or soil to be excluded whilst work proceeds in the cofferdam
- Soil conditions
- Water conditions, i.e. strength of flow, wave action, tide or groundwater range
- Availability of materials and plant
- Possible effect of the cofferdam construction on adjacent structures
- Possible methods of constructing the cofferdam
- Cost in comparison with other solutions.

The final choice may result in a combination of cofferdam types, eg sheet piling and ground stabilisation, or even in a change in mode of support, eg choosing a caisson rather than a cofferdam.

FIGURE 6.1
Cofferdams and other ground support systems

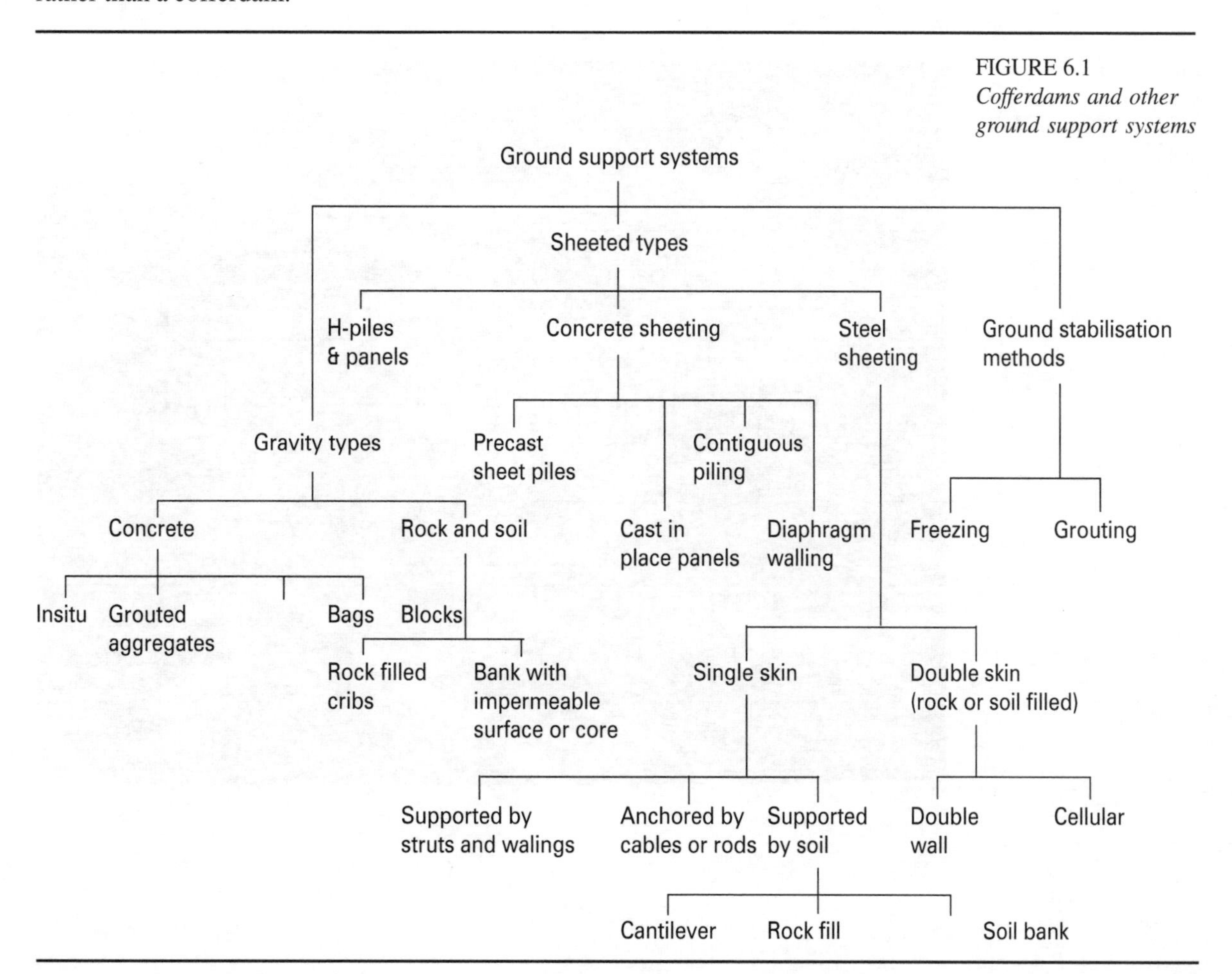

Gravity cofferdams

Earth and rock-fill cofferdams are the most common gravity type (Figure 6.1). They are enclosures formed by banks of soil or rock. They are very suitable for protecting large areas of excavation against flood waters and are best constructed in the dry during low-water periods. If the velocity of water is small and the head of pressure low, then earth fill is suitable; if the velocity is likely to be high, rock fill offers better resistance to scouring if founded on similar material.

The side slopes of the dam will depend upon the size and shape of stones; where rock of sufficient size is not available, a quantity of smaller rocks may be enclosed in wire-mesh nets or baskets (see Section 6.6 for 'Gabions'). The water at high tide or flood level may be controlled by sealing the dam with clay or other suitable fine material (Figure 6.2). Water that flows beneath the dam can be intercepted by a ditch and drained away.

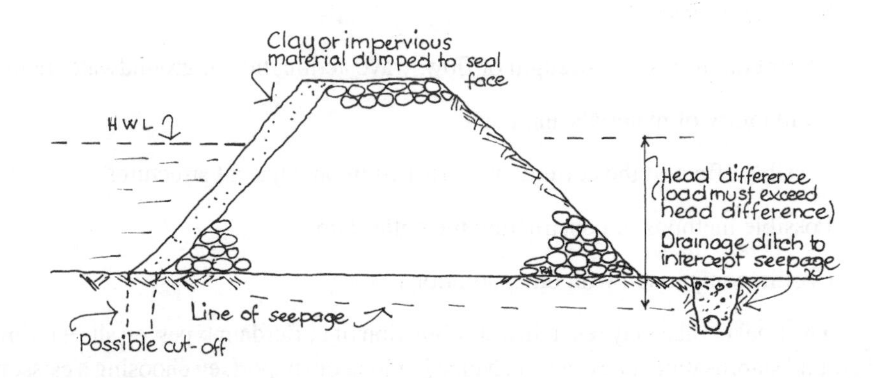

FIGURE 6.2
Rock-fill cofferdam

FIGURE 6.3
Contiguous piling used in cofferdam work
(Soil Mechanics Limited)

Sheeted cofferdams

Sheet piling is the main material used in the formation of cofferdams. A combination of H-beam piles and concrete planks is suitable for supporting deep excavations in waterlogged ground but only if ground water lowering methods are also employed inside and outside the cofferdam.

Concrete sheeting includes precast sheet piles, precast or cast in-situ panels or slabs, contiguous bored piles and conventional diaphragm walling (see Chapter 4 on Diaphragm Walling). The two most popular methods of concrete sheeting are contiguous piling (Figure 6.3) and diaphragm walling (Figure 6.4). It should be noted that although these forms of structure are permanent, in that they are not demolished or dismantled on completion of the works, they are not necessarily an integral part of the permanent works but can be designed as such.

Contiguous piling may be used where headroom or vibration prevents the driving of steel sheet piling. Where boulders would hamper the driving of sheet piling, special measures are necessary.

Diaphragm walling is particularly suitable for large cofferdams in weak or waterlogged ground, because great stresses involved can be resisted by increasing the thickness of the diaphragm.

Steel sheeting can be divided into two distinct types: single-skin, which has one or more vertical stages, and double-skin, which consists of two lines of sheet piles or circular cells of sheeting filled with rock or other material.

Single-skin sheet pile cofferdams. Steel sheet piling, of dimensions suitable to withstand the external pressures, is driven into an impermeable stratum and, if required, supported above the cantilevered end by struts, walings, or anchors. The amount and type of support will depend on the external pressures. Figures 6.5 to 6.8 show the various types of support. The construction of sheet steel piled cofferdams will be the same as that of other forms of sheet piling, which are fully described in Chapter 4.

FIGURE 6.4
Circular cofferdam using diaphragm walling for pump house 70 m diameter – walls 1 m thick and 30 m deep
(ICOS Engineering Limited)

FIGURE 6.5
Steel sheet cofferdams (Edmund Nuttall Limited)

FIGURE 6.6
Circular cofferdam supported by ring walings (Gleeson Civil Engineering Limited)

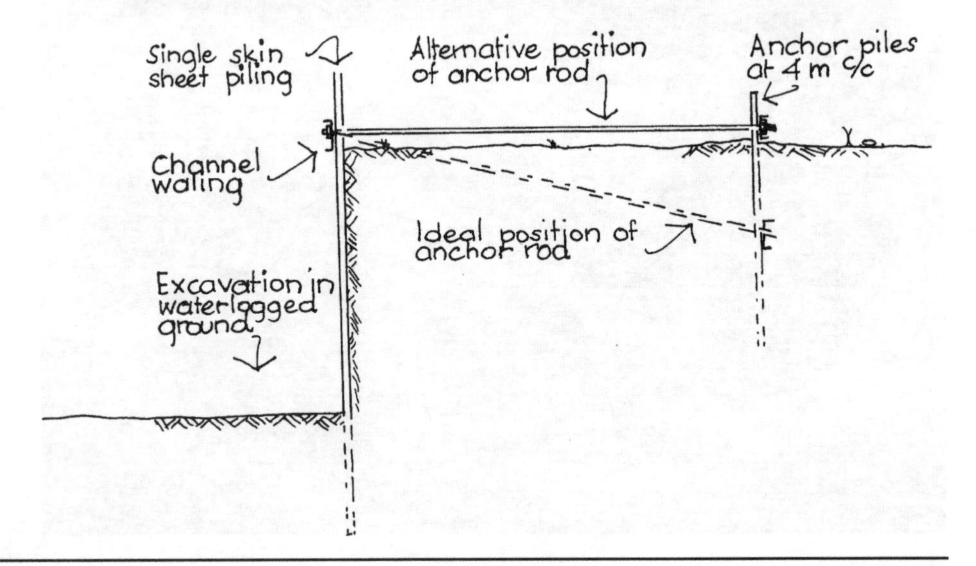

FIGURE 6.7
Single skin cofferdams anchored back by steel rods

Double-skin cofferdams. These are self-supporting gravity structures, of either the parallel-sided double-wall or the cellular cofferdam type (Figures 6.9 to 6.11). The stability of these dams depends on the fill material and on the arrangement of the sheet piling. The width of the dam should not normally be less than 0.8 times the height of retained water and/or soil. Cofferdams which exclude water must be provided with sluice gates to allow the works to be flooded (if the nature of the work inside the cofferdam allows flooding) to equalise the external pressures during storm conditions.

Cellular cofferdams are constructed from straight-web steel sheet piling (see Chapter 4). The piling is interlocked and driven to form cells; the fill material develops high circumferential tensile forces in the piling which straight-web piling is best able to resist. Cellular cofferdams are suitable for resisting the considerable head differences which are encountered in harbour and dock works. The piling does not have to penetrate the hard stratum for stability, although some penetration is necessary to prevent seepage under the dam. Cellular cofferdams can be used on irregular beds of rock when the sheet piles are cut to fit the rock profile.

Double-wall cofferdams consist of two parallel lines of sheet piling connected by walings and tie rods at one or more levels; the space between the walls is filled with material to give stability. The inner line of piling is designed as a retaining wall suitably keyed into solid strata, and the outer line as anchorage. Ordinary steel pile sections are preferable for these structures.

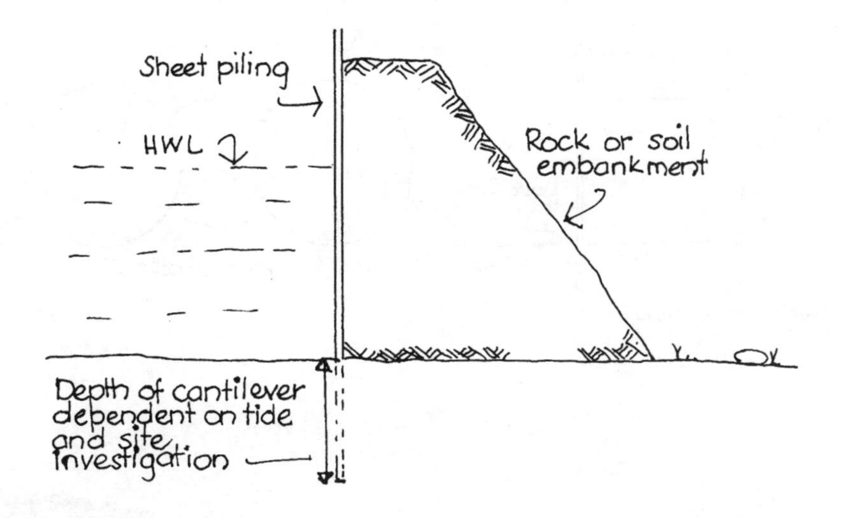

FIGURE 6.8
Single skin cofferdam supported by soil

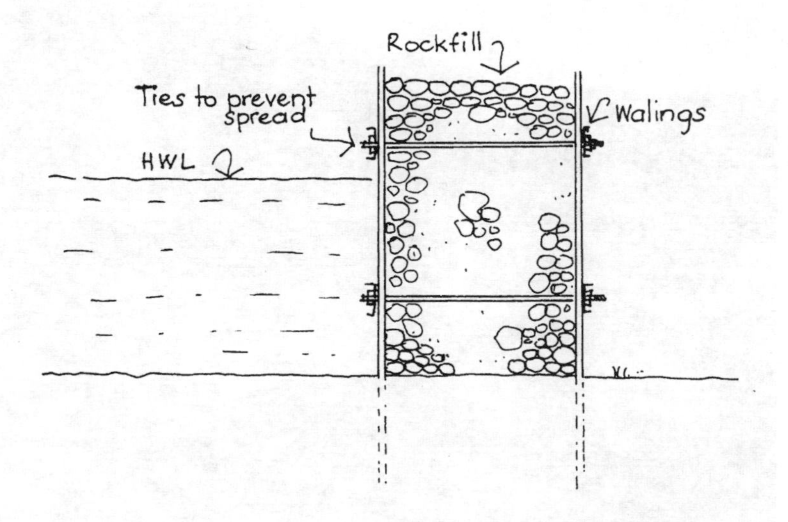

FIGURE 6.9
Double skin cofferdam (parallel sides)

Ground stabilisation

This may also be used in cofferdam construction, the practical methods being chemical or cementitious injection and freezing techniques. These methods of dealing with ground conditions are described in Chapter 3.

6.1.3 Economic factors

The factors for consideration in this section are the same as those mentioned in Section 6.1.2 under the heading of 'Selection of cofferdams'. The design for a cofferdam may be executed in various ways and in different materials, but a cost 'break-even' point can be established for the various solutions. The main economic limitations are size and depth of cofferdam; above a certain size the cost of shoring becomes prohibitive with thin wall dams, and thick diaphragm walling may well prove more economic.

A diaphragm wall may be sunk to depths to which steel sheet piling cannot be driven and so deep cut-off walls can best be formed by this method. However, for cofferdams of shallow depth, ie up to 10 metres, steel sheet piling is usually the least expensive. Earth and rock-fill are even cheaper than steel sheet piling for very large shallow cofferdams. The nature of the permanent works and method and sequence of excavating may affect the amount of strutting and shoring needed in the cofferdam and thus influence both the design and the cost. Safety is also of paramount importance, since many accidents have occurred in cofferdams.

FIGURE 6.10
Cellular cofferdam

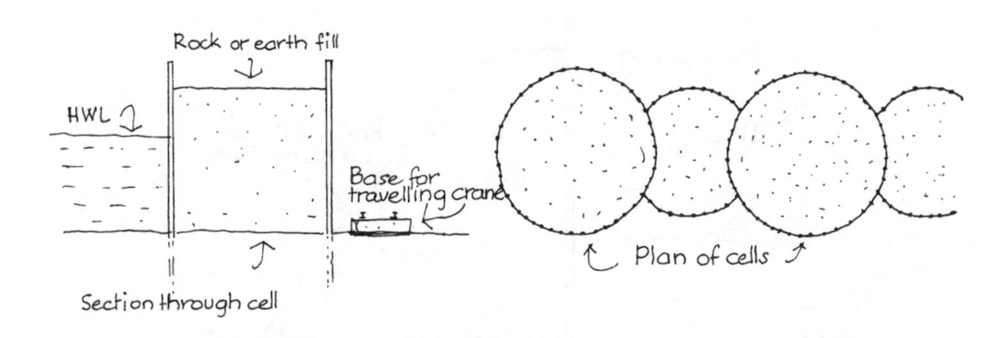

FIGURE 6.11
Cellular cofferdams for Leigth Harbour Development
(Edmund Nuttall Limited)

6.2 CAISSONS

6.2.1 Introduction

Caissons are structures which are sunk through ground or water to provide working space for the purpose of excavating and executing work at a prescribed depth, and which subsequently become an integral part of the permanent work. They may be constructed of steel, cast iron segments or reinforced concrete, or a combination of these materials. The plan area of the caisson in relation to that of the superstructure should be sufficient to provide for some deviation from its precise position during sinking.

Choice between cofferdams and caissons

The main difference between a cofferdam and a caisson is that a cofferdam is a temporary enclosure and a caisson is normally incorporated with the permanent works. The type of structure being erected within the enclosure will indicate whether a permanent enclosure is economically viable.

The chief factors influencing the choice between the two enclosures are ground conditions and the depth to which the work is to be taken. If the sinking of the enclosure requires compressed air working, a caisson may be used, although it is possible to fit cofferdams with air-decks. Generally, cofferdams are suitable for depths of up to 18 metres below high water level, while for greater depths caissons should be employed.

Design

The design of a caisson (usually contractor design) will be determined by several factors, such as proposed method of sinking, size of caisson and nature of the permanent works. The sides of the caisson should be free from bulges and constructed so as not to lose their shape during sinking; in some cases steel caissons are strengthened by insitu concrete between inner and outer walls to prevent buckling and to add weight to prevent floating. The bottom of a caisson is fitted with a shoe or cutting edge which projects beyond the face of the caisson: this allows the shoe to cleave a hole larger than the caisson and so reduces skin friction. Skin friction of between 10 kN/m^2 to 25 kN/m^2 can be assumed when the caisson is moving, although much greater friction has to be overcome to start movement. This friction can be greatly reduced by using water jets or bentonite as lubrication, additional kentledge or concrete being added to maintain movement.

6.2.2 Caisson types and forms of construction

There are four main types of caisson:

- Box caissons
- Open caissons
- Compressed air caissons
- Monoliths.

Box caissons are prefabricated boxes, usually in concrete, with sides and a bottom, which are set down on a prepared base. The box is then filled with concrete to form a massive foundation for a pier or similar structure (Figure 6.12). They are, however, unsuitable where foundations may be subject to erosion by fast moving water; this can be solved by setting the caisson on a piled foundation if the sub-strata permit the driving of piles; the piles also act as anchorage against buoyancy.

Excavation of the site is carried out by dredger or grab in normal conditions, ie gravel or mud bed, and a layer of crushed rock is levelled on the bed of the sea or river to receive the caisson. If the substratum is rock, it should be levelled off, by blasting, before laying the crushed rock base. Positioning and sinking caissons is covered in Section 6.2.3.

The box caisson may be concreted by one of three methods: tremie pipe, pump, or bottom-opening skip, the first being the most common method in recent years. Box caissons must be anchored or ballasted to prevent flotation before the concrete fill has been placed.

Open caissons are structures which are open at both top and bottom. They are suitable for foundations in waterways where the sub-stratum is soft clay or silt and therefore easily excavated by grab. They are also suitable for deep foundations in water where compressed air working would require air pressures above allowable limits. Open caissons are not suitable for sinking through ground containing obstructions, eg large boulders, unless the depth permits their removal by divers or by compressed air working.

The caisson (Figure 6.13) is sunk by grabbing the soil through the open wells. When the caisson has reached the desired depth, the bottom is plugged with a layer of concrete and the well is pumped dry. The foundation is completed by filling the well with concrete or hearting ballast. Care must be taken to ensure that the caisson is sufficiently loaded to resist flotation when it is pumped out.

FIGURE 6.12
Box caisson

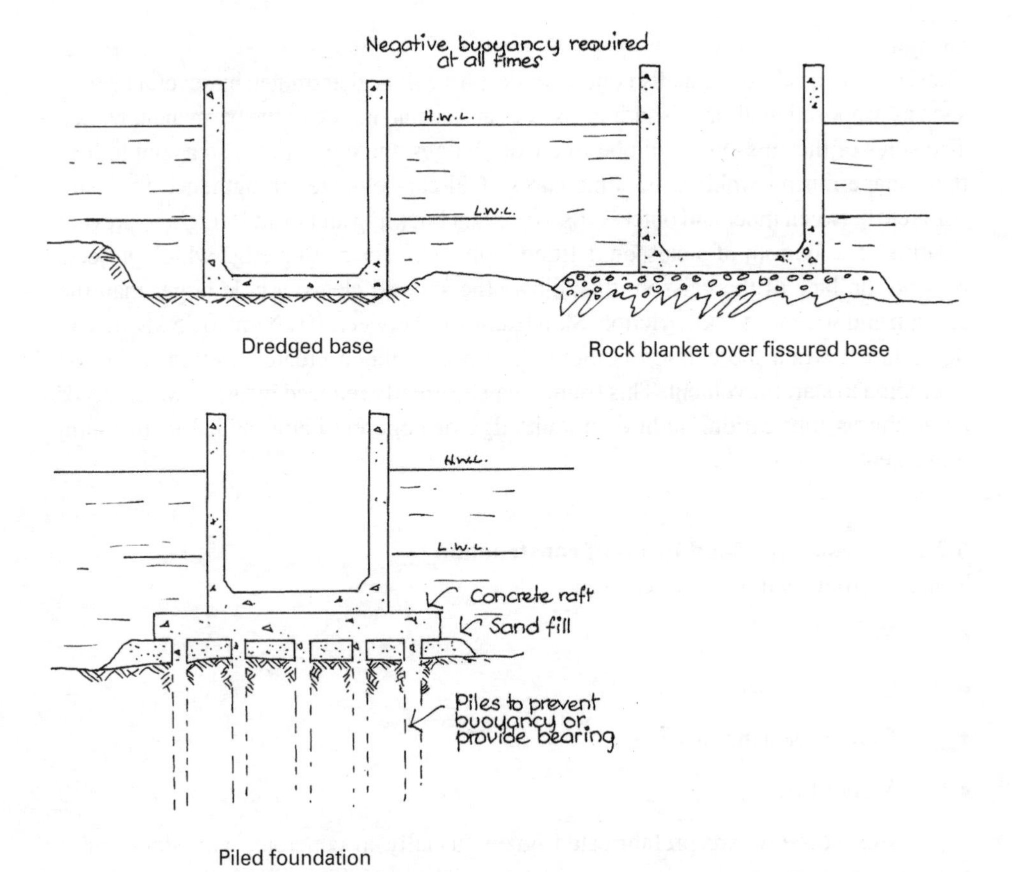

Compressed air caissons (Figure 6.14) are suitable for sinking foundations in troublesome ground, eg ground containing obstacles that require moving by hand. The caisson is positioned and sunk (see Section 6.2.3) and is then lowered into the soil as the soil is removed from within it. The soil is excavated by hand with the aid of pneumatic tools, and loaded into skips for hoisting up through the muck-lock. This method of sinking is very suitable for foundations which by other methods might result in the settlement of adjacent structures. The advantages include dry working conditions and accurate levelling and testing of the foundation bed, together with ideal conditions for the placing of concrete. However, the rate of sinking is very slow because the manual labour and the depth to which these caissons can be sunk in water are limited by the air pressures required.

When the roof of the working chamber is to form part of the structure, the concrete, as ballast filling, should be carefully placed to ensure that no void is formed against the roof. Grout pipes should be incorporated in the roof structure to facilitate the grouting of any spaces formed by shrinkage or by incomplete filling of concrete. Air locks should be located at a height that will not descend below the highest water level during sinking. If the depth of sinking is great, it may be necessary to extend the air shaft at various stages of sinking.

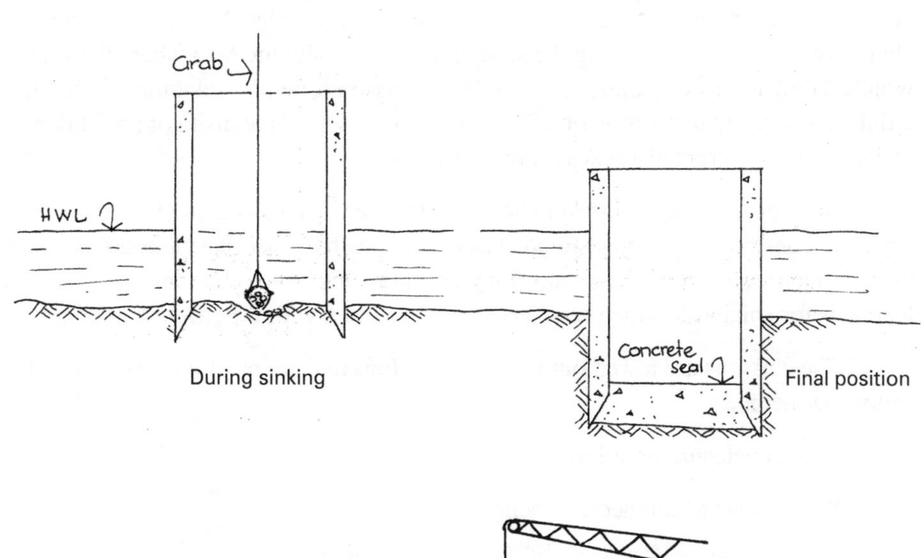

FIGURE 6.13
Open caisson

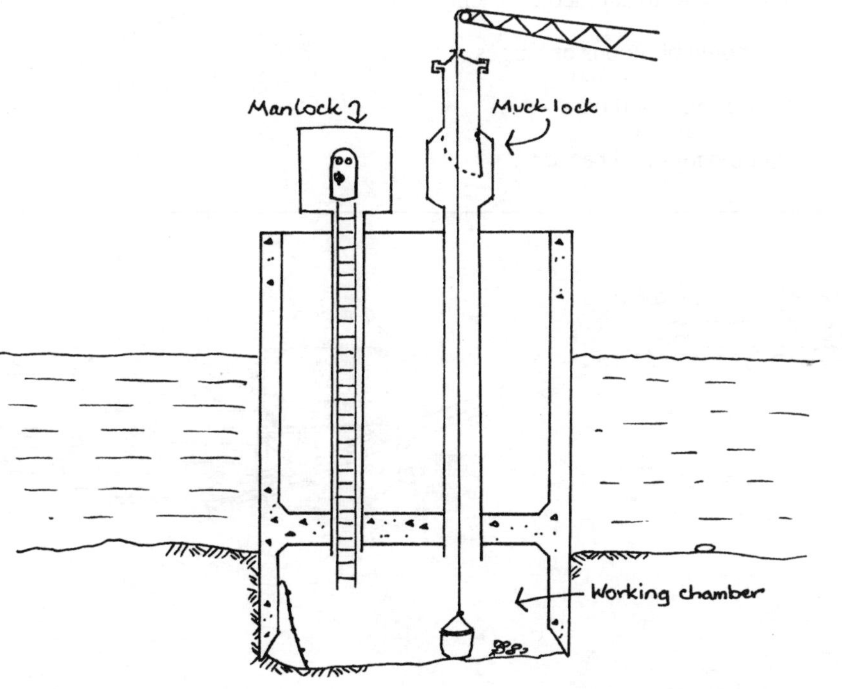

FIGURE 6.14
Compressed air caisson

In some cases it is possible to design a caisson which can be sunk partly by open grabbing and partly by compressed air. When caissons are to be sunk wholly by compressed air working, the working chamber can occupy the whole area of the caisson rather than a series of separate cells; this will allow freedom of movement for the workforce and materials and also facilitate easier control over sinking.

Monoliths are similar to open caissons but differ in that they are much heavier. The monolith consists of reinforced concrete walls of substantial thickness to provide sufficient weight to prevent overturning: for this reason they are often used for quay walls which have to resist great impact forces from ships coming in to berth. Their great weight makes them unsuitable for sinking through very soft deposits because it would be difficult to control the verticality of the structure.

With both open caissons and monoliths it is common practice to increase the height and weight of the caisson by casting, insitu, further sections as the structure sinks into the ground. Air decks can be fitted to monoliths and open caissons to permit compressed air working.

6.2.3 Positioning and sinking of caissons

If the caisson is to be founded in a river or sea, it may be partly constructed in dry dock and towed to the site. If this is the case, the shoe of the caisson, which may incorporate 10 to 30 metres of caisson wall, is constructed and then completed when in position. An alternative method of launching the shoe can be seen in Figure 6.15 where the shoe is winched down a slipway into the water. If the caisson is being sunk on land, the shoe will be constructed in position on a weak concrete base and the insitu or prefabricated walls will be constructed as excavation proceeds.

Box caissons are designed to be floated to the site and sunk in place, so normal dry dock construction is most suited. These caissons have permanent bottoms which provide the necessary seal for buoyancy and are sunk at low tide and anchored or loaded before high tide. Open caissons have to be fitted with watertight diaphragms.

The caissons are towed out to the site by tugs and are positioned by one of the following methods:

- Piling enclosure or dolphins
- Wire cables to submerged anchors
- Anchored pontoons or barges
- Sand island method
- Radio buoys and beacons.

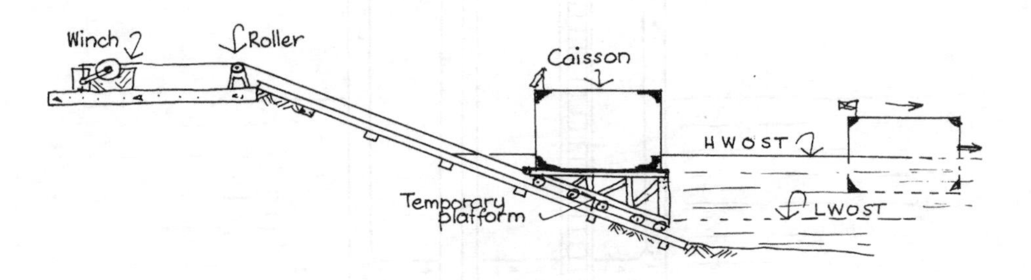

FIGURE 6.15
Launching caisson by means of slipway

Piling enclosures or dolphin support. Piling, in the form of H-sections or tube sections, is driven into the river or sea bed to form a three-sided enclosure. The piles are linked with welded beams to form fender rails, a further line of piles being driven outside the enclosure to provide support for a working platform. When the platform has been constructed (Figure 6.16), the caisson is towed into the enclosure and the opening in the enclosure is sealed with a connecting truss beam. The clearance between the fender rails and caisson will depend on the size of caisson and the proposed accuracy of sinking.

Dolphins, normally used for mooring ships in estuaries or rivers, are usually formed with raking piles and concrete platforms.

Wire cables to submerged anchors. Where the tides or currents are not very strong, the caisson may be positioned and lowered into place by winching from submerged anchors. The anchors must be heavy enough to withstand the winching operation: in some cases H-piles are driven down below the mud line to act as anchors.

Floating pontoons, barges or camels (large hollow steel floats) are the simplest forms of guide platform to construct in tidal water. The pontoons are anchored to form a three-sided enclosure (Figure 6.17) and are joined by a heavy truss to form the fourth side when the caisson is in position. The pontoons should be large enough to carry all the construction equipment needed to extend the caisson and excavate the soil.

Sand islands are used for sinking caissons in fast flowing water which would create difficulties for the anchorage of pontoons or for floating a caisson to the site. The island is formed over the proposed site with dredged sand or gravel; steel sheet piling

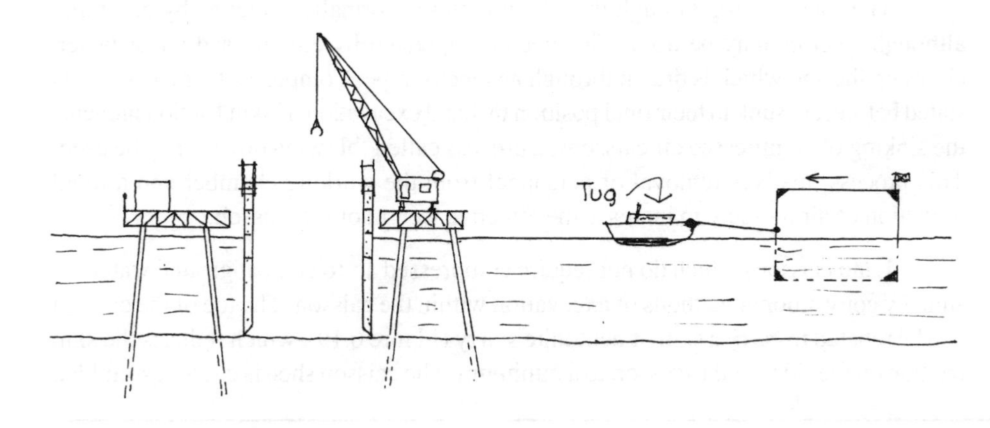

FIGURE 6.16
Temporary platform for positioning and sinking caisson

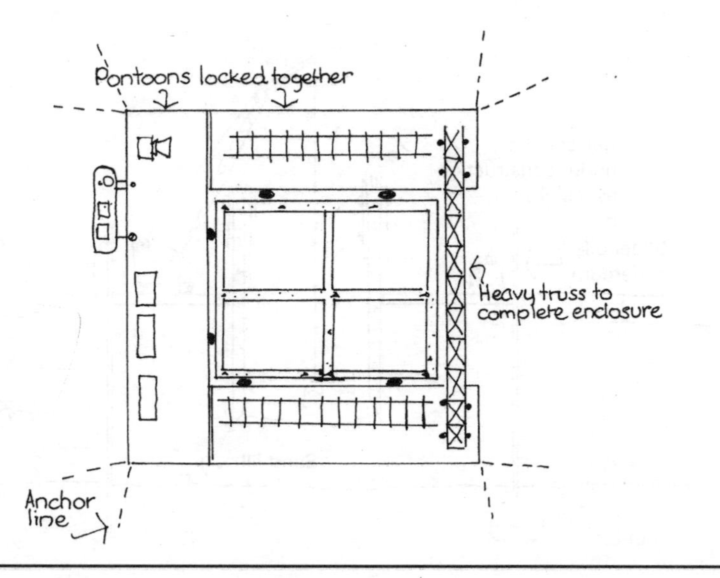

FIGURE 6.17
Caisson in position between anchored pontoons

may be used to contain the material. The caisson is constructed on the island and sunk through the sand to the bed. The disadvantage of this method of caisson positioning in rivers and estuaries is the possibility of bed and river bank scouring: the island (Figure 6.18) reduces the width of river and causes the flow of water to scour the adjacent banks and bed planes.

Radio buoys and beacons are used for accurate position and alignment of caissons for bridge piers and other works where caissons require alignment over great distances.

The sinking of caissons can be achieved by one of the following methods:

- Free sinking or jacking down between anchored pontoons
- Lowering by cranes from anchored pontoons
- Winching from submerged anchors
- Controlled sinking, using air domes, and ballasting within lateral control, usually radio beacons.

The initial rate of sinking will depend on the amount of insitu walling to be completed. The top of the caisson should maintain a minimum of 10 metres (in very rough conditions 25 metres) above high water level to receive the extension of the walls. The caisson sinking is accomplished by weighting, or in the case of compressed air caissons and open caissons which have false bottom plates, by controlled flooding. Large open caissons usually have double skin walls, which make them buoyant for towing into position: the cavity between the skins is then filled with concrete for the purpose of sinking.

The final sinking through the silt and mud is normally achieved by grabbing, although ejectors may be used. The ejectors, operated by compressed air or water, churn up the soil which is drawn through an ejector pipe. Compressed air caissons, as stated before, are sunk to their final position by hand excavation. If skin friction prevents the sinking of compressed air caissons, a process called 'blowing down' may be used. This process involves removal of personnel from the working chamber and a rapid reduction of air pressure to increase the effective weight of the caisson.

Land caissons which do not require compressed air to combat ground water are sunk by conventional methods of excavation within the caisson. The toe of the caisson can be rebated to carry a skin of bentonite slurry (Figure 6.19), which reduces the skin friction on the sides of the caisson to a minimum. The caisson shoe is cast or assembled

FIGURE 6.18
Sinking caisson through sand island

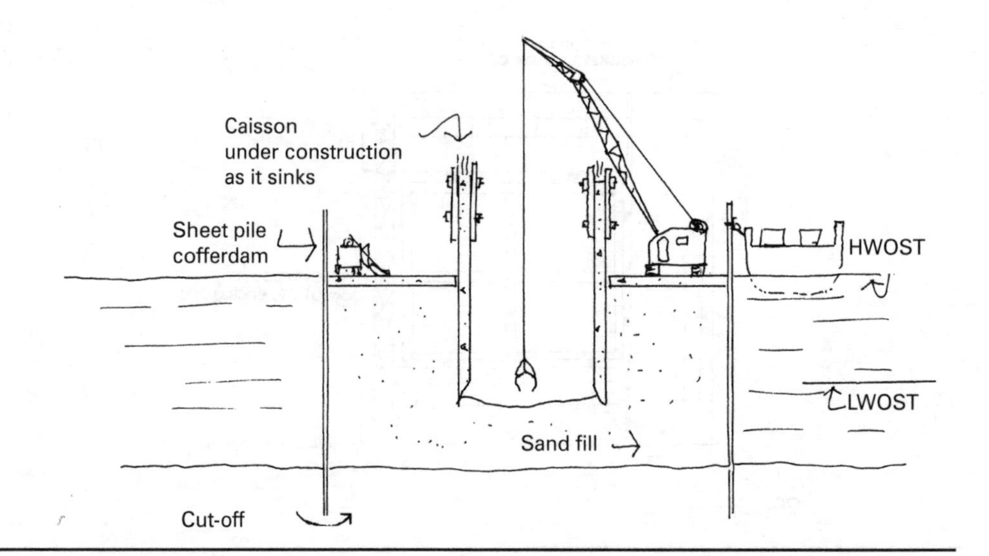

in position and excavators remove the ground under the toe of the shoe and allow the structure to move slowly into the ground. As the excavation and caisson construction continues, the annular space outside the caisson, created by the rebated shoe, is filled with bentonite from pipes carried down inside the wall and out through the shoe. The space may vary from 25 mm to 75 mm in width, depending on the depth of sinking and on ground conditions.

6.2.4 Sealing and filling caissons

All loose material should be removed from the bottom of the excavation and the excavation should be level or concave. In some cases it will be necessary for the haunching adjacent to the cutting edge to be removed by hand: this may have to be carried out by a diver.

When sealing open caissons and monoliths the first layer of concrete will have to be placed under water by tremie or bottom opening skip: on completion of the first layer the caisson may be pumped fully or partially dry for further filling, but if this procedure is followed the caisson must be sufficiently loaded to prevent flotation.

Compressed air caissons are concreted in dry conditions. The first layer of concrete, normally 600 mm thick but varying according to the span, seals the floor of the working chamber and is carefully vibrated under the cutting edge of the shoe. Subsequent layers of concrete are placed and vibrated until only a small space is left between the concrete and the roof of the working chamber. The space between concrete fill and the working chamber roof is then pressure grouted with a 1:1 cement:sand grout and left for three or four days before the air shafts are finally filled with concrete.

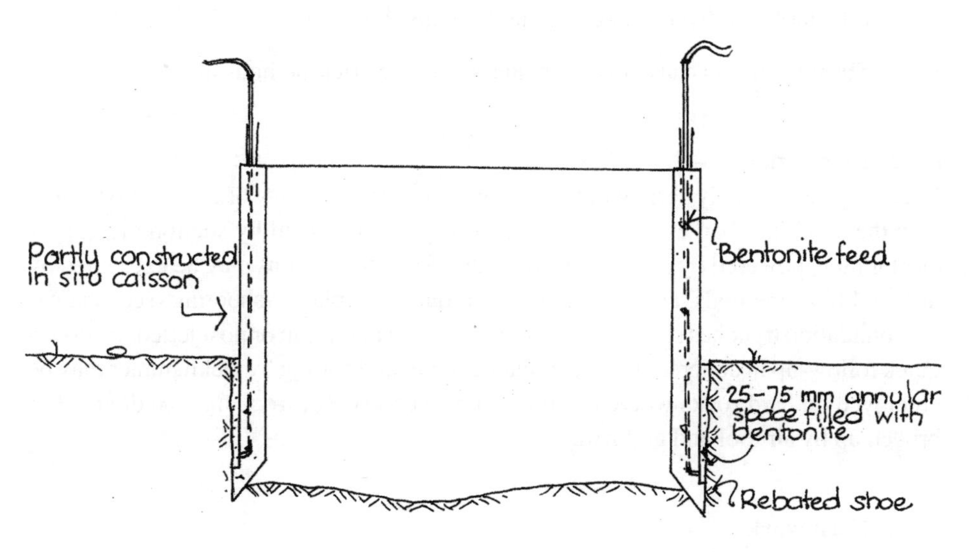

FIGURE 6.19
Sinking a caisson on land (with bentonite lubrication)

6.3 UNDERWATER FOUNDATION CONSTRUCTION

6.3.1 General considerations

Underwater foundations may have to be constructed in circumstances which prohibit the use of cofferdams and caissons; the reasons for this are usually the cost and feasibility of the works. Where small-scale or non-repetitive foundation work has to be carried out, the cost of constructing a caisson or cofferdam may prove too high to be a practical solution. If the stability of cofferdams or caissons is endangered by tides or 'blowing', the cost of 'making safe' may be excessive compared with other construction methods. Long foundation pads to receive precast concrete blocks for harbour wall construction are difficult and costly to form by caisson construction, but insitu methods have proved very successful in calm waters. However, the use of cofferdams in conjunction with underwater construction is common and must not necessarily be considered a separate form of construction.

Some of the problems or limiting factors in this work include:

- Transport of personnel and material; this proves to be difficult in anything but very calm waters.
- Lifting gear must be carefully positioned and pontoons securely sprung anchored to prevent excessive 'snatching' when lifting loads.
- Skin divers cannot work efficiently in currents over 2 knots or for long periods in depths of water exceeding 15 metres or in waters where visibility is virtually nil. Hard suit divers can carry out work involving torques, welding etc.
- Strong currents cause loss of materials and restrict methods of placing.

6.3.2 Excavation

Excavation is normally carried out by dredgers or grabs, which work off pontoons and load the spoil into barges. Gravel and sand may be excavated by suction dredger or ejector tube, powered by air or water-pressure. Rock formations may be excavated by drill and blast methods, and large boulders reduced by plaster-shooting (see Chapter 3). Foundation trenches should be blasted out in one operation or post jetted to allow a quick follow-up in concreting. Underwater drilling and cutting by jackhammer may be employed on hard rocks where blasting is unnecessary; soft rock may be dredged or broken up by air tool and grabbed.

6.3.3 Formwork

Formwork may be of a temporary or permanent nature. Temporary forms should be made of steel and designed to allow ease of placing by divers. Sand bags may be used to anchor the forms and prevent leakage of concrete under the toe of the form: if an improved seal is required a skirt of plastic sheeting around the forms, suitably anchored with sand bags, may be satisfactory. The effective pressure on the formwork is that due to the submerged weight of the concrete only, but forms can be designed to withstand pressures that occur in normal dry conditions: this provides robust forms which can cope with any underwater conditions. If the concrete foundation is limited in height the forms may consist of bagged cement or concrete placed in layers and left in position on completion. Higher lifts of concrete work may be supported by precast concrete blocks which can form part of the permanent work by using suitable ties: typical formwork sketches are shown in Figures 6.20 and 6.21.

6.3.4 Underwater concreting

When a mass of fresh concrete moves through water, or when water flows over the surface of concrete, some of the cement is washed out of the mix. Therefore as much of the concrete as possible should be kept out of contact with the water. This involves maintaining the concrete flow within the initial mass placed, allowing the initial mass to protect the fresh concrete.

To achieve the flow of concrete within the mass first placed, the mix must have a high degree of workability. This ensures that the leading edge of the concrete is kept moving forward under the pressure of additional material: a slump of 150 mm is normal. Extra cement over and above the normal requirements for the mix design must be added to offset loss through water, but too much cement will result in excessive laitance forming on the surface of the concrete.

Tolerances in underwater concreting

Dimensional tolerances can approach those for concreting in the dry, but only at great cost and considerable effort. The tolerances specified should be set with due regard to the requirements of the works and the particular conditions under which the work is being constructed. Screeded foundations for precast concrete harbour walls may be laid to a level of 25 mm to 100 mm depending on visibility. Screeding widths of up to 6 metres are possible in good conditions, but smaller widths are to be preferred.

Placing of concrete

This is achieved by three principal methods: tremie, underwater skip and pumping.

Placing by tremie. A tremie (Figure 6.22) is a steel tube suspended in the water from a crane, with a hopper fixed to the top end to receive the concrete. The tube must be watertight, smooth-bored and of adequate diameter for the size of aggregates being used: diameters of 200 mm and 250 mm are commonly used. The tremie is erected vertically over the area to be concreted with the lower end resting on the bottom. A travelling plug, formed from cement bags, foamed plastic or similar material, is placed in the pipe as a barrier between the concrete and water.

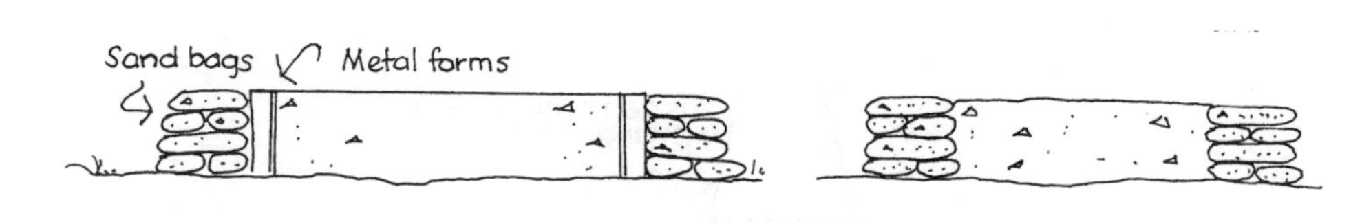

FIGURE 6.20 *Temporary formwork for underwater foundation (shallow water only)*

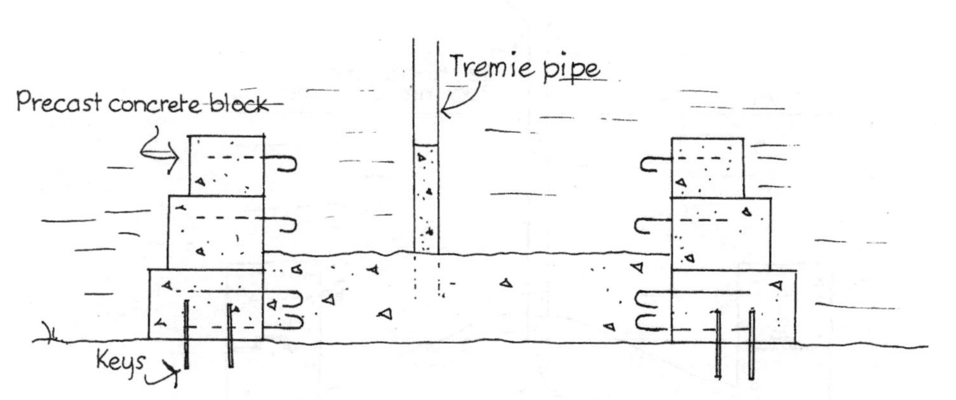

FIGURE 6.21 *Permanent formwork incorporated with foundation design*

The water in the pipe is displaced as the weight of concrete forces the plug to the bottom: some plugs float to the surface but if they remain in the concrete their presence is insignificant. After the pipe has been filled with concrete it is raised off the bottom to allow the concrete to flow. Thereafter, the flow should continue to feed the interior of the initial mass.

The rate of flow of concrete is controlled by raising and lowering the tremie, but care must be taken not to lift the tremie out of the mass. If the bottom of the tremie is lifted out of the mass, the seal will be broken and concrete will be weakened by water as it rushes out of the pipe; seals are often broken when attempting to clear blockages in the pipe. A broken seal may result in a damaged surface to the mass of concrete and consequent removal before work can recommence. Simultaneous placing through more than one tremie is recommended where the concrete cannot be placed from one position: one tremie will serve an area of about 30 m^2.

Placing by skip. Skips used for underwater concreting should be of the bottom opening type which can be operated automatically or manually. The skip should be equipped with a top cover consisting of two loose overlapping canvas flaps, which are kept in position by water pressure; the skip may also be fitted with a skirt to confine the concrete on release. The skip is filled with concrete and the canvas covers are put in position before gently lowering into the water: rapid lowering through the water may disturb the canvas covers and damage the concrete. On reaching the bottom the skip is gently emptied to minimise turbulence of the water around it.

The choice between placing by skip or by tremie rests on economics and on the plant and skills which are available. Placing by skip is the slower but more practical method for thin beds, whereas tremies are suitable for large concrete pours. The continual movement and placing involved with skips makes the concrete subject to a greater loss of strength or damage by exposure to water.

The pumping of concrete is discussed in Chapter 2 and therefore needs no further explanation here; it is used in underwater concreting when very large concrete pours are involved. Other methods, such as re-usable bottom-opening bags (toggle bags), may be used for very small concrete pours.

FIGURE 6.22
Underwater concreting using tremie pipe

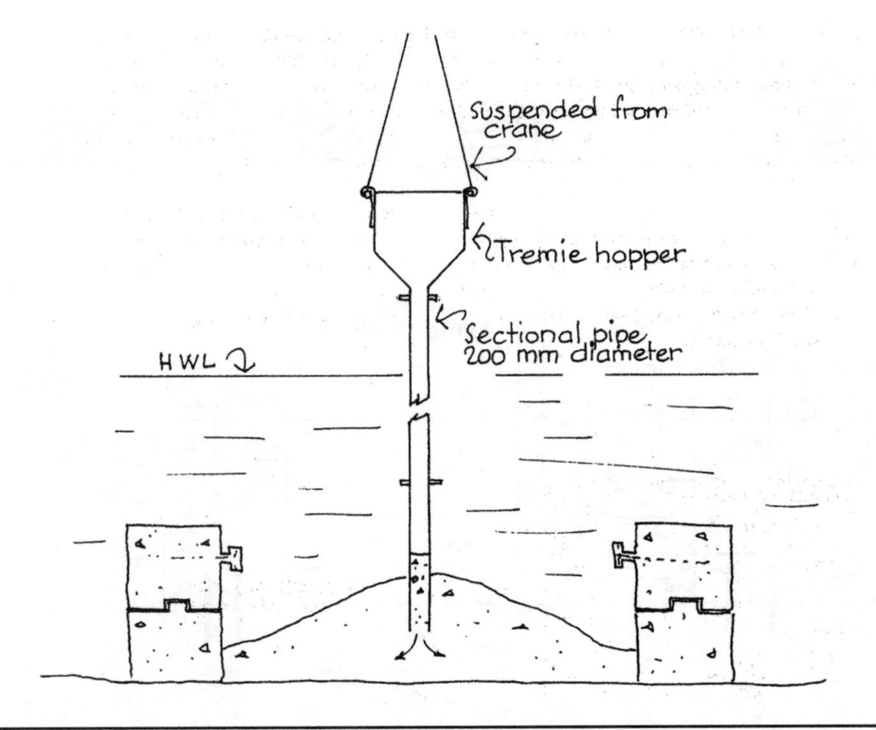

6.4 SEA WALLS, DOCKS, JETTIES AND OTHER MARINE STRUCTURES

6.4.1 Methods of construction

Sea walls are constructed to resist encroachment by the sea and are often incorporated into the construction of a promenade. They are constructed in various materials, ranging from masonry blocks and precast concrete units to insitu concrete. The design of the wall should minimise the effect of wave action and prevent the under-scouring of the foundation (Figure 6.23). Concrete blocks with projecting reinforcement (Figure 6.23a) may be used in conjunction with an insitu backing.

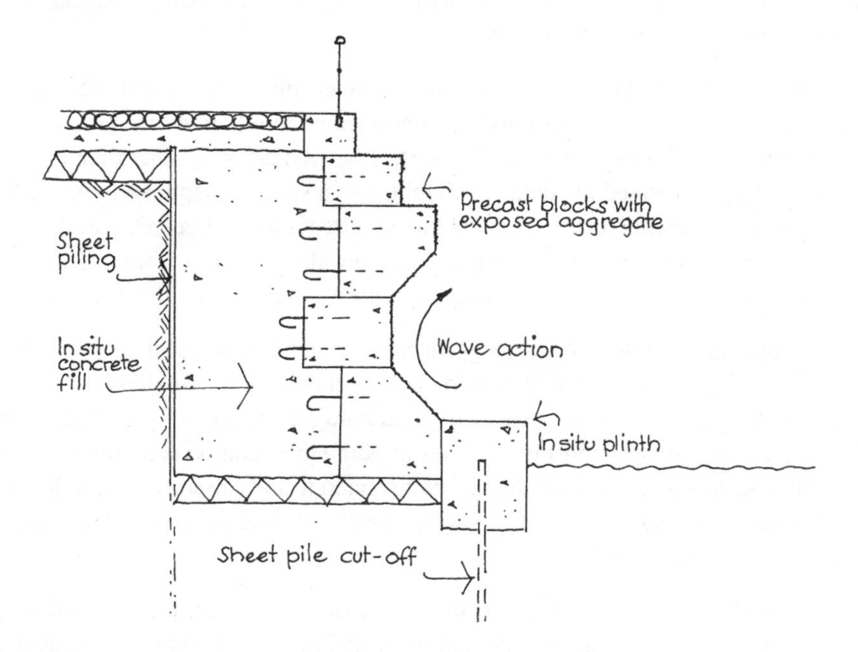

FIGURE 6.23(a)
Sea wall construction

FIGURE 6.23(b)
Alternative to sea wall – stepped revetment

Docks may be divided into two types: dry docks and wet docks.

Dry docks differ in design from wet docks in that they have to withstand hydrostatic pressure when the dock is emptied. The walls of the dock should be incorporated with the floor to form a rigid structure and so reduce uplift. If sufficient dead weight cannot be introduced into the construction to prevent buoyancy, a venting system must be provided to relieve the hydrostatic pressure; alternatively anchorage can be used.

The walls are normally massive insitu structures which are cast in large, deep trench excavations. The walls are cast in stages of up to 2 metres thick, with stepped construction joints (Figure 6.24). Longitudinal joints should be constructed at intervals of 15 metres; these joints are essentially construction joints with water bars and should be filled with bitumen or other sealing compound. The floor of the dock must, in addition to resisting uplift, be strong enough to distribute a ship's load without settlement or undue deflection. If the loads encountered are very high, piling support may be adopted, both for supporting a ship's load and resisting uplift. Lock entrance gates, normally of steel construction, may be pivoted or sliding.

Wet docks are large areas of water bounded by vertical solid walls against which vessels tie up. The walls must be impermeable to retain the water at high tide level. Locks are provided if entry to the dock is desired at times other than high tide. The walls may be formed by sinking monoliths to a suitable depth and joining them together with insitu concrete: a space of 2 to 3 metres is left between the monoliths to facilitate jointing and finishing. Alternatively, the walls may be constructed with deep diaphragms (Figure 6.25), decking being supported by cross-wall diaphragms.

Wharves are berths for shipping which may retain the surrounding soil or simply provide mooring facilities. Those constructed to retain soil are usually mass concrete walls constructed by means of caisson or diaphragm walling (Figure 6.25). Open wharves, which provide mooring facilities at both sides, can be constructed of piles with concrete decking. An alternative to insitu concrete and caisson construction is the use of precast concrete blocks, which may be dovetail-keyed and weigh anything from 10 tonnes to 30 tonnes each.

As with other large marine structures, caissons are also commonly used for this type of work and are particularly suited to construction of breakwaters and quay walls. The caisson fill may be rock, sand or concrete (Figure 6.26).

FIGURE 6.24
General view of lock wall construction showing stepped construction joints (Edmund Nuttall Limited)

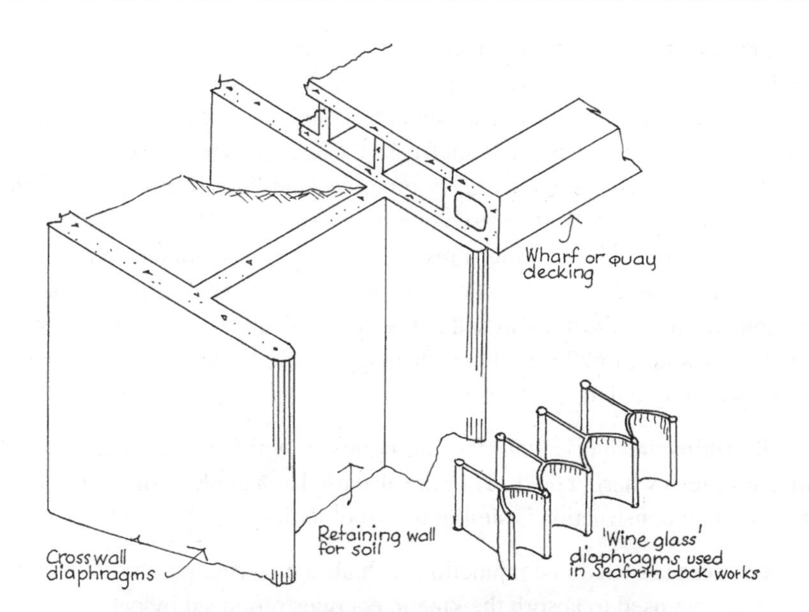

FIGURE 6.25
Diaphragm walling in dock and wharf construction

FIGURE 6.26(a)
Construction of quay wall using caissons ballasted with sand
(Royal Boskalis Westminster nv)

FIGURE 6.26(b)
General view of quay wall construction with side-cast stone dumping barge engaged in coastal protection activities
(Royal Boskalis Westminster nv)

Jetties jut out into the sea, usually at right angles to the shore line, although T-shaped and L-shaped jetties are not uncommon. They are open structures, usually of steel tubular or box piles with a heavy concrete deck (Figure 6.27). They may be used for off-loading heavy cargoes, in which case the deck may require extra bracing to the piles; alternatively the jetty may carry pipelines and light lifting gear which do not require heavy bracing. The jetty structure must be designed to withstand impact loads and 'bollard pulls' from berthing ships: this is usually accommodated by raking pile construction and fendering, the latter to avoid holing the ship. If the sea bed is a rock formation the piling construction will not be economically produced by normal driving methods; holes up to 600 mm diameter may be bored into the sea bed and steel or concrete piles grouted in.

Dolphins are individual mooring points to which vessels may be tied while waiting to enter a wharf or dock. They are also used as a guide to ships entering narrow harbours. Their construction is similar to that of jetties.

Fenders are used in conjunction with all the marine constructions mentioned above. They are used to absorb the kinetic energies produced by berthing vessels. To achieve the necessary absorption, they have to be flexible and may take the form of tubes or springs of metal, rope and plastic. Floating fenders, of rubber and timber, are used to distribute loads over many vertical fenders at the wharf side. If floating fenders are not used the load will normally be applied at deck level and the deck must be suitably braced and protected.

6.4.2 Other marine structures

Breakwaters or moles

Breakwaters and moles are constructed in the outer harbour area to dampen heavy waves and swell so as to provide easier entrance and exit of vessels. They may be constructed with concrete blocks, rock fill, or a combination of both (Figure 6.28). The choice of material will depend upon the conditions of the site, ie depth of water, foundation

FIGURE 6.27
Jetty under construction at Kingsnorth Power Station
(John Laing & Son Limited)

conditions, range of tides, availability of materials and the anticipated extent of fine weather during construction. Vertical-sided breakwaters are suitable for shallow waters up to 15 metres deep; working in depths above this proves difficult for divers who position the blocks.

Where blockwork is used a foundation is prepared by dredging the marine bed and laying a concrete base (see Section 6.3.4 'Underwater Concreting'). Blocks are lowered by cranes operated from pontoons, and are positioned by divers: the location of the blocks is made easier by a dovetailed jointing system.

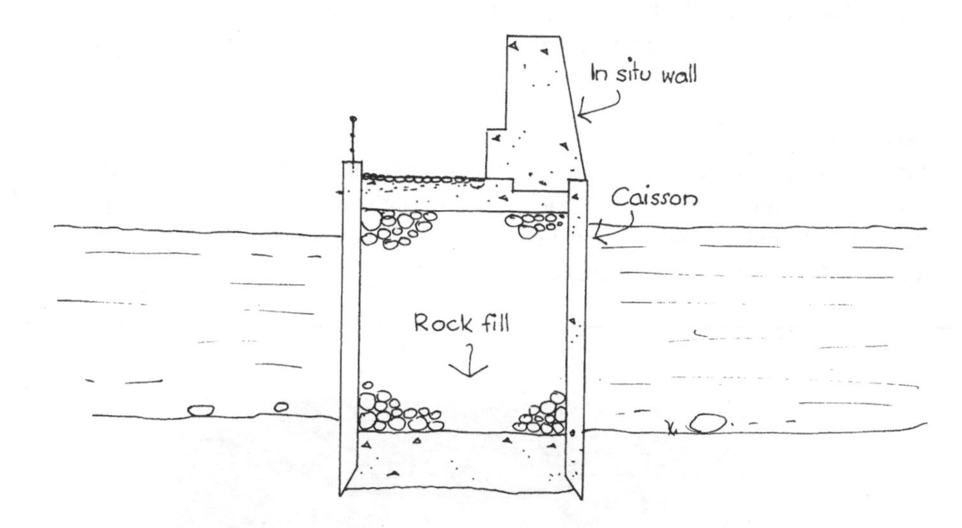

FIGURE 6.28(a) *Breakwater or mole construction*

FIGURE 6.28(b) *Rock-filled breakwater* *(Royal Boskalis Westminster nv)*

Rubble or rock-fill breakwaters are suitable for both shallow and deep waters. The rock fill should contain heavy stones, ranging from 1 to 5 tonnes each in weight, to prevent movement by wave action. The material is transported and placed by bottom-opening or side-tipping barges, some of which have a capacity of 600 m^3. Rubble breakwaters require protection against pounding of the deck area by heavy seas and this can be achieved by casting a concrete slab, or by grouting the top layer of rock or laying a precast interlocking deck. Above the water level, rock is loaded from barges by crane to provide the formation to the seaward side of the breakwater (Figure 6.29).

Composite breakwaters for very deep water consist of rock-fill and precast concrete block walls, the walls being taken to a depth of 5 to 10 metres below low water. The concrete blocks for breakwaters of this magnitude should be very heavy, averaging between 20 and 50 tonnes each.

Groynes

Groynes are small section walls of concrete or other suitable material which are built to protect or retain beach material. Steel sheet piling may be used suitably capped and backed with concrete: adequate penetration of the piles prevents underscouring of the structure by wave action (Figure 6.30).

FIGURE 6.29
Pontoon mounted crane unloading rock to protect the face of reclaimed land
(Royal Boskalis Westminster nv)

FIGURE 6.30
Groyne construction as part of coastal defence
(Royal Boskalis Westminster nv)

6.5 DREDGING AND RECLAMATION

6.5.1 General considerations

Dredging

Dredging is the process whereby sub-aqueous excavations are carried out by plant located above water level, and may be undertaken for the following reasons:

- To lower the bed level to permit the passage of ships.
- To obtain materials for use in land reclamation.
- To obtain aggregates which, after desalination, can be used in the manufacture of concrete.
- To obtain materials for use in the construction of roads and other civil engineering projects.
- To facilitate the construction of civil engineering works.

Reclamation

Reclamation is the process of depositing materials either in the sea or in low-lying swampy areas in such a way that useful areas of land are formed. Almost any type of material can be used for reclamation, depending on the use to which the land is to be put. This will range from agricultural land and land for light industrial uses, which can utilise materials which have low load-bearing capacities, to land for the construction of dock and harbour installations and power stations, which will require high quality incompressible materials (Figure 6.31).

FIGURE 6.31
General view of land reclamation
(Royal Boskalis Westminster nv)

General considerations in the planning of dredging operations
The one consideration that is common to all types of dredging is the consequence of removing the material.

There is little point in dredging in an estuary or harbour or cutting a navigable channel at sea if the void so formed is going to be filled by natural deposition of materials within a very short time. So the first step must be to carry out site investigations and hydrographic surveys to see if the project is feasible from practical, economical and ecological viewpoints.

Hydrographic surveys will indicate the characteristics of the seabed and the likely movement of material that will be caused by tides and currents after dredging has taken place. On more important schemes this will be supplemented by a study of scale models which can reproduce marine conditions with great accuracy.

Having established the feasibility of a project, the next stage is to determine the methods of execution, including the necessary plant requirements. Among the more important factors which affect this decision are:

- Location of site
- Dredging depth
- Type of material to be dredged
- Disposal of material.

Location of site. The selection of plant and equipment will depend to a large extent on the location of the site; equally the location of the site partially governs the selection of plant and equipment. The dredging of inland waterways such as rivers, canals and lakes will normally require small, easily controllable dredgers such as grabs, cutter suction dredgers or small bucket dredgers. It may even be necessary to use a dredger that can be transported overland. Dredging within harbours or within easy distance of the shore will be done with bucket, trailing suction dredgers or small cutter suction dredgers. Dredging in the open sea calls for large, well equipped trailing suction dredgers, either with self contained hoppers or serviced by fleets of barges with the necessary tugs, etc.

Dredging depth. Navigational requirements can require dredge depths varying from 2 or 3 metres in inland waterways to about 25 metres for very large vessel approach channels. Under-water pipeline trench excavation will usually require depths within this range also. Dredge depths of up to 60 metres are being presently achieved by the special suction dredgers used for land reclamation with sand in the Ijsselmeer and Maas/ Rhine Delta areas of Holland.

Type of material to be dredged. Almost any type of material can be dredged. Mud, silt and sand can be removed with bucket or suction dredgers or clam grabs. Harder materials such as clay have to be broken up by bucket or cutter suction dredgers before they can be removed. Hard rock, in some instances, can be excavated by purpose-made rock bucket dredgers, but more often than not will have to be broken up with explosives.

Disposal of material. The vast quantities of material which are removed during dredging operations have to be deposited somewhere else. If there is no area suitable for reclamation, or if the material itself is unsuitable for such a use, the cost of disposal can form a major part of the cost of the operation. Even though the volumes extracted during most inland dredging projects are relatively small, it may still be necessary to employ a fleet of lorries with special water-retaining bodies to transport the material to suitable disposal areas.

The disposal of material dredged from harbours or sea channels presents a greater problem. Depending on whether the material can be utilised or not, it is generally pumped inshore through a pipeline or dumped at sea, but care is required in the selection of the dumping ground. If it is too close inshore it may foul the beaches or silt up a navigation channel. At sea, it must be dumped away from shipping lanes, and in some instances consideration must be given to the possible effect that dumping may have on marine life.

General considerations in the planning of reclamation operations

As with dredging, the prime consideration must be the effect that the project will have on its surroundings.

With inland sites, factors which have to be taken into account include the stability of the surrounding land and the effect on existing watercourses. Proper provision must be made to ensure that the run-off from the area can pass round, through or under the area, even at times of flood.

Reclamation of coastal sites, especially on a large scale, can interrupt the natural process by which the foreshore remains stable. This may result in siltation of existing navigable waterways or the undermining of sea defences. It is often necessary to construct scale models with simulated tides and currents to study the effects that a reclamation scheme might have on its surroundings.

Having established the feasibility of the project, the planning stage will involve consideration of:

- Location of the site
- Type of material
- Transport of materials.

The location of the site. For inland sites the question of access is all-important. If necessary, road and/or rail access must be provided, although these will usually also be required as part of the permanent works when the site is developed. Coastal reclamation sites, in addition to the above, may require access from the sea, ie a navigable channel. The material from this channel may itself be used in the reclamation.

Type of material. For most coastal sites there is rarely any problem about the material to be used. Sand is not only the easiest and cheapest to extract but is also suitable for supporting most types of development; however, the haul distance can be an economic factor. With regard to inland sites, material availability will depend on the location of the site. Excavated material, rock, shale, and pulverised fuel ash can be used for reclamation. The uses to which the site is to be put may rule out some of the more compressible materials, but generally speaking it would be uneconomic to locate a reclamation site in an area where suitable material was not available.

Transport of materials will involve the mobilisation of large amounts of plant. The plant used will determine the rate at which reclamation can be carried out and a great deal of planning will be required to determine the best combination of equipment. The loading of transporting plant, the cycle time, the size of the vessels, the placing requirements, will all need to be carefully calculated, as it is their efficiency which will have the greatest effect on the economy of the scheme.

6.5.2 Plant and equipment

Dredging plant

Dredging plant may be broadly described under two main heads:

- Cutter suction dredgers
- Digging dredgers.

The various types of dredgers are shown in Figure 6.32. The following is a brief description of the more important types of dredging plant.

Suction dredger

This is by far the most common type of dredger in use today.

Trailer suction hopper dredgers (Figure 6.32) comprise a self-propelled hull containing a tank or hopper which is filled with sand, silt or some clays (depending on the drag head used and the hopper design) by one or two suction pipes which usually trail alongside the vessel with the suction head dragging along the bed.

The diameter of the suction pipes will be between 400 mm and 1300 mm with a maximum dredging depth of about 35 metres, and the hopper capacity will be up to 3000 m^3 for the smaller vessel and up to 10 000 m^3 or larger for the deep-sea type. The suction pipes are raised and lowered by crane or winch. A 500 mm diameter pipe will deliver approximately 650 m^3 of sand into the hopper per hour (this varies with the type of material).

The trailer suction hopper dredger has great mobility and loads itself while under way. As soon as the hopper is full, the ship heads for its unloading point, where the spoil will be dumped.

Discharge is either by bottom dumping, which involves the opening of hydraulically operated doors or valves in the bottom of the hopper, or by a further series of suction pipes which can discharge over the side into barges if the vessel cannot reach the dumping ground itself.

Cutter suction dredgers (Figure 6.32). Whereas the trailer suction dredger can work in sand, silt and light clay conditions, the cutter suction dredger is designed to break up and remove firm and more cohesive materials, including soft rock. The dredger has a central well which houses a rigid ladder which can be raised and lowered by means of a crane. The ladder carries a suction pipe and, at the end of the ladder, a revolving cutting head with teeth which bite into the firm material.

To increase the thrust on the cutting head and to enable the dredger to 'walk' ahead into the face, 'spuds' are driven into the sea bed: these are anchor legs which can be raised or lowered hydraulically. Cutter suction dredgers can operate up to a maximum depth of about 40 metres and the material broken up is sucked up the arm and discharged through a floating pipeline or occasionally into barges.

Reclamation dredgers are simple suction dredgers which can be stationed remote from or close to the reclamation area, depending on whether a booster station is used, and transfer sand or silt to barges or via a pipe-line to its final location. Water jets remix sand in the barge which is being unloaded, in order to form a pumpable mixture. The reclamation dredger is essentially an inshore craft and will need to be towed to a safe location during storm conditions.

The bucket wheel dredger (Figure 6.32) is another type of cutter suction dredger which has a wheel cutter instead of a crown cutter. The wheel cutter consists of closely spaced bottomless buckets which cut through the material to prepare it for suction action. The suction pipe and wheel cutter form an integral unit.

FIGURE 6.32
Types of dredgers
(Royal Boskalis Westminster nv)

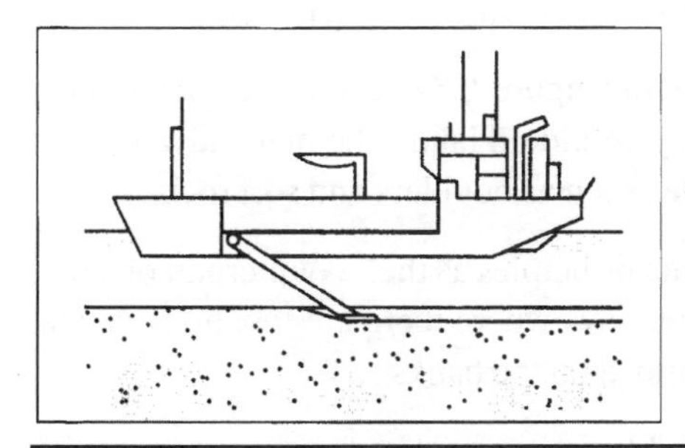

The **Trailing Suction Hopper Dredger** is able to load its own hold with the aid of centrifugal pump(s). Loading takes place with the ship under way.
Discharge is normally via a bottom dumping arrangement or occasionally by pump discharge, usually to the shore.

The main advantages of this type of vessel are a relative immunity to adverse weather and sea conditions, its operational independence, an ability to transport soil over long distances, and little interference to other shipping traffic.

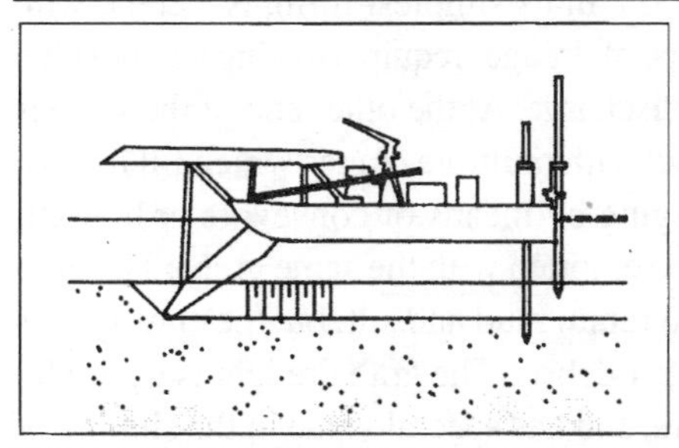

Dredging with a **Cutter Suction Dredger** takes place with the vessel moored by means of spuds and/or anchors and combines a powerful cutting action with suction. Dredged material can be discharged into barges or, more commonly, pumped via a pipeline to the shore for disposal or for land reclamation.

The main advantages are the ability to dredge a very wide range of materials, including soft rocks, and to convey by pumping the dredged material to the disposal or reclamation area. The vessel can operate in shallow water and produce a fairly uniform bottom level.

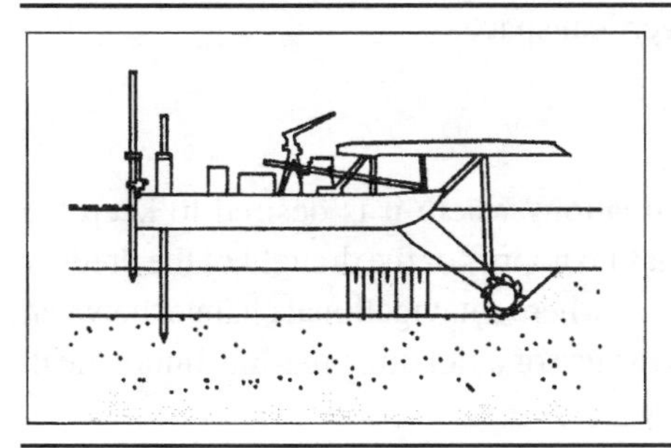

The **Bucket Wheel Dredger** is essentially a cutter suction dredger fitted with a different cutting device. Instead of a crown cutter, it employs a wheel cutter comprising closely spaced bottomless buckets arranged around the circumference of a circular assembly, within which the suction intake to the dredging pump is located.

The main adcantages, depending on soil characteristics and local circumstances, are: reduced spillage and reduction of over-dredging.

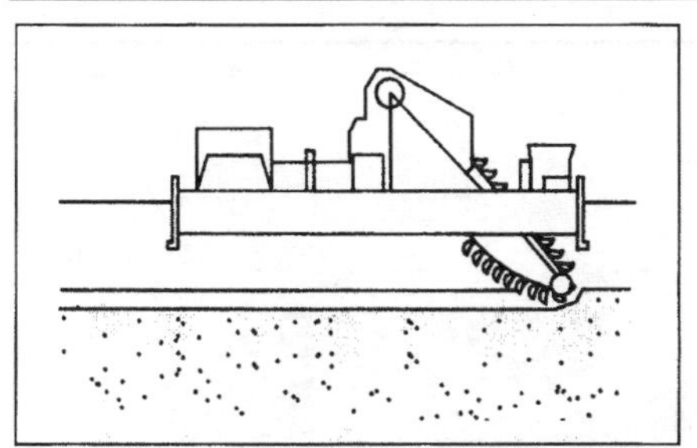

The **Bucket Chain Dredger** uses a continuous chain of buckets to scoop material from the bottom and raise it above water. The buckets are inverted as they pass over the top tumbler, causing their contents to be discharged by gravity on to chutes which convey the spoil into barges alongside.

In mining applications the dredger can feed directly to an attached processing plant. Positioning and movement are achieved by means of six winches and anchors.

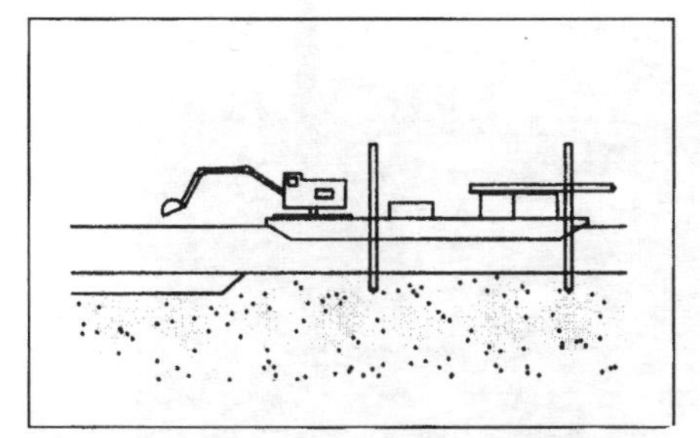

The **Backhoe Dredger** is based on the common land-based backhoe excavator which is mounted at one end of a spud-rigged pontoon.

The backhoe's main advantage is its ability to dredge a wide range of materials, including debris, boulders, stiff clay and soft, weathered or fractured rock.

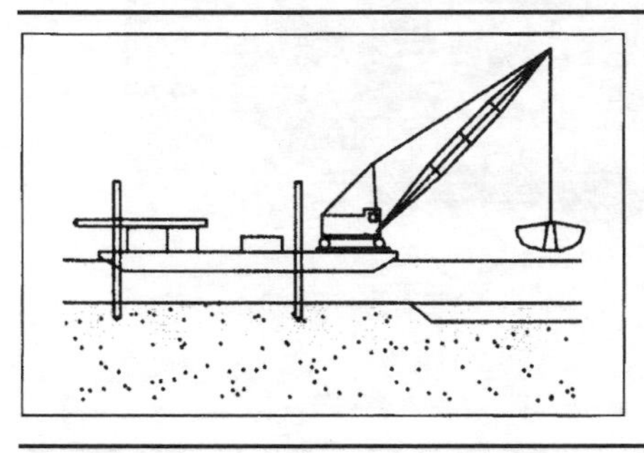

The **Floating Grab Dredger** employs a grab crane mounted on the deck of an anchored or spud-rigged pontoon. Material is loaded into independently operated hopper barges.

The ability to dredge in difficult soils without being seriously affected by debris or occasional small boulders is the main advantage of this type of dredger.

Digging dredgers

Bucker chain dredgers (Figure 6.32) are the most common, in which a continuous chain of buckets scoop material from the bottom and raise it above the water level. The spoil is discharged on to chutes as the buckets invert over the top tumbler.

Bucket dredgers or backhoe dredgers (Figure 6.32) can be employed on a mobile craft. In principle they operate as they would on land. The main advantage is that the machine can cope with large obstacles such as boulders and soft rock.

Dragline dredgers operate on the same principles as their counterpart on land. Their uses are limited to shallow inland waterways where there are few bridges, and the material dredged from the bed can be dumped on the banks.

The **floating grab dredger** (Figure 6.32), in its simplest form, is a crane with grab attachment mounted on a pontoon. This type of dredger requires towing into position and a barge in attendance to receive the grab discharge. At the other end of the scale is the grab hopper dredger, a self-propelled vessel with multiple cranes which fill its own hopper. The hopper itself can be self-discharging by means of conveyors or bottom-discharging through opening doors, or it can be emptied with the same grabs. The grab itself will generally be of the clamshell type to retain mud and silt, but the claw type is useful for picking up large rock pieces and other debris. The grab dredger is generally used on inland waterways, docks and harbours, and has special value in the sinking of caissons where it can operate in a confined vertical space.

Ancillary dredging equipment

Barges are used to transport materials in situations where it is desired to keep the dredger working full time, where the dredge has no hopper or the draught of the dredger is too great for it to reach the dumping ground, or where a navigational channel between the dredger and spoil reclamation area makes the use of a floating pipe-line impractical.

FIGURE 6.33
Barge with tractor mounted grab off-loading rock
(Royal Boskalis Westminster nv)

Barges can either be dumb, in which case they will be manoeuvred by tugs and towed in strings, or self-propelled. Bottom-dump barges are self-discharging. This can be achieved by having hydraulically or chain-operated doors in the bottom of the hopper, which can be opened when the barge is over the required dumping area.

Another method of bottom dumping is the **split discharge barge**, which comprises two halves pivoted about hinge points at the top of the superstructure fore and aft. Hydraulic rams force the two halves apart and the sand discharges through the bottom.

Inverting barges have also been designed which can be overturned by filling and emptying water tanks in the hull, thus dropping the load.

Well barges are used for transporting material from suction dredgers and bucket dredgers to be unloaded by reclamation dredgers.

Barges have also been designed to transport stone of various sizes. Small stones up to 225 mm in size can be discharged by means of conveyors and hydraulic shovels, while larger rock pieces are transported on trays which can be lifted off bodily at the unloading point. Alternatively bottom-opening or tipping barges can be used (see Figure 6.33).

Booster pumps and pipework. Booster pumps are used in delivery pipelines from reclamation dredgers, suction dredgers and cutter suction dredgers, where the combination of pumping distance and material grain size make it necessary to have more horsepower in order to pump the material at an economic production rate through a given size of pipeline. Such booster pumps are nowadays frequently mounted within the dredger (some dredgers have three- or four-stage pumping), while flexibility of planning is maintained by the availability of separate booster stations which may be either skid-mounted for use on land or installed in self-contained dumb pontoons which can be coupled to a pipeline when required.

6.5.3 Construction and materials

Dredging and reclamation with sand

A typical large scale land reclamation project will involve the following:

- Site establishment and mobilisation.
- Dredging of a stockpit (or dump harbour).
- Construction of sea walls or bunds.
- Pumping sand behind sea walls or bunds.
- Stabilisation of surface.

Site establishment and mobilisation. In addition to the normal requirements, site establishment may involve:

- the bringing of road and rail access to the site to enable vast quantities of materials to be transported economically;
- the provision of a high voltage electricity supply;
- the installation of radio and radar communication systems;
- the installation of conveyor and cableway systems if these are required;
- the construction of an independent plant yard and offices which may be in use for more than five years;
- and in remote areas the construction of a camp for the workforce.

Mobilisation will often entail the movement of dredgers from various parts of the world, and the timing of these movements will be crucial to the start of the operation.

Dredging of a stockpit. Having set up in the site, the first operation will be the construction of a stockpit and any necessary channels. In simple terms this is a pit in the sea bed, located centrally about the area to be reclaimed, into which the dredgers or barges dump their load. A cutter suction dredger (or dredgers), sitting in the middle of the harbour, sucks up the sand which has been previously dumped and pumps it into its final position.

The location and size of the stockpit will depend entirely on individual situations. It should be as close inshore as possible to minimise the pumping distance and afford protection from storms, yet this will be balanced to a certain extent by the need to shorten the length of the approach channel, which might require almost continuous maintenance dredging. Alternatively, the reclaimed material may be loaded into dump or self-propelled well-barges, which travel to a reclamation dredger which then unloads them and pumps the material ashore. This obviates the need for a stockpit, but is a method generally used only in more sheltered waters.

Construction of sea walls. When sand is placed by artificial means on top of an existing beach, the natural action of the sea may tend to wash all the 'foreign' sand away, though this is not always the case. Thus in a coastal reclamation scheme it may be necessary to protect the reclaimed area by sea walls. Sea wall construction is discussed in greater depth elsewhere in this chapter and will vary greatly with the availability of material and marine conditions. The design shown in Figure 6.35 provides a satisfactory solution for most locations by varying the size of the rock pieces.

FIGURE 6.34
Use of stockpit in land reclamation

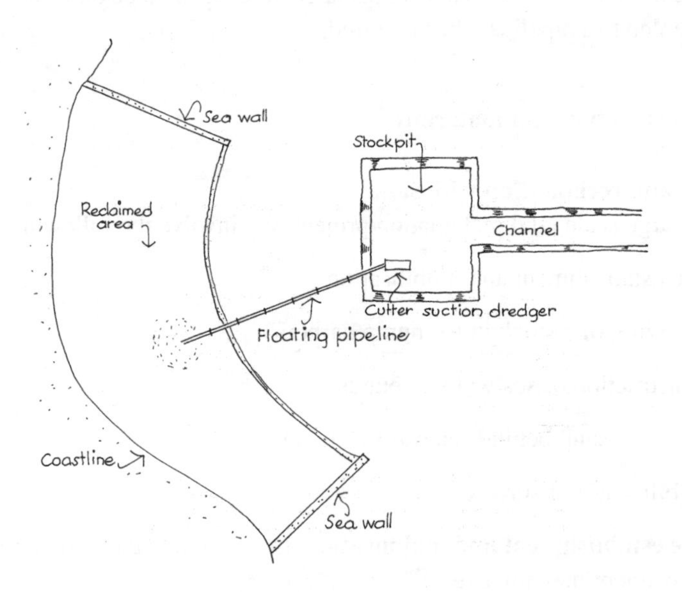

FIGURE 6.35
Section through sea wall suitable for protecting reclaimed area

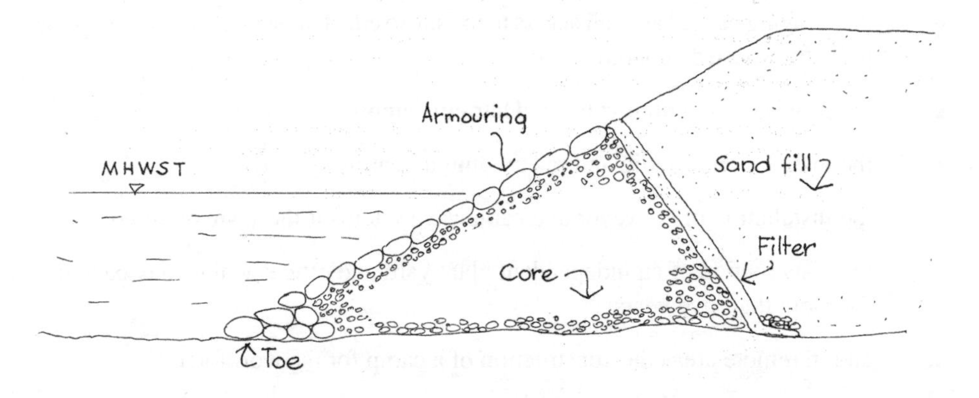

The construction of the sea wall will vary according to the depth of water. If at some state of the tide there is at least 3 metres of water, it will be possible to place the lower materials by special bottom-dump barges. Elsewhere the material will normally be placed by transporting from the land and bulldozing into place.

The main components of this type of sea wall are (see Figure 6.35):

- Toe – formed of heavy rock pieces of precast concrete units strong enough to withstand and break up the predicted forces acting upon the foot of the wall.
- Core – a graded rock heavy enough to withstand normal seas until the armouring is placed.
- Armouring – formed of large rock pieces laid on the face of the wall where wave action can be anticipated; smaller rock pieces may be used by grouting between the stones with a bitumen binder.
- Filters – graded gravel layers or man-made fibre sheeting which prevent the sand behind the wall from getting leeched through the wall.
- Sand – above the level of the highest predicted storm conditions, the wall can be formed of sand with a thin flexible facing if wave run-up can be anticipated.

The construction of the sea wall must necessarily be carried out in advance of the reclamation, and will be the critical factor in determining the speed at which reclamation takes place. Construction time will be affected by tidal working, storms, the necessity for armouring to follow fairly closely behind the placing of the core and the particular restrictions of the working area. A great deal of planning is needed to enable this sort of operation to be carried out successfully.

Pumping sand behind sea walls. Once a protecting arm of sea wall is constructed, it will be practicable to start pumping sand behind the wall. The dredgers, which may be up to twenty miles away, will deposit their loads either directly or by means of barges into the stockpit. The cutter suction dredgers suck up the sand and pump it through floating pipelines to its final position behind the sea wall. The sea water pumped with the sand runs off through specially constructed temporary sluices and the sand forms a level surface which is finally trimmed by bulldozer and grader.

Stabilisation of the surface. The final operation is the stabilisation of the sand surface. Large areas of dry sand at an exposed coastal location will be prone to sand storms and sand losses of appreciable magnitude. In the short term, a weak bitumen emulsion sprayed on to the sand will bind the surface, while special grasses are established.

Dredging and disposal of materials other than sand

The dredging of materials other than sand will be carried out during the maintenance of inland waterways and harbours, the construction or enlargement of port facilities or the formation of shipping channels at sea. The actual dredging technique adopted will depend on the characteristics of the material dredged, but plant is available to deal with most conditions that will be met. The only exception to this is likely to be hard rock, which may have to be broken up with underwater explosive before removal by bucket or grab.

Reclamation with materials other than sand

As previously stated, reclamation with materials other than sand will generally be carried out at inland sites in areas where the materials themselves are the by-product of other activities, eg fuel ash from power generation.

The sequence of events will be as follows:

- excavate from stockpile and load;
- transport;
- tip and spread;
- compact.

The first three of these, having been fully discussed in previous chapters, require no further comment. Compaction requirements will vary according to the use to which the land will be put. Whilst agricultural land will require little or no compaction, great care must be taken over the compaction of fill materials when the land is to be developed. The various types of compaction plant are fully described in Chapter 2, and the efficient execution of reclamation works will depend on the selection of sufficient machines of the right type to spread and compact the material as it is delivered. Stock-piling of material adds a further operation to the cycle and this will obviously increase the cost of the project.

FIGURE 6.36
A gauging weir and lock under construction, showing earth dam
(The Dredging & Construction Company Limited)

6.6 CANALS AND RIVER WORKS

6.6.1 Introduction

The various river and water authorities own or manage over 3000 km of canal and river navigation in England, Wales and Scotland, together with associated works. Apart from the constant upkeep of locks and bridges, an extensive programme of dredging and river bank protection is maintained.

Dredging operations are usually carried out by the authorities with their own labour and equipment. Normally dredging operations are undertaken by bucket and grab dredgers, with supporting disposal units.

Larger works, such as bridge building or reconstruction, lock and weir construction and similar works, are carried out by civil engineering firms under contract. A gauging weir and lock under construction can be seen in Figure 6.36. The photograph shows the river diversion necessary for such work, which was successfully carried out by the use of an earth-fill cofferdam.

6.6.2 River-bank protection

River-bank protection or works to control the course of rivers are sometimes called 'training works'. The work involved has two objectives: firstly to prevent damage by erosion, and secondly to improve the discharge capacity of a river channel or to reduce the natural deterioration of such channels.

Bank protection may take one of several forms, such as:

- Mattresses and geotextiles
- Gabions
- Steel sheet piling
- Stone pitching.

Mattresses (fascines), geogrids and geotextiles are particularly valuable for slope protection work on river banks. Originally they were made from willow branches to the size required, today they are made from polypropylene twine or polyester yarns. '**Geogrids**' is the collective term for net-shaped synthetic fabrics used in geotechnical engineering. One specific geogrid is Fortrac. The material is formed into grid pattern sheets, with a mesh size of 20mm × 20mm; this material forms the retaining grid for soil and rock fill (Figure 6.37). The material is produced in rolls 3.7 m wide, with up to 200 m in a roll. The retaining wall is built in layers, the face of each layer being formed by a shutter anchored to the lower formed layer by mild steel reinforcing bars. When the layer of fill has been compacted, the free end of the mat is pulled over the face back and retained by further fill material. This gives stability to the wall as it is built.

One specific geotextile is the Enkamat, a flexible nylon mat. The mat (Figure 6.38) is used to prevent erosion of embankments and is fixed with nylon or steel pins as the embankment is rebuilt. Alternatively steel pins can be used to anchor the mat. This material allows vegetation to grow quickly and therefore has environmental advantages. Mats vary in thickness, up to 20mm, and can be filled with bitumen-bound gravel, or contain pre-grown turf for areas requiring immediate protection.

FIGURE 6.37(a)
Fortrac mattresses used to retain embankment
(MMG Civil Engineering Systems Limited)

FIGURE 37(b)
Fortrac wall in various stages of construction
(MMG Civil Engineering Systems Limited)

FIGURE 6.38(a)
Enkamat being used to protect banks in the Norfolk Broads
(MMG Civil Engineering Systems Limited)

FIGURE 38(b)
During and after the application of Enkamat to a flood relief channel
(MMG Civil Engineering Systems Limited)

Gabions (Figure 6.39) are essentially wire boxes filled with small stones to form large blocks. There are two basic types, one consisting of a frame of woven wire covered with PVC and the other of frames fabricated from welded high tensile steel mesh.

Gabions are delivered in a folded state and assembled on site, being filled with rock, broken concrete or boulders, and then closed at the top. The whole structure is wired together and lifted into position prior to further wiring to adjacent and underlying gabions. There is no need for any drainage and the boxes are flexible enough to take up settlement without damage.

River walls should be of heavy construction if the current is strong or if the bank has to be contained. Protection of the gabion wall against scouring action can be achieved by the use of a flexible apron. The apron consists of a mattress 0.5 metres thick, formed with small gabions. As the bed is scoured away under it, the apron folds down, eventually forming a curtain wall which stops further undermining.

FIGURE 6.39(a)
Gabions used for river bank protection
(River and Sea Gabions Limited)

(i) Gabions shortly after construction

(ii) The same scene two years later, showing vegetation

Steel sheet piling (Figure 6.40) has been used for bank revetment and protection for many years and its construction is covered fully in Chapter 4, section 4.1.

Stone training work and precast concrete units may also be used to counter erosion. The stone work, in the form of pitching of substantial size, is dumped along the river bank and allowed to sink into the silt. The pitching is quickly bound together with silt and forms a strong bank; however, the use of pitching or concrete units is not as aesthetically pleasing as some other forms of revetment.

FIGURE 39(b)
Anchored gabions (Platipus Anchors Limited)

FIGURE 6.40
Steel sheet piling in river bank protection

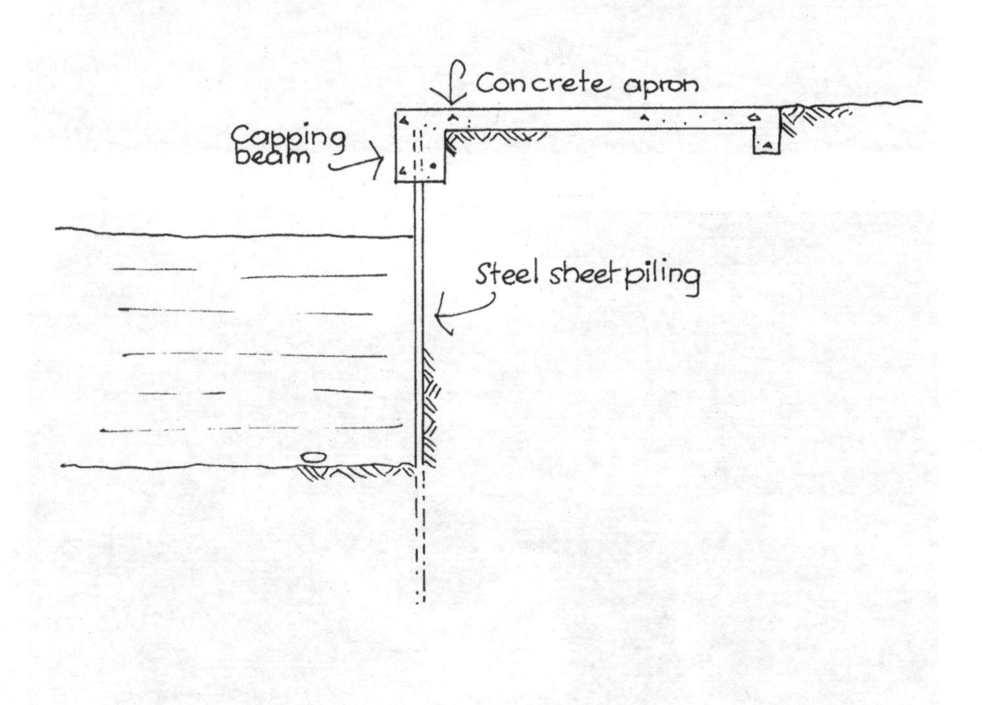

6.7 RESERVOIRS AND LIQUID-RETAINING TANKS

6.7.1 General considerations

Reservoirs includes two basic forms. Those for the catchment of crude water are referred to as 'open impounding reservoirs', and those used in the distribution of water are referred to as 'service reservoirs'. The latter are also known as 'covered' service reservoirs because they must be enclosed to protect the water.

Liquid-retaining tanks are generally used by industrial or public utility undertakings for the storage of liquid gas, oil products, acids, alkalis and other liquids.

Reservoirs are normally constructed in an elevated position in order to eliminate or reduce the amount of pumping. The ground condition at such levels is normally free from ground water, and the structure, if formed in the ground, can take a very simple form. If, however, the reservoir has to be constructed above ground to give the required head of water, the construction may be very complex. Most gravity schemes consist of a dam across a valley to impound a natural stream. They are all similar in principle in that they all impound the water and have overflow weirs, draw-off points and some form of cut-off walling or other construction to intercept any stray water which may flow under the dam. In addition, of course, they have treatment plants to improve the quality of the water.

Liquid retaining tanks must be designed with the following considerations in mind:

- type of liquid to be stored;
- temperature of liquid;
- size of tank;
- position of tank (ie whether in the ground or above ground on columns);
- type of tank (open or closed, lined or unlined);
- shape of tank (circular or rectangular).

These and other factors will have an influence on the construction and specification of the work to be done.

6.7.2 Construction

Reservoirs which are designed to impound water – normally dams – can be constructed of earth, rock, concrete, or any combination of these materials. The geological factors, together with the size of the reservoir, will determine the choice; earthen embankments are suitable for clay sites and concrete dams for rocky ground. The earthen embankment must have a watertight face, or a centre core which can be achieved by puddled clay or concrete, although bituminous grout may be used in conjunction with a layer of coarse material (Figure 6.41). Concrete dams may also include cut-off walls or grout-curtains to prevent seepage of water, depending on the geological conditions.

Reservoirs for the distribution of water and those for other liquids are constructed using the same basic design principles, the only difference being that water tanks must be covered, and this involves the construction of a roof structure. The reservoir or tank, when empty, must be designed to prevent flotation or damage to the floor if constructed in waterlogged ground. Consideration must be given to the type of material to be used if constructed in ground which contains injurious chemicals or which is subject to settlement or movement. If tanks are to be constructed in concrete, consideration must be given to the risk of cracking as the concrete cools, especially in thick sections.

While overall contraction may be adequately dealt with by the provision of movement joints, the risk of cracking due to differential contraction may require special attention. Construction joints and movements joints can be seen in Figure 6.42. These joints must be provided in the construction if cracking is to be controlled; the spacing of them will depend on the length and thickness of the wall, the cement which is used and its temperature during construction.

Sliding joints, for the purpose of accommodating expansion or contraction in the walls, are shown in Figure 6.43 (g to i); such joints are suitable for tanks containing hot liquids where the jointing material must not soften unduly with the heat.

Circular tanks may be designed as simple cantilever structures or ring tension structures in which ring reinforcement restricts the outward deflection of the wall. Rectangular tanks will normally be designed as simple slabs cantilevered from the floor, the corners providing suitable restraints to deflection. For this reason there should be no vertical construction joint at the junction of two straight walls.

FIGURE 6.41
Section through earth dam with clay centre and cut-off wall

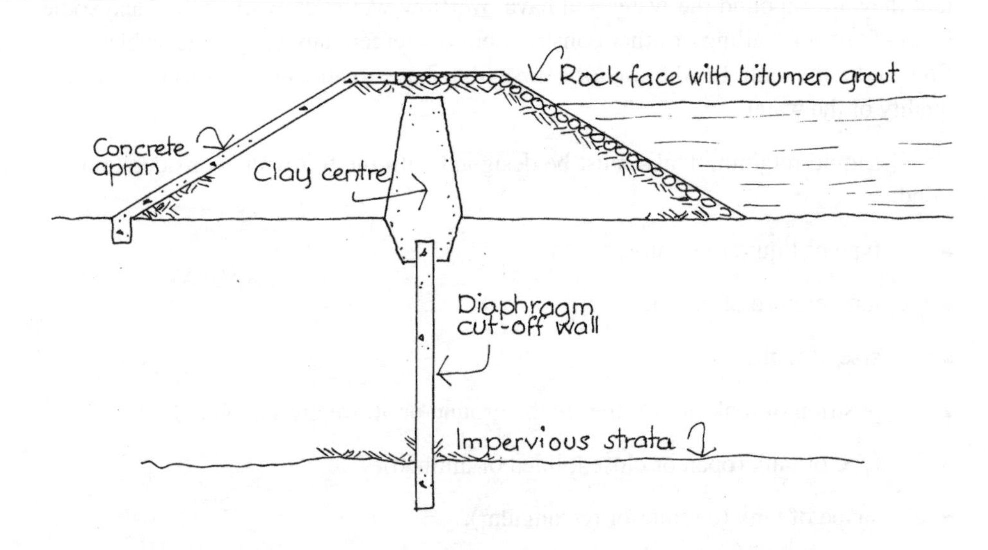

FIGURE 6.42
Joints in concrete walls to accommodate movement

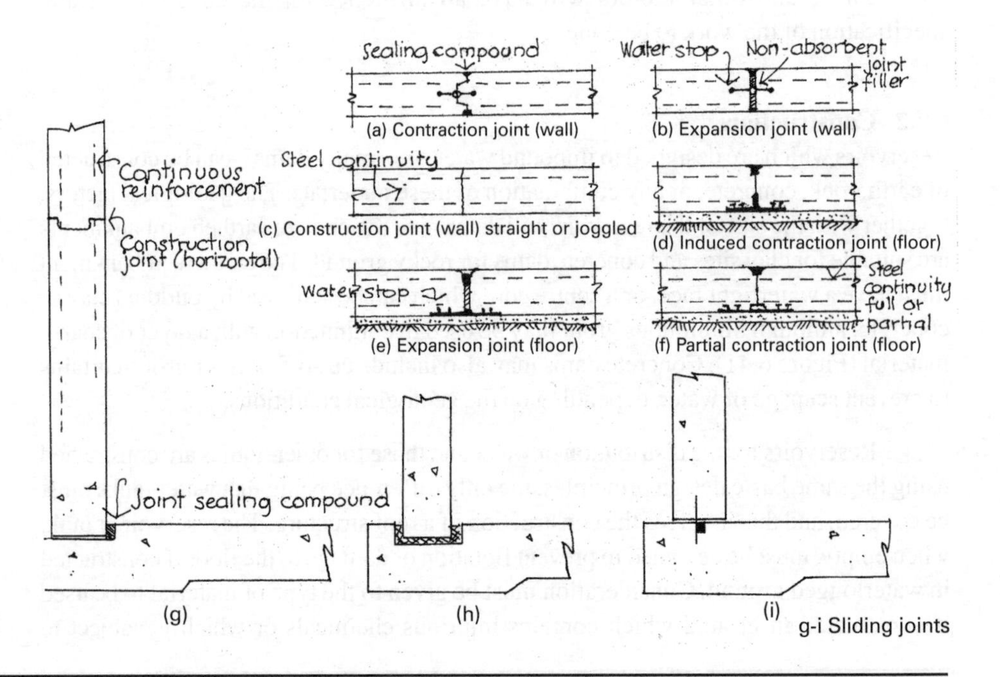

Tanks that have roof structures which are connected to the top of the walls may be subject to excessive pressures due to the movement of the roof. Where the tank is surrounded by soil, which can follow the contraction movements, a layer of compressible material, usually foamed plastic, should be placed between the tank walls and the ground (Figure 6.43 (a)); this will allow for wall movement caused by expansion of the roof slab. Alternatively, the roof slab can rest on a sliding joint (Figure 6.43 (b)), the joint being of bituminous material, stainless steel, multi-layer rubber or synthetic fabric. If sliding joints are not acceptable, the roof may be cantilevered from columns (Figure 6.46 (c)), allowing roof and walls to move freely.

Floors to reservoirs and tanks must be constructed carefully to reduce shrinkage. Laying a floor in alternate bays or in 'chess-board' squares, with seven-day intervals between the placing of alternate bays, will eliminate 50 percent of the primary cooling shrinkage but will not control the hardening shrinkage.

Two schools of thought prevail in the control of shrinkage: the first allows the slab to move freely on a sliding layer of synthetic material, with adequate expansion joints (Figure 6.44a); and the second is to restrain the slab on a rough concrete sub-floor, thereby restricting shrinkage cracks to a size that will not permit leakage. In the latter case (Figure 6.44b), reinforcing bars give continuity at the joints. Some floors have been laid successfully in two or three layers, each one separated by a waterproof membrane, usually synthetic polymer sheeting. The upper layers cover the joints in the lower layers.

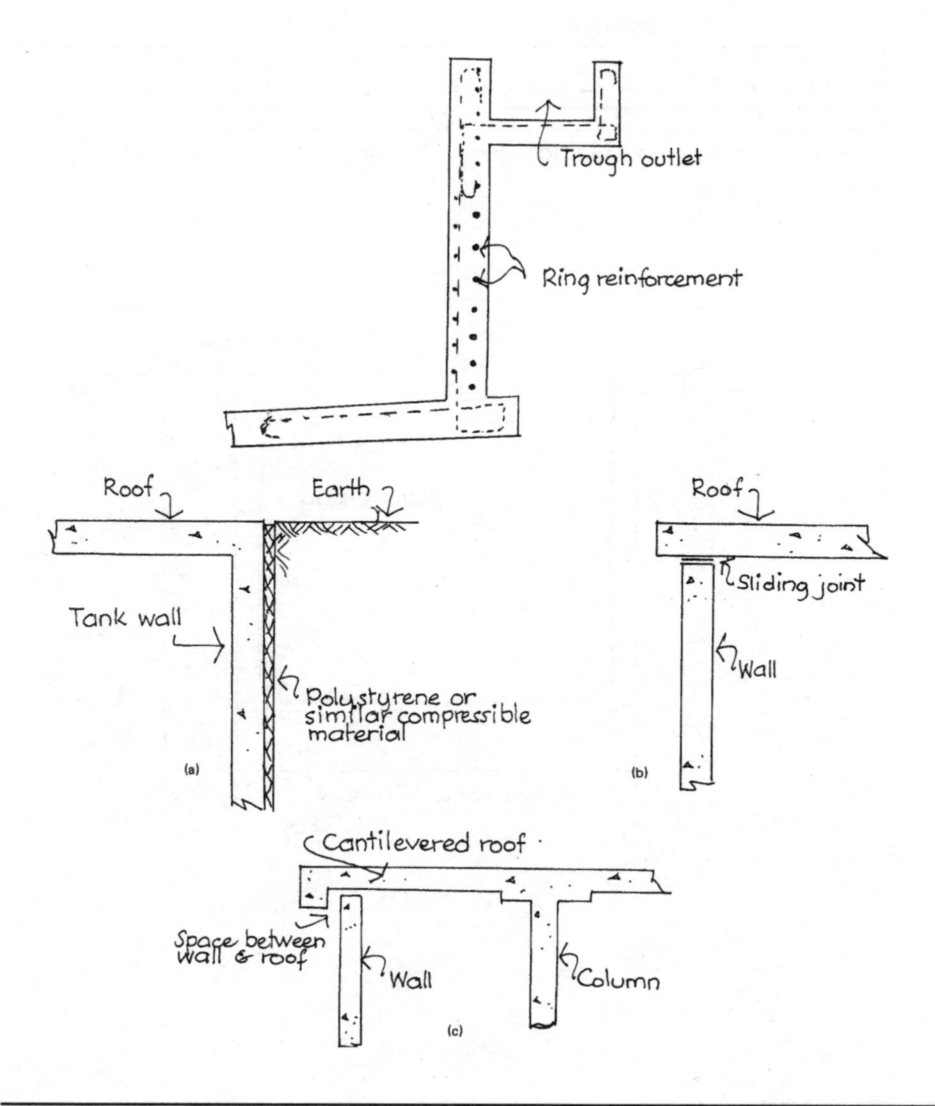

FIGURE 6.43
Provision for expansion and contraction of tanks and tank roofs

Floors that are suspended, ie on columns or on piles, are designed as normal concrete suspended floors. Tanks below ground should have a roof screeded to falls and covered with granular material to permit drainage of ground water.

In-ground storage tanks

In-ground storage tanks or reservoirs are used for the storage of liquid natural gas (LNG). The tanks may be lined or unlined, depending on the nature of the soil and the height of the water table. The ground is frozen by the freezing method described in Chapter 3, and when a wall strong enough to resist the outer soil pressures has been frozen, the tank is excavated. Since natural gas liquifies at –160°C the temperatures must be kept below this or the gas will 'boil off'. The most suitable soil for unlined tanks of this nature is stiff clay or marl; however, it is recommended that tanks are lined.

Metal-lined tanks with concrete walls and floors (Figure 6.45) are the most practical in construction. The ground is excavated and a concrete tank is cast, to which a metal lining is fixed. The problem of casting concrete against the frozen earth can be overcome by using an efficient insulatory material between the concrete and the ground; the reinforcement should be of nickel steel or other non-brittle steel. Steel with 9 percent nickel can be used in temperatures down to –200 °C. The metal lining is normally nickel-steel or aluminium alloy.

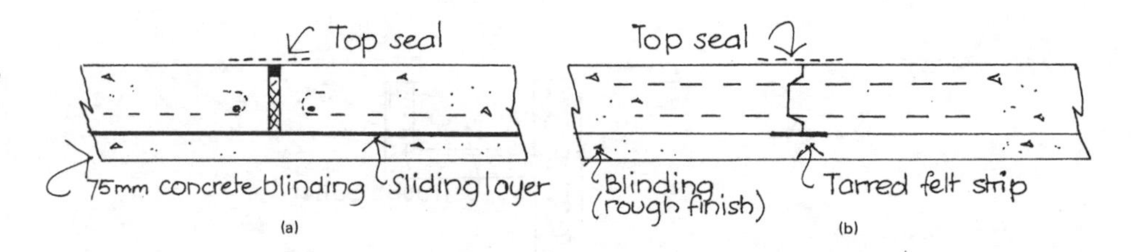

FIGURE 6.44
Different treatments to floor slab to counteract movement

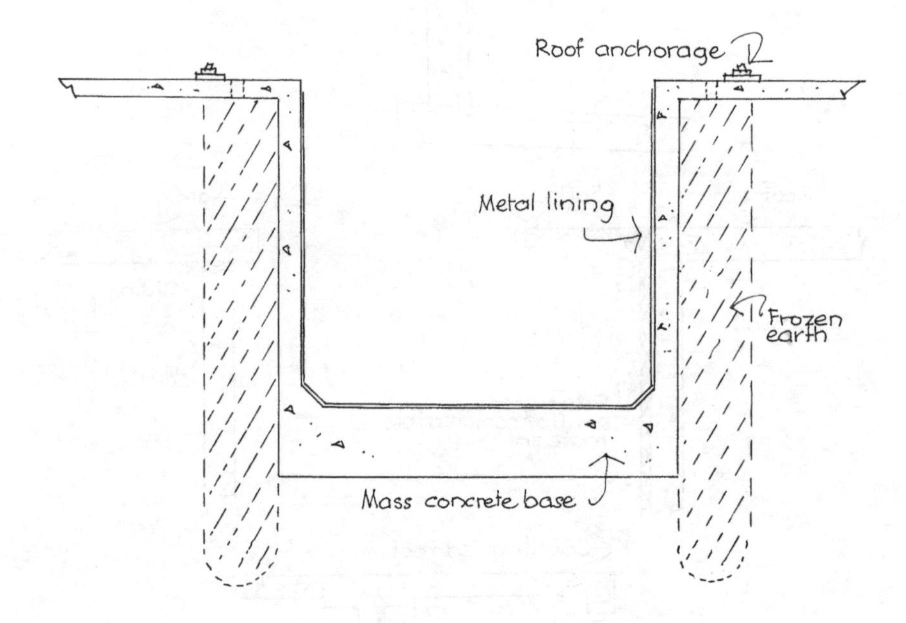

FIGURE 6.45
In-ground storage tank for liquid natural gas

6.7.3 Waterproofing

In considering methods of waterproofing concrete structures, certain major factors must be satisfied: stresses in the concrete and steel must be kept within reasonable limits; the anchorage between steel and concrete must also be adequate; and shrinkage and temperature stresses must be considered.

It should be noted that fine hair cracks can be reduced by using deformed bars in lieu of plain round bars, which highlights the fact that anchorage between steel and concrete can be improved.

Integral waterproofing is not favoured by many engineers or authorities, although manufacturers claim great success with their products. A possible risk in the use of admixtures is that of bad workmanship – there is a tendency to rely only on the admixture and so to overlook strict control of concrete batching, mixing and placing. The object of most integral waterproofers is to fill the pores in the concrete, thus making a denser and less permeable material; some admixtures obtain a similar result by chemical action on the cement.

Waterproof linings include asphalt, renderings, plastic sheeting and paints. The oldest, perhaps, is asphalt tanking, which requires no explanation in this book. Waterproof renderings containing a patent waterproofing additive are suitable for tanks which remain filled most of the time. If waterproof renderings are allowed to dry out, and remain too dry for a long time, they may become subject to cracking and spalling.

Geomembranes, which include high and low density polyethylene, Butyl, EPDM, Hypalon synthetic rubbers and PVC, are suitable for a wide variety of liquid retaining structures, such as reservoirs, settlement lagoons, ornamental lakes, landfill linings, canals and dams. Figure 6.46 shows a membrane being used in a reservoir where the membrane is laid over a carpet of sand 100 mm thick, and jointed by vulcanising. The peripheral joints can be formed by sealing the sheeting to a perimeter skirt, which in turn is sealed to a water-bar in the toe of the reservoir wall. Applications are also illustrated for effluent treatment plants (Figure 6.47), and industrial settlement lagoons (Figure 4.48).

FIGURE 6.46
Butyl sheeting being laid on sand base to provide floor of reservoir
(Butyl Products Limited)

6.7.4 Special linings

Many liquids can be stored in concrete tanks, but there are some groups, especially acids and sulphates, which are extremely harmful to concrete. Concrete must be protected from such liquids and this is done either by lining the tank with a material that will resist corrosion or, alternatively, by lining the tank with a material that can be replaced at suitable intervals of time. The latter method of protection is known as 'sacrificial linings'. Protection linings include silica-of-soda solutions, glass linings, acid-resisting asphalt, wax linings, lead linings, and other soft metal linings. The lining is chosen to resist the corrosive action of the particular liquid.

Sacrificial linings, on the other hand, are linings which do not resist the corrosive action of the liquid but allow an attack to take place over a period of time until the tank requires emptying and the lining renewing. The lining may be rendered in cement mortar or some other relatively cheap form of cover which can be 'sacrificed' to prevent serious attack on the concrete.

FIGURE 6.47
Membrane used in a reed bed for effluent treatment
(SGS Geosystems Limited)

FIGURE 6.48
Membrane used in an industrial settlement lagoon
(SGS Geosystems Limited)

CHAPTER 7

Roadworks, Bridges, Subways and Airfield Construction

7.1 ROADWORKS

7.1.1 Earthworks

Road construction can be divided into two distinct phases: first the earthworks, and second the construction of the pavement which overlies the earthworks.

The earthworks are concerned with the preparation of the soil to bring it to the correct levels, gradients, profiles, and strength required. The finished level of such earthworks is referred to as the **formation level** and the soil immediately below that level is known as the **subgrade**. As discussed in Chapter 1, it is essential that the earthworks are preceded by testing of the soil; this will establish the properties of the soil and assist in the design of an economic subgrade.

The formation of embankments is discussed in Chapter 3. It should be noted that although 'fill' materials are defined as 'suitable or unsuitable', a material that may be suitable for the construction of embankments in a dry state may nevertheless be unsuitable in a wet state. The type of plant used in compaction is also important. So it can be seen that the suitability of fill material depends not only on its physical properties but also on the conditions in which it is to be used and on the methods used for compaction.

Since most earthworks are formed by cut-and-fill the relevant considerations are as follows:

- Determining side slopes for cuttings and embankments
- Treatment of compressible subsoil.

The side slopes of non-cohesive soils are governed by the natural angle of repose of the material, and this can be calculated from site and laboratory tests. Typical ratios for side slopes in clay embankments range from 1 in 2 to 1 in 4, but the actual ratio must be determined from tests; it is therefore difficult, and in some cases unwise, to produce tables of side-slope ratios without full knowledge of the materials being used (see Chapter 3 – Slopes in embankments).

Compressible subsoil may be removed or stabilised, depending on its quality. Peat and similar materials should be excavated and replaced with granular fill; other methods of removal include bog blasting, overloading and jetting, which are described in Chapter 3. When the nature of the subsoil material is such that the cost of full or partial excavation cannot be justified and consolidation is likely, or if the embankment itself consists of compacted clay of a very high moisture content, then sand wicks or sand drains may be used. Sand wicks (see Chapter 3) or sand drains are sand-filled bore holes; the sand wick is a sand-filled stocking which is lowered into a borehole. The sand wick decreases the length of the drainage path which the water has to travel and so dissipates pore water pressure and gives greater stability to the soil (Figure 7.1).

Subgrade strength

The strength of the subgrade determines the thickness of the pavement needed and should therefore be made as high as possible. The desired subgrade strength can be achieved by:

- Removing poor material in cuttings and replacing with selected fill.
- Ensuring compaction of the subgrade to a high dry density.
- Providing adequate subsoil drainage.
- Avoiding the use of materials subject to frost damage.

The suitability of materials and their compaction have already been discussed.

Subsoil drainage must be provided to deal with:

- Seepage through pavement and verges
- Seepage from higher ground
- Seasonal rise and fall of the water table.

Seepage through pavements is difficult to eliminate, since joints deteriorate over a period of time, and surface materials, which are subject to a very wide range of temperatures, eventually crack, allowing water to penetrate. One satisfactory solution is to apply hot tar or bitumen as a sealing coat to the subgrade; this has the dual purpose of shedding the water to a side drain (Figure 7.2) and also of protecting the subgrade during construction. Moisture seeping through verges and moving into the subgrade, causing swelling and shrinkage of the subgrade, may be intercepted by a subsoil drain situated between the verge and the carriageway, or, in the case of motorways, between the verge and hard shoulder (Figure 7.2).

Seepage from higher ground occurs when a layer of permeable soil overlies an impermeable strata. The water can be intercepted by a cut-off drain, which may be incorporated with the drain at the verge if the impermeable layer is less than 1.2 metres below the surface. Where the impermeable layer is at a lower level, the cut-off drain

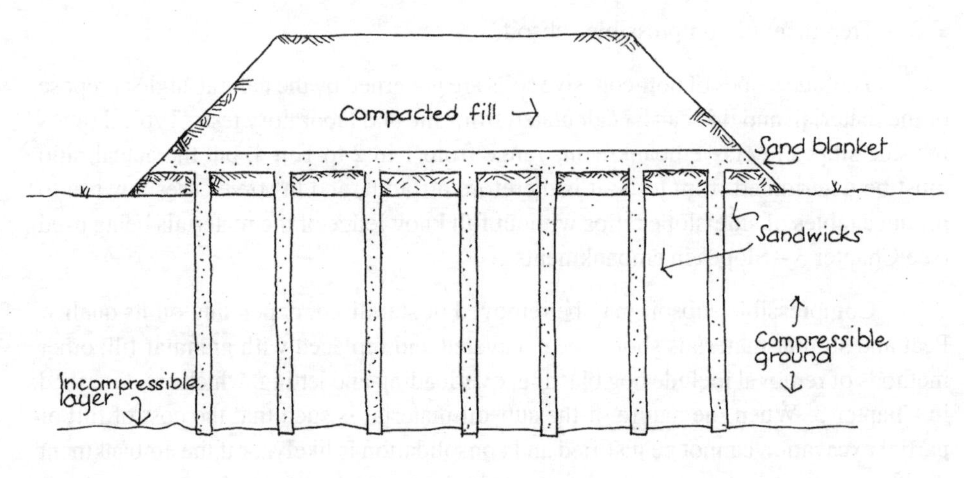

FIGURE 7.1
Sand wicks under road embankments

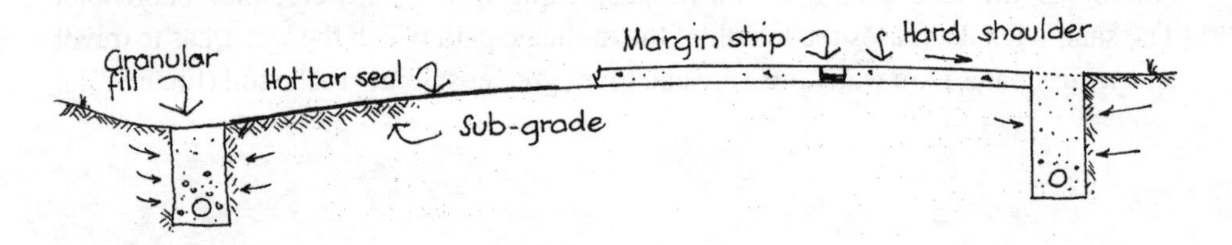

FIGURE 7.2
Use of sub-soil drains to intercept ground water

should be taken to a depth that will keep the seepage water at least 1.2 metres below formation level.

Seasonal fluctuation in the water table should be controlled by drains, as shown in Figure 7.2, so as to keep the water table 1.2 to 1.5 metres below the formation level. Frost damage to the subgrade can be expected with certain types of soil which draw and hold moisture in the soil pores. Limestone gravels are likely to be susceptible if the average saturation moisture content of the material exceeds 2 percent. All crushed chalks are frost susceptible; the magnitude of frost heave increases linearly with the saturation moisture content of the chalk.

The final strength of the subgrade is assessed on the California Bearing Ratio (CBR) scale. The correlation between CBR value and soil type is shown in Table 7.1; this table is helpful when calculating the sub-base thickness. The surface treatment of the formation will vary from simple rolling to coating with bituminous materials. If the formation is not immediately covered with sub-base material it should be protected by an impermeable plastic membrane (500 gauge) having laps of 300 mm at the joints. Rock-fill or rock cuttings should be blinded with fine material before the protection layer is placed

Soil stabilisation

If the natural properties of the subgrade do not possess the strength required to support and distribute the proposed loading, the subgrade strength may be increased by soil stabilisation.

Stabilisation may be achieved by various agents, which include cement, lime, bitumen and chemicals.

Cement stabilisation is achieved by mixing cement with pulverised soil to form a material which, when compacted and allowed to harden, possesses appreciable strength. The material to be stabilised should not contain sulphates in excess of 1%. The thickness of layer to be stabilised should not be less than 75 mm when compacted, and should be compacted in layers of up to 200 mm thick at one pass.

Bituminous materials such as cut-back bitumen, and chemicals, have all been used for soil stabilisation in hot dry climates, but their use is outside the scope of this book.

TABLE 7.1 **Estimated laboratory CBR values for British soils compacted at the natural moisture content**

Type of soil	Plasticity index (percent)	CPR (percent)	
		Depth of water-table below formation level	
		More than 600 mm	*600 mm or less*
Heavy clay	70	2	1
	60	2	1.5
	50	2.5	2
	40	3	2
Silty clay	30	5	3
Sandy clay	20	6	4
	10	7	5
Silt	–	2	1
Sand (poorly graded)	non-plastic	20	10
Sand (well graded)	non-plastic	40	15
Well-graded sandy gravel	non-plastic	60	20

Pavement design

Pavement design is a specialised topic that cannot be adequately covered here. However, it is important to note that pavement design depends on the relationship between pavement thickness, subgrade strength and the axle loading likely to occur. For design purposes the number of standard axles has to estimated, this is done in terms of millions (msa); for example a road might have to be designed for 13 msa or 65 msa. A standard equivalent single-wheel is used, which in the US is taken as 8200 kg, while in the UK the loading is limited to 10 tonnes for wheels at the end of an axle or 11 tonnes where wheels are spaced along the axle. The number of standard axles involved, and hence the design of the pavement, is determined from tables which convert the number of commercial vehicles using a road daily into standard axles. In the UK pavements must comply with the Specification for Highway Works (MCHW1).

7.1.2 Flexible and flexible composite pavements

Flexible pavements (Figure 7.3) consist of a layered system of materials which distribute the wheel-loads to the sub-grade. The thicknesses of individual layers must be such as to distribute the loads without permanent deformation of the material, thereby presenting an uneven running surface. Flexible composite pavements consist of upper surfacing and roadbase material bound with bituminous binder and the lower roadbase of cement bound material (CBM); these are discussed later.

Sub-base

The sub-base is the construction layer that is placed over the subgrade after the subgrade has been waterproofed. Various materials can be used but often the specification will call for granular materials with a minimum crushing value and specific grading. The material is spread by machine and compacted by heavy rollers, in layers not exceeding 150 mm.

The required thickness of sub-base is determined from the cumulative number of standard axles to be carried and the CBR of the subgrade; this can be obtained from published design charts. For major roadworks the sub-base can be formed with stabilised materials, cement-bound materials or lean concrete. Free-draining materials, such as quarry overburden or crushed rock, should be used in preference to materials containing large amounts of fines.

Roadbase materials

Any material that remains stable in water, is unaffected by frost, and has a satisfactory CBR value when compacted, can be used for roadbase construction. In the UK specific materials are permitted, details of which are found in the *Design Manual for Roads and Bridges*, Volume 7: 'Pavement Design and Maintenance' (1994).

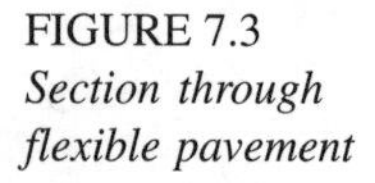
FIGURE 7.3
Section through flexible pavement

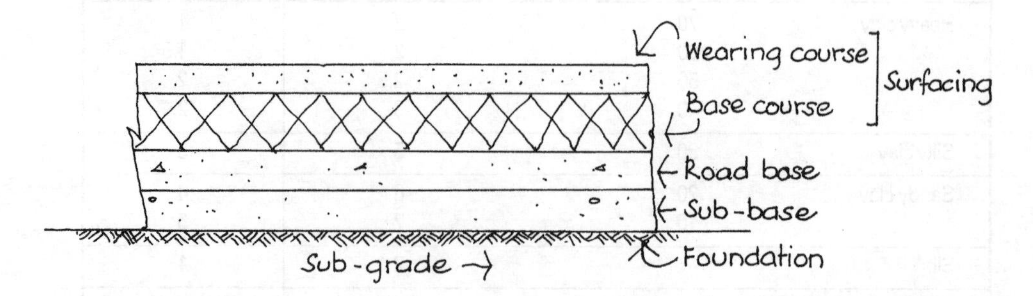

Permitted roadbase materials include

- Dense Bitumen Macadam (DBM)
- Hot Rolled Asphalt (HRA)
- Dense Tar Macadam (DTM)
- DBM + 50 Penetration Bitumen (DBM50)
- Heavy Duty Macadam (HDM).

DBM is a continuously graded material of relatively low binder content (100 penetration grade). It is resistant to deformation but has less resistance to fatigue cracking than other standard materials.

HRA is a gap-graded material: the gaps between the larger pieces of aggregate are filled with binder and some form of filler, such as sand, with high binder content (50 penetration grade). It is easy to lay and compact and has high resistance to fatigue cracking. However, it is susceptible to deformation under heavy traffic.

HDM is a continuously graded material with a greater percentage of fines than DBM and also has a 50 penetration grade binder. It is less easy to compact than DBM but has the highest stiffness and good resistance to fatigue cracking and deformation.

DBM50 has the same composition as DBM but because of the higher grade binder it is superior to DBM.

DTM is similar to DBM but uses tar binder, grade C50 or C54.

Bituminous basecourse materials should contain crushed rock or slag coarse aggregate, unless local experience indicates successful use of gravel.

Cement Bound Material (CBM) includes soil cement, cement bound granular materials, and lean concrete. The aggregate may be an all-in aggregate having a maximum nominal size of 40 mm and not less than 20 mm. The water/cement ratio is determined by the compaction requirements but is about 6 percent of the dry weight of the materials. The thickness will range between 150 and 250 mm. This material is used mainly in flexible composite designs.

Outside the UK a range of other materials are still used for roadbases, including:

- **Dry-bound macadam**. Normally supplied for base construction in two sizes: a single-sized coarse aggregate, nominally 38 mm to 50 mm in size, and a fine graded material, 4 mm to dust, for blinding the surface of the coarse aggregate. The coarse material is laid in a 75 to 100 mm layer and compacted; this is followed by a 25 mm layer of fine material and compaction is continued.

- **Wet-mix macadam**. Graded crushed rock or crushed slag aggregate normally with 2-5 percent water; an added provision is that the percentage of material passing a No 200 sieve may be increased to 10 percent. The material is laid in layers 75 to 150 mm in thickness by a paver or similar machine, and compacted with either a heavy smooth-wheeled roller or by vibrating roller until the required density is achieved.

Recycled bituminous material

The Transport and Road Research Laboratory have carried out tests with recycled roadbase and basecourse materials and found that the use of recycled bituminous material has the potential to reduce costs as well as save energy and natural resources. In the trials Dense Bitumen Macadam (DBM) containing 50 percent reclaimed material, and Hot Rolled Asphalt (HRA) containing up to 60 percent of reclaim materials were used. The trials demonstrated that recycled materials, produced in off-site mixers and laid with conventional plant, could meet the requirements of the relevant British Standard Specifications. The development offers other environmental benefits, such as:

- Reduced aggregate extraction
- Less need for tipping sites
- Reduced need for importing bitumen.

Surfacing courses

The surfacing layer of a flexible pavement is subjected to a great intensity of stress and must consist of high quality material capable of resisting such stress. It must also provide an impermeable weathering which protects both surface and base materials. In addition to these properties it must provide a high resistance to skidding.

The types of surfacing material used are numerous, but they are similar in that each consists of an aggregate bound together by tar or bitumen; the difference lies mainly in the type, viscosity and proportion of binder used, and the type and grading of the aggregate. Road tar, which is one of the main binders, is obtained from crude tar, and bitumen, the other main binder, is produced from crude petroleum or found in natural deposits mixed with mineral aggregate. Asphalt is a mixture of bitumen and mineral matter.

Wearing course

The wearing course is the upper layer of bituminous material and is usually denser and stronger than the lower layer or base course. The thickness depends upon the specification of the material to be used and the amount of wear expected. The main wearing materials are Hot Rolled Asphalt (HRA); Dense Bitumen Macadam (DBM); Dense Tar Macadam (DTM) and Porous Asphalt (PA). In the UK the wearing course on trunk roads has to be either HRA or PA, which are precisely specified. PA is an open graded material designed to enable rapid drainage of surface water from the road, thereby reducing spray. PA also reduces tyre noise. A separate basecourse layer is optional under the HRA wearing course but should always be provided under the PA wearing course.

Base course

Base course materials, which should be of a minimum thickness of 60 mm, may consist of HRA, DBM, DTM or Porous Asphalt (PA). The wearing course should be laid on the base course as soon as possible, normally within three days. With the exception of compressed natural rock asphalt and mastic asphalt, laying is carried out by machine. The base course is shaped with the appropriate crossfalls and gradients before the wearing coat is laid.

Recycled wearing courses

As with recycled base course, it has been found that recycled wearing courses can be less expensive than conventional wearing courses, and have particular applicability in wearing course replacement. Two process are used, repaving and remixing.

In the case of **repaving** the process restores surfaces that are in sound structural condition by bonding a thin overlay to the pre-heated, scarified and re-profiled road surface. The existing wearing course is reused in combination with a thin layer of new material.

With the **remixing** process a purpose-built machine pre-heats and scarifies the surface before augering the material into a pugmill mixer, where it is blended with freshly mixed new material. This recycled mix is placed evenly on the heated surface to form the replacement wearing course.

7.1.3 Rigid pavements

A rigid pavement consists essentially of a concrete slab resting on a thin granular base. The loads and stresses are distributed over a wide area of subgrade by the rigidity and strength of the pavement. The pavement may be Unreinforced Concrete (URC), Jointed Reinforced Concrete (JRC) or Continuously Reinforced Concrete (CRCP). The concrete slab should be Pavement Quality Concrete (PQC), manufactured, laid and cured in accordance with the Specification for Highway Works (MCHW1).

Sub-base

The function of the sub-base is to assist drainage, to protect the subgrade against frost, and, in the case of fine-grained soils, to prevent pumping (the ejection of water and silt through joints or cracks caused by the downward movement of the slab due to heavy wheel loads).

The materials used are usually granular, eg crushed rock, crushed slag, crushed concrete, natural sand, gravels or well-burnt non-plastic shale. The materials should be graded. The thickness of the sub-base depends on the type of subgrade and should follow the recommendation of standard design tables. If the subgrade is susceptible to frost, the total thickness of sub-base and concrete slab should be a minimum of 450 mm. After the pavement slab has been designed, the thickness of the sub-base should be increased, if necessary, to gain a total pavement thickness of 450 mm.

Concrete slab construction

When the sub-base has been prepared, it is common practice to provide an anti-friction membrane over the sub-base before laying the concrete slab. This layer is normally polythene sheeting, which performs the extra function of preventing grout loss from freshly laid concrete. Figure 7.4 shows preparation prior to the placing of the road slab. Figure 7.5 shows a typical road form in position to receive the concrete slab.

The slab is normally placed by a concreting train which runs on a heavy duty road form to prevent deflection. The form is bedded in position on the base at least 24 hours before concreting the slab. The concrete train usually includes hopper units which feed the concrete on to the base via a conveyor belt. Alternatively, this operation may be carried out by a screw-type spreader. Concrete is laid to the level of the fabric reinforcement and, following the placing of the fabric, a second spreader and compactor unit completes the slab. The top layer of concrete is placed with a surcharge of up to 25 percent of the slab thickness to gain maximum compaction; the actual surcharge will depend upon workability of the mix.

TABLE 7.2 **Classification of subgrades for concrete roads and minimum thicknesses of sub-base required**

Type of subgrade	*Definition*	*Minimum thickness of sub-base required*
Weak	All subgrades of CBR value 2 per cent or less	150
Normal	Subgrades other than those defined by the other categories	80 mm
Very stable	All subgrades of CBR value 15 per cent or more This category includes undistributed foundations of old roads	0

Air-entrainment of the concrete should be specified for either the full depth of slab or for at least the top 50 mm. Air-entrainment increases the resistance of concrete to frost damage and to the destructive action of de-icing salts. The materials used for air-entrainment produces minute bubbles in the hardened concrete which prevent saturation by capillarity and so relieve the stresses which otherwise occur when pore water freezes. Since air-entrainment has the disadvantage of weakening the concrete, the volume of air should be restricted to 4.5 percent by volume of the concrete.

An alternative method of laying the slab is by slip-form paver. This machine, which requires no side forms, is mounted on crawler tracks and is capable of laying pavements at speeds in excess of 2 m per minute. The concrete slab is moulded to the required thickness and extruded at the rear of the machine. Line and level is achieved by automatic sensing probes at each corner of the machine; the probes straddle a length of tensioned wire which has been set to the correct gradient. Variations in the level of the base are detected by the probes and the machine level is corrected automatically by hydraulic jacks.

Reinforcement

Reinforcement may be either steel fabric or bar reinforcement, the latter being deformed and spaced at centres of not more than 150 mm. The diameter of bar to use in lieu of fabric can be established from design tables. Concrete cover to the reinforcement should be 60 mm, unless slabs are less than 150 mm in thickness, in which case 50 mm cover is required. The reinforcement should terminate at least 40 mm and not more than 80 mm from the edge of the slab and from all joints except longitudinal joints.

FIGURE 7.4
Slab preparation prior to concrete laying

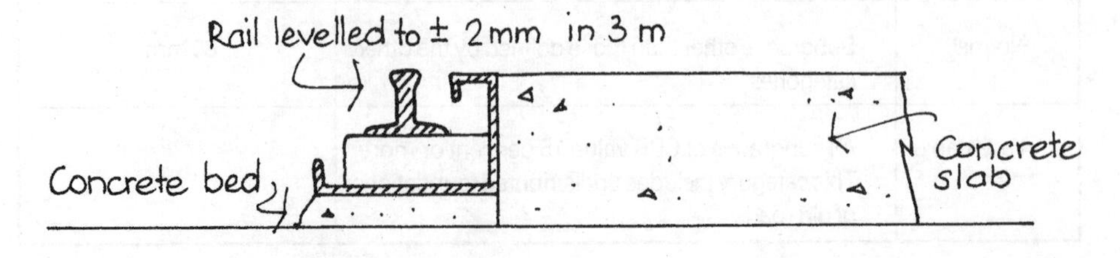

FIGURE 7.5
Road form in position

Where two- or three-lane carriageways are constructed in one operation, reinforcing mats with 8 mm transverse wires at 200 mm centres may be used to span the longitudinal joint in place of bars; if so used, the 8 mm wires must span at least 500 mm either side of the longitudinal joints. Where three-lane carriageways are constructed in two widths, and where each slab is wider than 4.5 metres, special transverse reinforcement, 600 mm longer than a third of the slab width, should be placed centrally in each slab.

Joint construction

Joints are formed in concrete slabs for the purpose of allowing and controlling movement; the movements include expansion, contraction and warping. The spacing of joints depends on the amount of reinforcement used, which in turn depends on the proposed traffic intensity; slab thickness; frictional restraint of the subgrade; and the temperature at which the concrete is placed. Expansion joints, however, may be replaced at the discretion of the engineer by contraction joints when the slab is constructed during summer months. Maximum spacings for expansion joints range from 25 to 27 metres in jointed reinforced concrete (JRC) slabs and from 40 metres (for slabs up to 230 mm thick) to 60 metres (for slabs over 230 mm thick) in unreinforced concrete (URC). Maximum spacing of contraction joints ranges from 12 to 24 metres in reinforced slabs and from 4 metres to 5 metres in unreinforced slabs (for slabs under or over 230 respectively).

Joint-filling and sealing

There are two types of material used in joints: a filler which separates the slabs, and a sealing compound which fills the top 25 mm of the joint, thus resisting the entry of water and grit. Materials suitable for joint filling are impregnated fibre board; cork; sheet bitumen; and rubber.

Joint sealing compound must have good adhesion to concrete, extensibility without fracture, resistance to flow in hot weather, and durability. There is no perfect solution to all these requirements but some adequate solutions are:

- Straight-run bitumen
- Resinous compounds
- Rubber-bituminous compounds.

The last has superseded most other types.

Expansion joints (Figure 7.6) must prevent unrestrained horizontal movements of the slabs; to accommodate the movement a compressible material 25 mm thick should be provided between the slab faces. The compressible material must be protected against the ingress of grit by filling the upper part of the joint with sealing compound to a level 5 mm below the surface of the slab. An alternative material to sealing compound is a preformed neoprene compression sealing strip.

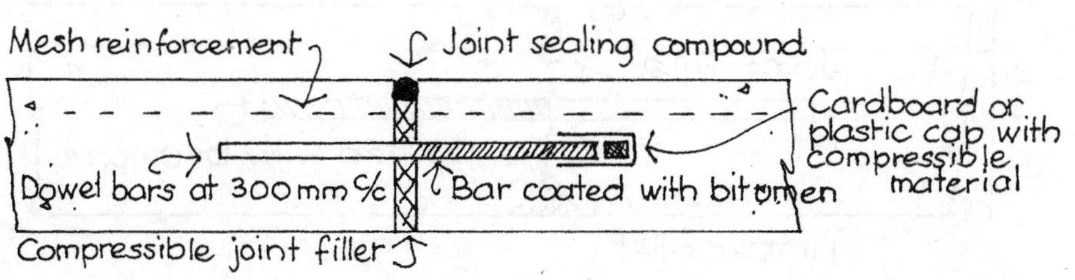

FIGURE 7.6
Expansion joint

To prevent movement of the slabs, and at the same time to ensure load-transfer, a system of dowel bars is introduced. The dowel bars are positioned at mid-depth of the slab at centres of 300 mm; the diameter of the bar varies with the thickness of slab, but usually ranges from 20 mm to 30 mm. Free movement of the slab is achieved by providing a plastic sleeve 100 mm long to one end of the dowel; this sleeve should contain a 25 mm pad of compressible material. In addition, the free end of the dowel (the one fitted with the sleeve) should be coated with a bond-breaking compound.

Longitudinal joints should have tie-bars 12 mm in diameter by 1 metre long at 600 mm centres; the bars should be fully bonded. Alternatively, a mat of reinforcement fabric may be used in lieu of bars. Difficulties arise when tie-bars are misaligned or when bars require bending aside to allow easier construction; these difficulties may be overcome by casting threaded couplings into the first slab and connecting the bars for the second slab as required (Figure 7.7). In both cases, where dowel bars are used they must be carefully supported by cradles or cages to ensure correct alignment.

Contraction joints

These joints are similar in construction (Figure 7.8) to expansion joints except that the filler material and dowel bar sleeves are omitted. Dowel bars are used to transfer loads across the joints and one half of each bar is coated with a bond-breaking material to allow contraction to take place. In addition, the interface of the slabs may be coated with bitumen before the second slab is cast.

For continuous construction 'dummy' joints may be formed (Figure 7.9) by creating planes of weakness at the required spacing. A plane of weakness is induced

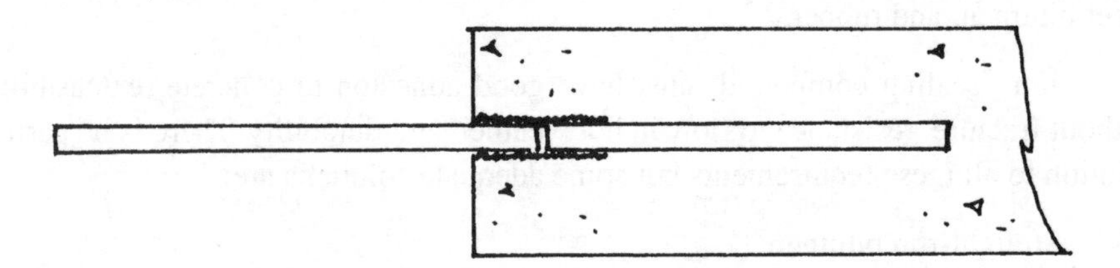

FIGURE 7.7
Threaded couplings for longitudinal joints

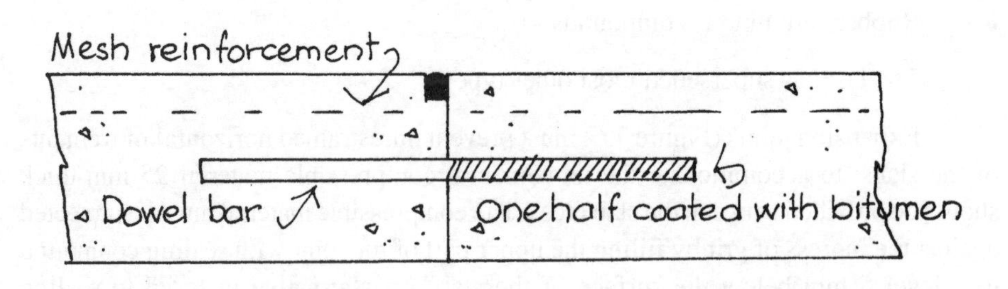

FIGURE 7.8
Contraction joint

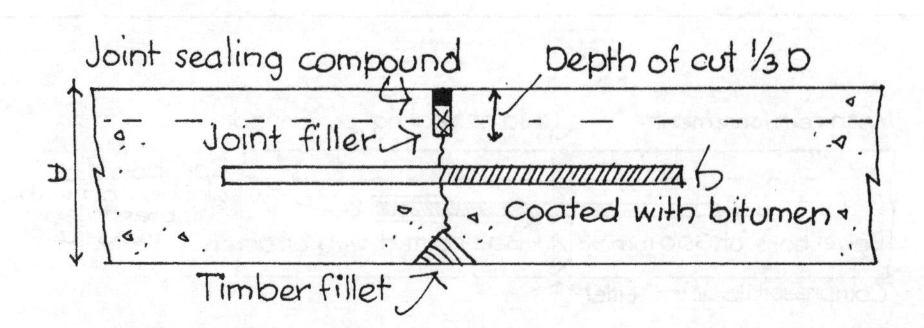

FIGURE 7.9
Dummy contraction joint

by means of a timber, plastic or steel fillet, fixed to the surface of the base and cast into the slab. An upper groove is formed by a vibrating plate while the concrete is plastic, or one may be cut with a saw; it is then filled and sealed with a suitable compound. The fillet may be eliminated if the groove is cut to a depth of not less than one-third of the depth of the slab, but in that case the concrete should be made with crushed stone aggregate.

Warping joints

Transverse joints are needed in unreinforced concrete slabs to relieve stresses of restraint due to contraction and warping; warping is caused by vertical temperature gradients within the slab, and stresses caused by these may be higher than those caused by contraction. For that reason 'warping' or hinged joints may be used in lieu of the normal contraction joints. These joints, sometimes referred to as tied warping joints, consist essentially of a contraction joint with a special arrangement of reinforcement (Figure 7.10).

Slab finish

On completion, the surface of the slab may be textured by brushing with a wire broom at right angles to the centre line of the carriageway. This gives a better skidding resistance and a uniform appearance. The slab should be cured immediately after brush treatment by spraying with a curing compound.

Pre-stressed concrete slabs may be constructed in lieu of traditional reinforced slabs, giving greater slab lengths without joints; slab lengths of up to 300 metres have been constructed without joints. The disadvantages of this form of construction are lack of continuity in the pouring of concrete, due to jacking; extra cost of supervision during stressing; and problems of maintenance if roads should require cutting for services at a later date.

7.1.4 Rigid composite pavements

Rigid composite pavements employ Continuously Reinforced Concrete Pavements (CRCP) with bituminous surfacing. This combination of concrete slab and bituminous surfacing may be justified because of its trouble-free performance (Figure 7.11). It has the advantages of both flexible and rigid pavements in that it gives good riding

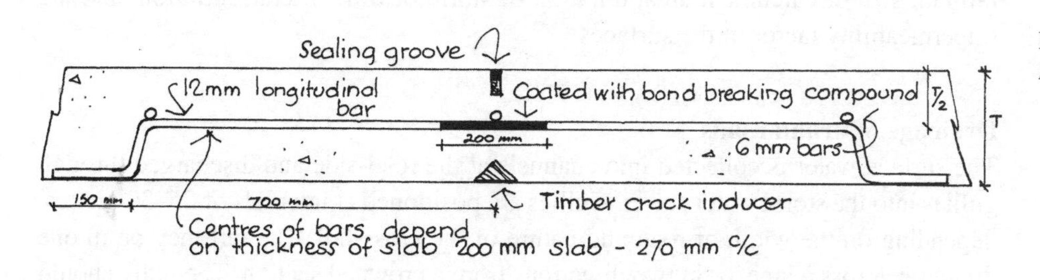

FIGURE 7.10
Detail of warping joint

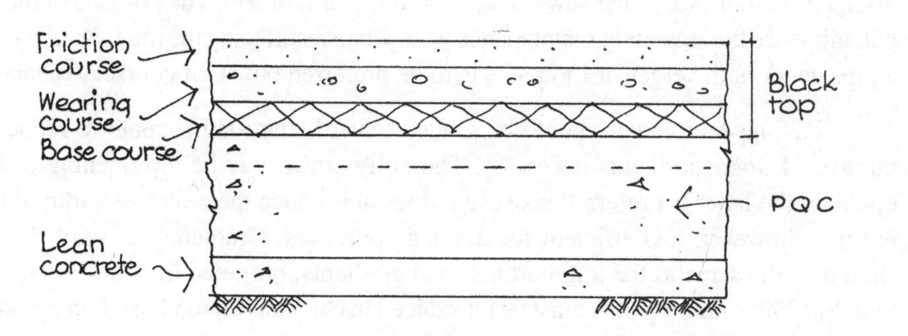

FIGURE 7.11
Rigid composite pavement

quality and resists settlement, particularly in roads which are extensively disturbed by excavation for service trenches. The subgrade, drainage and sub-base follow closely that for normal reinforced concrete bases. Alternatively, PQC concrete may be used. The concrete does not require air-entrainment. The thickness of the concrete base varies from 140 mm to 250 mm, depending on the subgrade and traffic intensity.

Reinforcement should be long mesh, not lighter than 5.5 kg/m^2, or longitudinal deformed bar reinforcement of cross-sectional area not less than 650 mm^2 per metre width of road. This heavy reinforcement is recommended in view of the absence of transverse joints. The cover for reinforcement and terminating distances from the slab edges are the same as those for rigid pavements, and the recommendations for reinforcement when constructing two- or three-lane carriageways are also the same as for rigid pavements.

Joints

It is recommended that no transverse joints be introduced except for unavoidable construction joints. Where construction joints are necessary the reinforcement should be allowed to project 700 mm beyond the end of a day's work, and when work is resumed the reinforcement should be overlapped the full 700 mm to minimise joint movement. Longitudinal joints should be provided so that the slabs do not exceed 4.5 metres in width; alternatively, reinforcing mats having transverse wires of 8 mm at 200 mm centres may be used to span the joint. When longitudinal joints are provided they should have tie-bars as described in the section on rigid pavements.

Surfacing

The surfacing of a heavy composite pavement should be similar to that for trunk roads, with either Hot Rolled Asphalt or Porous Asphalt as the wearing course.

7.1.5 Surface water drainage

The provision of adequate drainage facilities is essential in any pavement design. Drainage facilities must cope with water from the carriageways, hard shoulders, and footpath or cycle paths, as well as dealing with water from verges and adjacent catchment areas.

The design of surface water drainage is outside the scope of this book, but the reader should be aware that the design will depend on factors such as intensity of rainfall, size of catchment area, duration of storm or time of concentration, and the impermeability factor of the surfaces.

Drainage of urban roads

The surface water is collected into channels at the road-side and discharged through gullies into the storm water sewer. Gullies are positioned at intervals of 25–30 metres, depending on the width of road and nature of the cross-fall. The fall may be in one direction across a lane, or in two directions from a crowned section. The gully should discharge to a surface-water sewer under the verge or footpath. The position of the pipe will influence the amount of maintenance work when resurfacing the road; for this reason a pipe under both verges or footpaths is to be preferred but is, of course, expensive.

Pavings and verges should be graded towards the road channel to reduce the number of drainage points necessary. The gully cover may be top opening or side-opening; the latter is preferred because it does not reduce the effective width of road, but it is, however, less efficient for drainage purposes. Gradients for road channels should be the same as the longitudinal road gradients, provided that they are not less than 1 in 250; summit points must be introduced in channels on roads of flatter gradient.

Drainage of roads in open country

If the road is a minor road, the drainage may be achieved by simple openings or channels which feed into ditches or french drains. Most main roads, however, have a system of gullies and piped sewers. If the road has a hard shoulder the kerb is normally kept flush with the road surface and a precast concrete channel is placed at the outside edge of the shoulder to catch water. The channel discharges into gullies and then into a piped sewer or open channel, depending on the elevation of the road and the type of subgrade. An alternative method of retaining water along the edge of the hard shoulder until it flows into the gullies is to provide a raised concrete or mastic asphalt edging 75 to 100 mm high.

Road camber

The road camber or cross-fall should be designed to cope with heavy water run-off during a storm; if the cross-fall is insufficient to cope with heavy rainfalls, there will be a danger of skidding or aquaplaning on the water. The standard cross-fall for roads can be taken as 1:40. To eliminate problems that occur with water lying on a flatter road surface, however, the camber of individual roads will vary.

Super-elevation

When a vehicle travels round a bend, the horizontal centrifugal force tends to overturn the vehicle or cause a sideways movement, depending on the height and speed of the vehicle. This force can be controlled by banking the road surface at the bend (known as super-elevation). The degree of super-elevation depends on factors such as the type of road surface; speed limits; and radius of bend.

7.1.6 Hard shoulder, kerbs, footpaths and verges

Hard shoulders are continuous strips of hard standing alongside motorways and other major roads, on to which vehicles may drive during emergencies. They are normally 3 metres wide and constructed in similar materials to the main carriageway. The surfacing material consists of 50 mm bitumen macadam or tarmacadam with chippings to give a contrasting colour to that of the carriageway. Alternatively, a concrete hard shoulder with exposed aggregate may be used.

Coloured surfacing, consisting of bituminous-coated sand, has also been used to provide a distinctive hard shoulder.

Consideration must be given to whether the shoulder may at some time be incorporated into a carriageway as part of a road-widening scheme. If this is likely to occur, the shoulder should be constructed to fulfil the functions of the main pavement. The shoulder is normally separated from the carriageway by a 300 mm wide flush marginal strip contrasting in colour with the road surface. Alternatively a white ribbed line can be applied to act as a sounding band for the wheels of vehicles.

Kerbs (Figure 7.12) are used to contain the road construction and to define the limits of the carriageway. They may be constructed of concrete or asphalt, vertical or splayed in section, or sometimes level with the carriageway. Concrete kerbs may be either precast or insitu, the latter being laid with an automatic kerbing machine. Asphalt kerbs are also laid with automatic kerbing machines and should be laid as soon as practicable after the completion of the road surface. Vertical kerbs are only necessary where a footpath adjoins the carriageway or where the kerb is used to prevent a vehicle leaving the carriageway at a particular point. For most purposes the top of the kerb should be 100 mm above the road surface; if kerbs are placed too high, they induce 'kerb shyness' which effectively reduces the width of the carriageway. Flush or level kerbs should be preferred where their use is practicable.

Footpaths and verges

Footpaths may be constructed with concrete slabs or flexible paving. If concrete slabs are used they should be bedded using dabs of lime mortar on a base of ashes or similar material (Figure 7.12). Slabs should be bonded to resist movement. Flexible surfacing conforming to the appropriate BS should be laid to falls on a sub-base of granular material or other suitable material.

Verges should be provided alongside all classified roads in rural areas. They should have a width of 3.5 metres, of which the 1.2 metres adjacent to the carriageway should be free from obstacles. Verges separating cycle tracks from roads should be 2 metres wide; those separating footpaths from roads should be 1 metre wide.

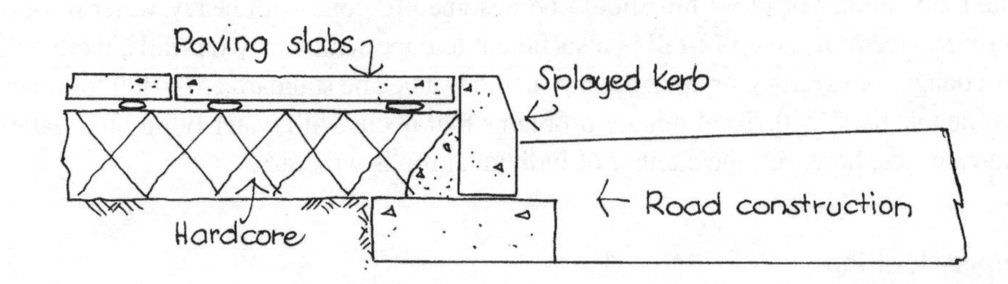

FIGURE 7.12
Section through kerb and adjoining pavings

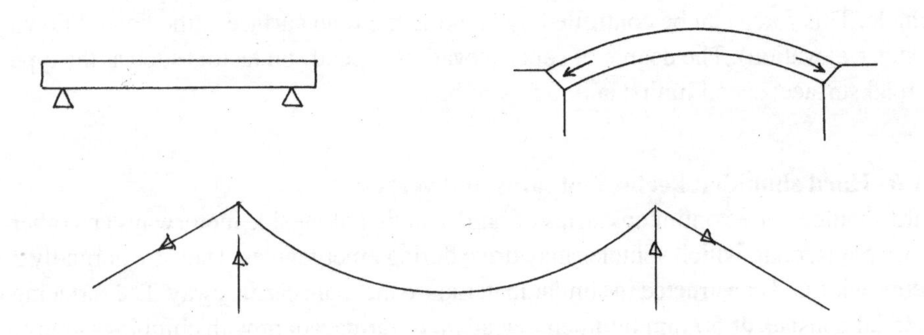

FIGURE 7.13
Basic principles in bridge design

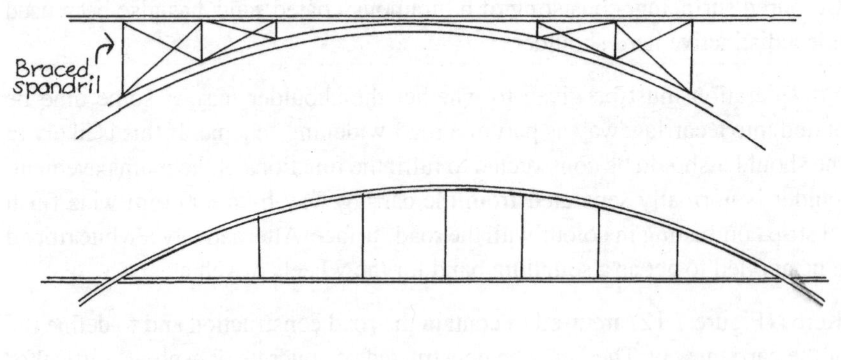

FIGURE 7.14
Arch bridge

(a) Above: Road above arch

(b) Below: Road below arch

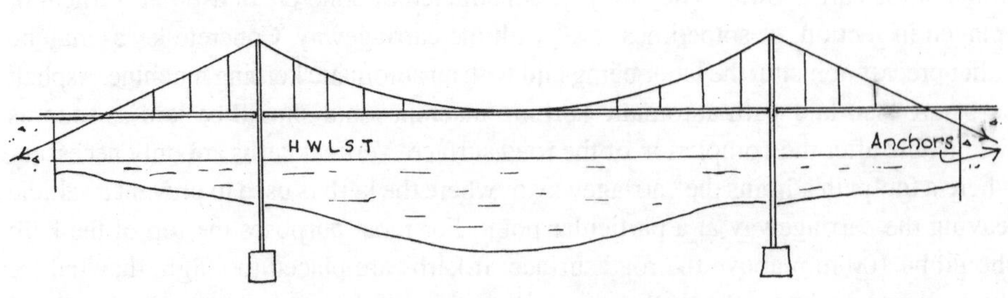

FIGURE 7.15
General principles of construction for a suspension bridge

7.2 BRIDGES

7.2.1 Types of construction

There are three basic types of bridge, depending on the form of the load-bearing structure: flat, convex and concave (Figure 7.13). They are better known as beam, arch and suspension-type bridges.

Beam bridges can be divided into two main groups: simple beam and cantilever. The simple beam bridge transmits the loads vertically through piers or abutments and is horizontally self-supporting. The cantilever beam transmits its loads through piers which are normally central to the beam; however, when the cantilever is loaded the beam exerts great pressures on the opposite end connection.

There are many variations in beam design, ranging from steel truss design to pre-stressed concrete units. The simple beam bridge is very economical for spans of up to 50 metres. The cantilever beam provides a means of producing much greater spans: bridges with spans of up to 540 metres have been constructed.

The **arch bridge** can carry greater loads than the beam bridge because the load-carrying member, the arch, is in a state of compression throughout. This makes the design suited to materials which are weak in tension. The arch supports the traffic either above or below the main structural form (Figure 7.14). The first type of support (that above the arch) is suitable for use where the bridge crosses a gorge and where rock or very hard material is present to resist the thrust. The arch may be hinged or pinned at the ends to eliminate any movement on the foundation. The spandrel of the bridges may be braced or open. Roads supported below the arch, by hangers, create a different form of structure in that the deck acts as a tie to the arch and therefore produces a bridge which can be used on foundations that would not resist the thrust of the first arch type. This type of bridge has been used for spans of 496 metres in steel and 305 metres in concrete. The rise/span factor is between 0.15 and 0.25.

The problem inherent in long arch bridges is one of erection; if intermediate piers are impracticable then it often follows that falsework (temporary support) will be impracticable. This means complicated methods of erection, usually involving designing the bridge to withstand stresses which, after erection, it will never have to bear.

A **suspension bridge** consists of a cable-hung decking supported by towers. The general layout (Figure 7.15) comprises a central suspended span with side spans; the latter may take the form of a simply supported beam over short spans. The towers are secured by main cables which are continuous between anchorages. The foundations of the towers are constructed by caisson or cofferdam methods (see Chapter 6) and the cable anchorages or foundations are taken through anchorage tunnels to suitable ground.

The deck of the suspension bridge must be stiffened to prevent undue deflection and to provide aero-dynamic stability. This is achieved either by introducing a continuous truss alongside or below the deck, or by designing the deck as an aerofoil. In either case models of the design must be tested in a wind tunnel to ensure stability and to achieve the most economic design.

The supporting towers of many large span suspension bridges are constructed in a cellular design, in either steel or concrete. Steel box towers were used for the first Severn Bridge between England and Wales. In addition to the new development for suspension towers, the Severn bridge has a 'torsion box' deck (Figure 7.16) which has been designed as an aerofoil. The box-deck or box-girder construction is now a very popular construction method used in all forms of bridge design.

Moveable bridges

These bridges are better known for their function, eg swing, bascule (or drawbridge) or vertical lift, than for their basic construction. They are in fact the same as, or variations of, the three basic forms of bridge. The need for moveable bridges arises from the demand for greater headroom than a normal fixed bridge can economically provide.

Swing bridges are used for spanning wide openings. They pivot on a central pier, and so have the disadvantage of reducing the actual navigational channel by half. Their advantage is that the whole structure is balanced on the central pier when the bridge is open, thereby reducing the amount of support mechanism. The bridge is a cantilever type which acts as a continuous girder when in the closed position.

Bascule or **pivoted cantilever bridges** are basically drawbridges which are operated by means of a counter-weight behind the pivot point. The Tower Bridge, London, works on this principle.

Vertical lift bridges consist of simple beams or girders which are supported and raised by cables from high towers. The cables, counter-weighted to balance the dead load of the bridge, pass over sheaves at the top of the tower. Spans up to 167 metres have been achieved by this method, which has proved to be the simplest form of movable bridge, though not aesthetically the best.

Choice of bridge system

The first important factor is the clear span required. If it is a long span, ie over 300 metres, steel construction is the most likely solution. However, concrete arch bridges are being developed for increasingly longer spans. The steel construction may be cantilever girder, arch or suspension. The cantilever form has great advantages in erection, since the cantilever arms can be built without formwork and the centre sections

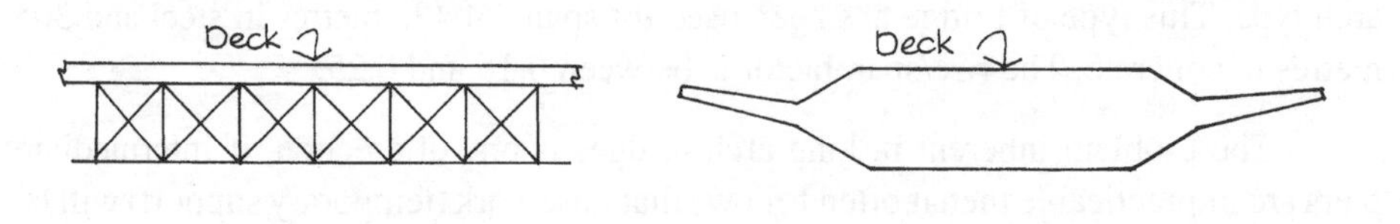

FIGURE 7.16
Methods of achieving stability of deck to suspension bridge
(a) Left: Lattice beam below deck
(b) Right: Torsion box as aerofoil

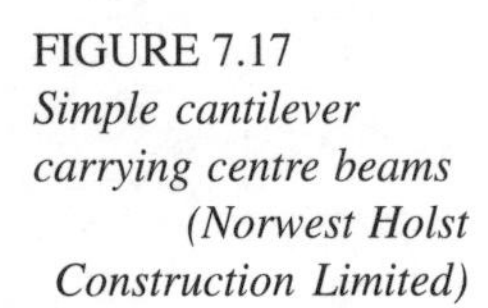

FIGURE 7.17
Simple cantilever carrying centre beams (Norwest Holst Construction Limited)

of the span can be floated out and lifted into position. The steel arch form is very suitable for spans of up to 500 metres, but is difficult to construct: the arch can be built out as a cantilever, but requires extensive back-anchoring which is very costly. Suspension bridges are the best known form for spans of over 600 metres. The high-tensile strength of cable wires produces a very economic design solution when compared with other forms of support.

In general, for a given span and given load the main girder weight decreases in the following order: cantilever, arch, suspension. However, foundation costs can be quite the opposite: foundations for cantilevers are simple by comparison with the other forms, owing to their vertical loading, and are therefore normally the cheapest. Foundations for suspension bridges are usually very extensive and costly. Small span bridges, up to 300 metres span, may be formed in steel, concrete or other material suitable for the span and load. In particular, steel box-girder construction and pre-stressed concrete box-girders have produced economic solutions in recent years. Many modern road bridges, especially 'overbridges' (carrying a minor road over a major road), have been formed as simple cantilevers carrying centre beams (Figure 7.17). The most economical solution, depending on span and aesthetics, has proved to be a combination of insitu concrete and pre-stressed units.

7.2.2 Foundation techniques and bridge construction

The construction of bridge foundations on land presents little difficulty compared with those in water. If the ground is water-bearing there may be a need for cofferdams or in some cases even caissons, although the latter is a rare occurrence on land. The foundations are taken down to bedrock unless this is uneconomical, in which case a piled foundation will be necessary.

Foundations in water present various constructional problems which include protection of the workforce, transporting and placing concrete, and reduction in working time (if affected by tides). The technique of forming the foundation will normally involve the use of cofferdams or caissons. If piles are used they are usually capped off to receive a pier foundation; alternatively, they are prepared to receive cross-beams which support the main bridge construction.

Beam and girder bridges may be constructed in steel or concrete. The beam may take the form of a lattice a solid beam, or a hollow box-section; in addition a steel structure may have plate girders.

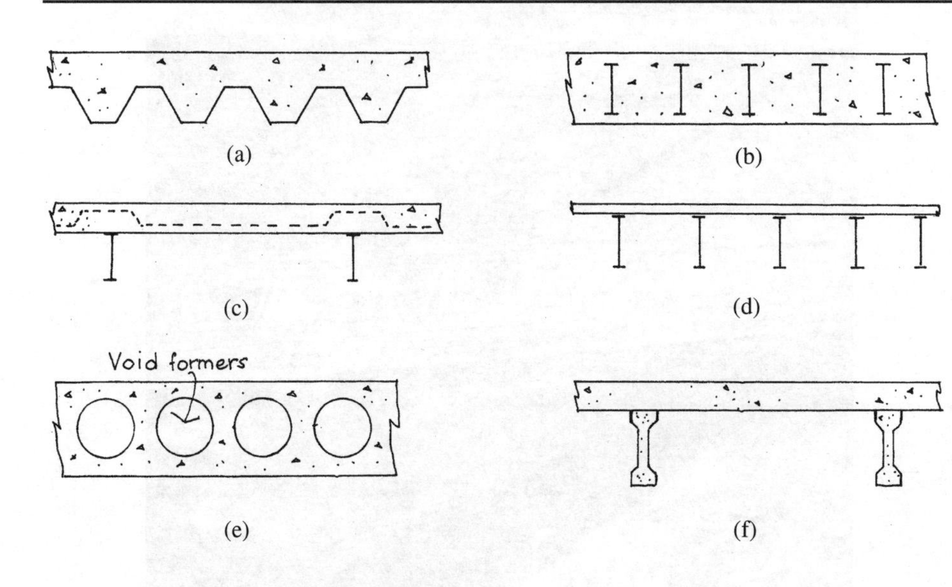

FIGURE 7.18
Decking to bridges

(a)-(d) Decking to steel bridges:
(a) Steel trough filled with concrete
(b) Steel beams encased with concrete
(c) RC slab on steel beams
(d) Steel deck on steel beams

(e)-(f) Decking to concrete bridges:
(e) Integral bridge deck with polystyrene formers
(f) Beam and slab deck

The piers are first constructed and the beams or girders are lifted on to them by cranes or hydraulic jacking. Cranes may be positioned on falsework cradled around the piers, or they may operate from ground level. When the bridge spans water, a gantry may be built either on piles at just above water level or between the piers at high level. The beam sections are floated into position by pontoon prior to lifting. If the spans are not too great, the beam sections can be 'rolled out' from the access point and lowered into position by crane. This method of construction is used for box-girder construction.

After placing the main girders, the bridge deck is positioned or cast insitu. Decking for steel bridges can take one of many forms: Figure 7.18 shows typical deck sections. Decking for concrete bridges may either be integral with the beam design or take the form of cross-beams and slab construction.

The design of an insitu bridge may vary, but for all forms of construction the one common factor is the falsework required: this ranges from very heavy towers, supporting steel beams for high bridges, to a forest of heavy scaffolding for low bridges. Various systems of falsework are shown in Figures 7.19 to 7.21.

If the construction involves cantilever beams, the anchorage or anchor arm is completed first, to prevent movement of the pier when the cantilever arm is constructed. Alternatively, both arms may be constructed simultaneously.

Erection of bridges

Erection of bridges varies a great deal from the erection of buildings in that temporary structures and braces are more widely used. If, however, the bridge is to be constructed on land there may be no need for temporary support such as trestles, although temporary

FIGURE 7.19
Falsework to high level bridge
(John Laing & Son Limited)

bracing may be required. If the bridge is to be erected over water there is the problem of positioning the crane and materials prior to lifting. Materials are normally floated out to the lifting position, either on barges or on pontoons; in some instances, eg box-girder construction, the unit may be towed into position by barge.

Where the space between bridge supports can be effectively utilised, a system of trestles may be used. This will be suitable as a means of temporary support for girder and truss construction.

The method of construction for bridges will vary with the type of structure and length of span. Members in plate-girder and trussed-girder bridges of up to 50 metres span can be lifted by mobile cranes without much difficulty. The main girders are first positioned, followed by the lifting and placing of the cross beams which support the deck. The main girders may be of a length and weight that require two cranes, one at either end, to lift the member on to the prepared seating. Truss-constructions can be supported on trestles or on a rigid platform which is designed to carry both bridge members and cranage. Long plate-girders or trussed-girders may be hauled across an opening with the aid of rollers and temporary trestles.

FIGURE 7.20
Formwork and falsework for bridge construction low level
(R.M. Douglas Construction Limited)

FIGURE 7.21
Special beam support to formwork over canal
(Norwest Holst Construction Limited)

The crane, situated on temporary trestles possibly at mid-span, hauls the member on to rollers on the trestle and completes the operation from a new position on the opposite bank. The operation is termed 'launching a girder' and it is very suitable for erection over deep ravines and gullies. An alternative method of positioning long beam-girders is cantilever launching. The beam is launched from one bank, having sufficient ballast to counteract any overturning force. If the span is excessive for a complete cantilever operation, a temporary trestle can be constructed, approximately one-third of the distance across the span; the trestle is capped with rollers to facilitate the movement of the girder.

In both methods of cantilevering, the beam is hauled across the span by means of a winch (Figure 7.22). Where more than one span is necessary for the bridge construction, the first span can be constructed by launching methods and the remainder by true cantilever construction. The first span acts as a counterbalance and working platform while the other spans are being erected.

Erection by means of barges is a suitable method for small bridges, which have to be constructed over waterways which cannot be closed. The barges or pontoons are equipped with suitably braced trestles on to which the bridge structure, usually a trussed-girder, is placed. They float the bridge into position, where it is lowered on to the bridge bearings by means of jacks; alternatively the barges may be lowered in the water by flooding the ballast tanks.

In all of these methods the actual girder or truss has to be lowered into position because the rollers or trestles are positioned so as to deliver the girder over its bearing. The operation of finally positioning the girder or truss is known as 'jacking down': the girder is fitted with jacking cleats to facilitate this operation.

Arch bridge construction

Steel arch bridges are erected by one of two methods; either by cantilevering the ribs out from the sides of the span, or by supporting the arch by means of trestles until it

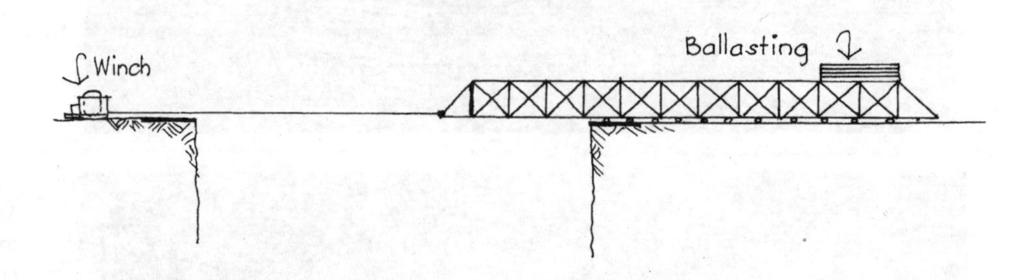

FIGURE 7.22
Launching a girder bridge

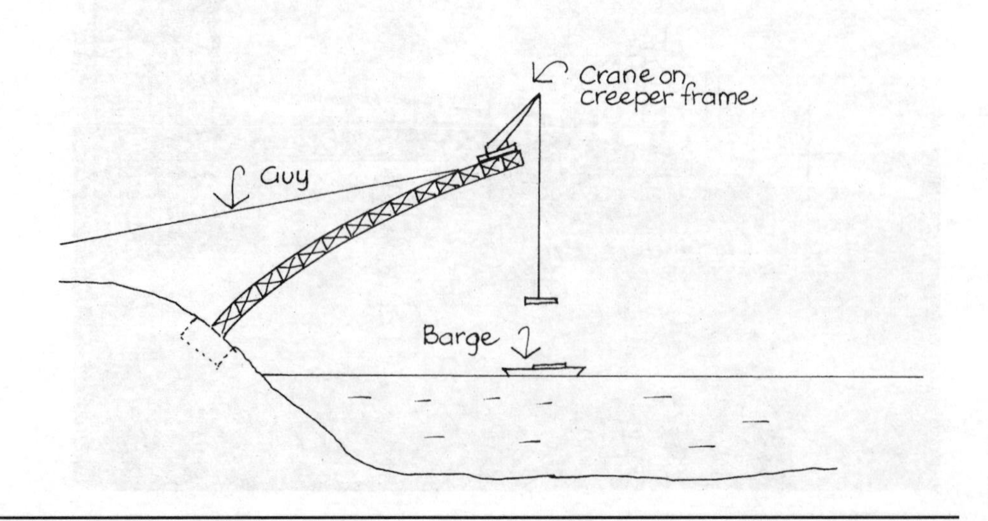

FIGURE 7.23
Erection of steel arch by 'creeper' frame

has been constructed. The second method is suitable for low arch bridges, where trestles can be constructed economically. The first method is suitable for bridges over deep ravines or rivers which cannot be spanned with trestles. The construction is achieved by means of a crane on a 'creeper' frame; a 'creeper' frame is a special unit on which the crane is mounted, which is hauled along the arch as it is constructed (Figure 7.23). The arches are held by tension bars until the bridge is complete (see Chapter 8, Figure 8.22).

For very large spans the arch is cantilevered in the same manner but is built out in sections, from the basic cantilevered structure, by means of a creeping crane. The crane is mounted on top of the arch and lifts sections of rib from barges or pontoons below. The work proceeds from both abutments, and the final gap, which may range from 50 mm to 100 mm, may be closed by 'fleeting' (lowering both arch ribs until they meet). The bridge deck is then constructed from both ends, which consist of beams suitably decked, hung from the arch.

Concrete arch bridges, if pre-stressed, may be constructed in a similar manner to steel arches. Normal reinforced arches are unable to withstand cantilever stresses and are therefore lifted on to falsework which must be erected from the river bed. In situ arches may be cast in the same manner.

One further method of erecting arch bridges, which may be classed as a modification to the method described above, is cableway erection. Arches are formed from each bank and cantilevered by tension cables as before, but the bridge sections are placed by cableway. This method relieves the guy ropes of the stress created by the 'creeping' crane. The cableway has the extra advantage of being able to pick up bridge sections from the bank, thus eliminating the cost of floating craft.

Suspension bridge construction

This involves four main operations: construction of the anchorage foundations; construction of the towers; spinning the main cables; and constructing the bridge deck. The anchorage foundations are usually taken deep into a hill-side and tied back to rock by means of piles or rock-anchors. The towers may be of steel or insitu concrete

FIGURE 7.24
Laying the wires in the saddle at the top of the tower–temporary catwalk in background (William Tribe Limited)

(constructed by slip-form methods) founded on massive concrete bases. The bases are usually constructed within a cofferdam or caisson. Climbing cranes may be used for the handling of steel and concrete.

The towers of suspension bridges are erected on specially formed foundations, usually caisson foundations. They may be erected by any form of climbing crane, but both tower cranes and Scotch derricks are commonly used. The cable housings are constructed at the same time; these may take the form of massive concrete blocks positioned in the water near each bank or they may consist of heavy anchor blocks cast deep into the hillside or river bank.

The suspension cables are built up from multi-strands of high tensile wire. The wire is carried from one anchorage point to the other by means of a grooved wheel which is fixed to a continuous wire cable. The endless cable carrying the wheel is driven backwards and forwards (spinning the cable), the wheel carrying the high tensile wire. On completion of one haul the wire is lifted off the wheel and fixed to the suspension shoe. The wheel then returns, carrying back another double strand of wire. Several wheels may be employed, laying as many as twelve wires on a double pass. The wires are all checked for sag and then bound together to form the main cable; anti-corrosion paint is applied to the cable during binding. The work of laying and binding the wires is carried out from a temporary catwalk (Figure 7.24) which is suspended just below the cable. The road deck may be erected by cantilevering the sections out from each end or by lifting the sections from pontoons on the water below. Alternatively, if the sections are hollow they may be floated into position (as for the Severn Bridge) and lifted.

The work is carried out from a temporary hanging walkway which is suspended from the pilot cable. Once the main cables have been constructed, a series of hangers is fastened to them by means of special brackets, starting at the centre of the span and working back to both towers. The erection of the bridge deck is then commenced from

FIGURE 7.25
Wires hydraulically clamped together to form main suspension cable
(William Tribe Limited)

both towers to maintain equilibrium during erection (Figures 7.26 and 7.27). If the cable housings are constructed 'off-shore', approach spans will have to be constructed at both ends. These can be constructed by traditional pier and beam construction.

Figures 7.28 to 7.31 show general views of the bridge construction in the Second Severn Crossing.

FIGURE 7.26
Construction of bridge deck – Severn Bridge
(William Tribe Limited)

FIGURE 7.27
Hoisting bridge deck sections – Severn Bridge
(William Tribe Limited)

FIGURE 7.28
Aerial view of the Second Severn Crossing
(Laing GTM)

FIGURE 7.29
Construction of concrete caissons for bridge piers – Second Severn Crossing
(Laing GTM)

FIGURE 7.30
Lowering concrete caissons into position – Second Severn Crossing
(Laing GTM)

FIGURE 7.31
General view showing bridge towers, decking and intermediate bridge piers – Second Severn Crossing
(Laing GTM)

7.2.3 Bridge bearings and expansion joints

Bridge bearings

Bridge bearings are made of either metal or a flexible material such as rubber or plastic laminates. Where spans exceed 15 metres an allowance must be made for angular deflection at the supports. At one end the span should have a roller, rocker or other efficient expansion bearing. Some provision must be made to prevent uplift at the bearing point; this can be achieved by a sliding lug or key. Metal bearings for small-span bridges may consist of steel blocks resting directly on the concrete or steel bases. Expansion can be accommodated by setting the steel block on phosphor-bronze strips, which are fixed to the supporting base. Where bridges are supported entirely on elastomeric bearings, it is often necessary to provide 'guide bearings' to prevent excessive lateral movement. Guide bearings do not carry load: they simply resist movement in one horizontal direction.

Metal bearings for large spans may be of the rocker or roller type (Figure 7.32); the lower saddle may be fixed or rest on phosphor-bronze strips or some other form of expansion base. Leaf bearings are used to suit applications where complete reversals of vertical loading can occur or where very large rotations are required about a single axis.

Rubber and plastic bearings may be used as alternative means of allowing movement. Rubber bearings consist of alternate layers of steel and rubber, which allow movement in both vertical and horizontal directions. Plastic laminates, Polytetrafluoroethylene (PTFE) by name, have been used in lieu of rubber: this plastic

FIGURE 7.32
Bridge bearings

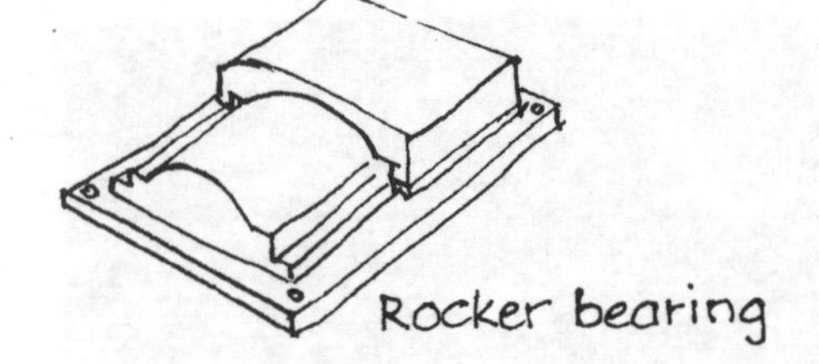

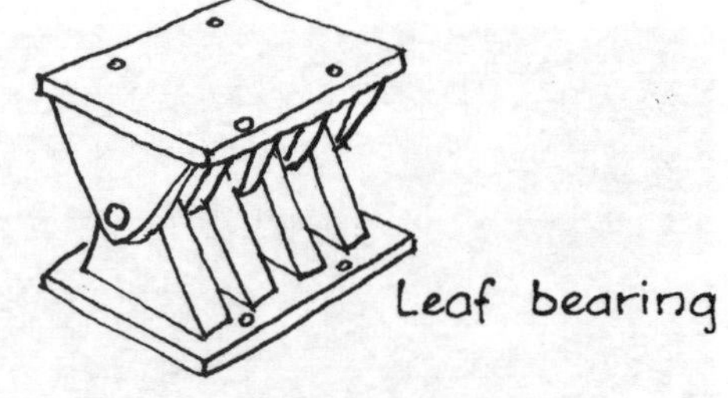

FIGURE 7.33
Comb-type expansion joint
(John Laing Construction Limited)

is highly resistant to adverse weather and is ideal for situations where low resistance to sliding is all that is required. The bearings may be located on the piers engaging the bridge beams by means of projecting studs.

Where necessary, resilient pads may be placed between the bridge bearing and the concrete pier, thus equalising the pressure under the bearing and preventing damage to the concrete by excessive stress concentrations.

Expansion joints

There are approximately 60,000 steel or concrete highway bridges in the UK, most of which have expansion joints to accommodate movement in the bridge deck. Expansion joints for bridges and elevated motorways can be formed in various materials and may function in different ways. The two most common ways to accommodate expansion are by an interlocking comb-joint or by one of the many flexible sealing elements, such as Asphaltic Plug Joints (AJPs); Elastomeric in Metal Runners (EMRs); and Reinforced Elastomeric Joint (REJs). The interlocking comb joint (Figure 7.33), which is used on continuous span bridge-work, is bolted to the concrete or steel deck at intervals which will accommodate maximum expansion. Other types of expansion joints (Figure 7.34) are more widely used and have their own characteristics and weaknesses, which give an average service life of between 7 and 11 years, depending on materials and method of installation.

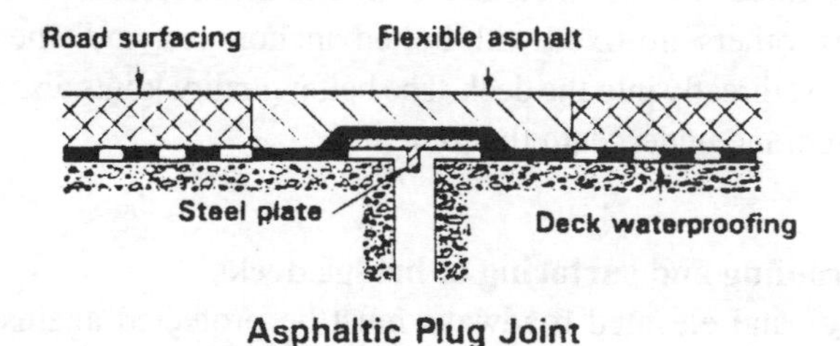

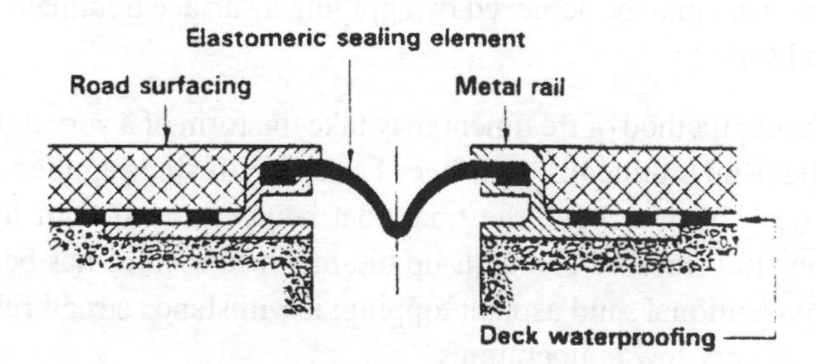

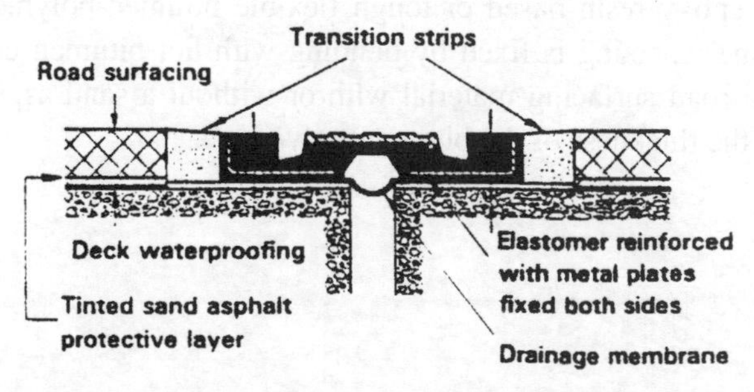

FIGURE 7.34
Common types of expansion joint
(Department of Transport – Transport Research Laboratory)

Asphaltic Plug joints (Figure 7.34) are essentially buried joints with a strip of surfacing over the joint to accommodate the movement. APJs are suitable for movement up to 40 mm and have replaced other types of buried expansion joint. APJs are inexpensive, provide a good running surface and can be installed in a few hours, minimising delays to traffic. However, they do require regular monitoring and maintenance because their life-span is very variable. Failure to maintain this type of joint on a regular basis could lead to high maintenance costs to deck ends, due to penetration of chloride-laden water, causing corrosion of steel reinforcement.

EMRs are joints formed with elastomeric material fixed in metal runners (Figure 7.34). In principle they provide a flexible cover over a designed expansion gap in the bridge deck, and are therefore simple in concept and potentially trouble-free if installed correctly. However, research has revealed some incompatibility between the movement capacity of the elastomeric material and the actual deck movement, resulting in overstretching of the material and dislocation of the elastomeric insert. Furthermore, the build-up of debris between the runners may exacerbate corrosion of the metal rails. EMRs are normally installed prior to surfacing and care must be taken to ensure that the metal rails do not stand proud of the surface finish.

REJs consist of elastomer reinforced with metal plates on both sides of the joint (Figure 7.34). The joint is set between transition strips which provide the finish between joint and road surfacing. This type of joint is normally fixed after the road surface has been completed and should be installed slightly below the road surface. If this is not done the joint will be subject to wheel impact damage. The method of fixing varies; some are fixed with resin anchors, which may result in debonding in heavily trafficked routes; others are fixed with bolted anchorages, and where these are used they should be cast directly into the deck. The better quality joints also have a secondary waterproof membrane underneath the joint.

7.2.4 Waterproofing and surfacing of bridge decks

Concrete bridges and elevated roadways must be protected against the deleterious effects of water and de-icing chemicals, which may cause corrosion of the reinforcement. This may be achieved by applying a surface treatment, or a membrane under the road surface.

The former method of treatment may take the form of a very thin coat of epoxy-resin or of a built-up bitumenized surface. The epoxy-resin is applied to the surface of the finished deck in two coats, the final coat being sprayed with fine aggregate to produce a non-skid surface. The built-up bitumenized surface has been developed to replace the conventional sand asphalt topping: it withstands structural movement and stays flexible at very low temperatures.

Sandwich membranes may be applied as an alternative method of waterproofing. They may be epoxy resin based or tough flexible bitumen-polymer sheeting. The bitumen-polymer sheeting is fixed by bonding with hot bitumen compound and is protected by a road surfacing material with or without a sand asphalt under-layer, depending on the thickness of the bitumen-polymer.

7.3 SUBWAYS

7.3.1 Methods of construction

The constructional system for an underpass or subway may be one of three methods commonly used:

- Precast concrete units
- Thrust-bored units
- Insitu concrete.

Precast concrete units are available as standard units; specially designed units may also be employed. Standard units may be supplied as complete box-like, open-ended sections; portal frame segments, which are located on a predetermined concrete slab; or separate units for the walls and roof. The box units (Figure 7.35) are assembled on a concrete bed and packed to the correct levels before winching together. The joints between the units are formed by means of a pre-formed sealant strip in a socket and spigot joint. The units are connected together by bolted connection plates in the floor and roof, or alternatively they may be pre-stressed by the Macalloy bar system (see Chapter 8). With pre-stressed portal frame units, the lower waterproofing membrane, which may be asphalt or neoprene sheeting, is placed on the concrete slab, and continuous granolithic concrete bearing pads, 300 mm wide and 25 mm deep, are laid on top of the membrane. After the pads have reached the required strength, the units are placed in position, the pad being lubricated with a graphite paste to reduce friction during stressing.

FIGURE 7.35
Precast concrete subway units being prepared for assembly (Johnson Construction Limited)

Wall and roof unit systems (Figure 7.36) consists of precast units for the walls and roof of the subway. The wall units are placed in position and the insitu floor is cast, using the units as shuttering. This is followed by placing the roof units and pouring the insitu loading slab, the thickness of which depends on the loading requirements. The wall units have plain joints without seal, which gives the advantage of dimensional flexibility.

The method of construction in thrust boring (Figure 7.37) has been fully described in Chapter 5, and consideration of details of inserting the units can be limited here to aspects concerning location. Normal methods involve the formation of a driving pit and the construction of a thrust block.

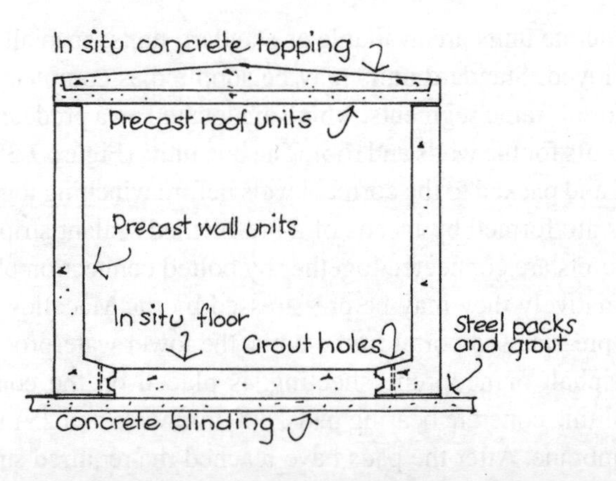

FIGURE 7.36
Wall and roof units in subway construction

FIGURE 7.37(a)
Units being driven in thrust-bored subway (M & H Tunnelling Limited)

FIGURE 7.37(b)
View inside subway being assembled (Johnson Construction Limited)

The units for jacking may be circular or rectangular, the latter being suitable for the majority of subways. Lubrication may be necessary during thrust-boring and this is usually achieved with a suspension of bentonite. The jointing of units to prevent the ingress of water is an expensive item. The units must have direct edge contact to transmit the thrust load, which eliminates the normal pre-formed sealing strip. The jointing method used must allow edge contact for jacking, yet be designed to receive a sealing compound from the inner face. The latter can be accommodated by forming a rebated joint which can be filled with mortar prior to the application of sealant. Insitu concrete subways need no explanation, the construction following the pattern for any underground concrete construction.

7.3.2 Methods of waterproofing

Waterproofing to subways may be achieved by applying mastic asphalt, bituminous or neoprene sheeting, or painted membrane.

Mastic asphalt should be laid in two coats and conform to the relevant Code of Practice. Joints in the layers should be staggered at least 150 mm. Bituminous sheeting may be applied in two or three layers, each layer being fully bedded in hot bitumen. The concrete surface must be carefully primed with cut-back bitumen or similar primer, allowed to dry before applying the first layer of sheeting. Joints between sheeting should be lapped with end laps of at least 150 mm and side laps of at least 100 mm.

Pre-formed sheeting membranes such as 'bituthene' – a self-adhesive bituminous polythene – have become increasingly popular in recent years. They require the same preliminary work as other membranes but are laid more quickly and can be applied easily (Figure 7.38). Tar or bituminous painting should consist of two coats of hot-applied tar or two coats of cut-back bitumen. Painted membranes, used on pre-cast units, must be compatible with the sealants used for jointing. On completion, the waterproof membrane must be protected against puncture. This is normally achieved by constructing a concrete-block skin against the membrane prior to backfilling with granular material.

7.3.3 Lighting, finishes and drainage

Lighting for subways may be incorporated in the walls or roof units. These should be recessed to give maximum protection against vandalism, and all light fittings should

FIGURE 7.38
Bituthene 1000 used for tanking subway (W.R.Grace Limited)

be of toughened glass. Finishes to subways may vary to suit criteria such as location and degree of use, level of vandalism anticipated, and types or quality of finish on adjacent buildings. For industrial and agricultural use a plain concrete finish may suffice, but subways and underpasses in urban areas should be bright and attractive, otherwise the subway may be avoided by a large proportion of potential users. Attractive surface finishes can be achieved by using mosaics, tiling and fine white aggregate rendering which can be applied by spray gun. Drainage to subways, which may be necessary in storm conditions, can be achieved by normal falls, screeded floor, gullies or continuous side channels. These channels or gullies will empty direct into the storm sewer, if it is low enough, or alternatively a sump and pump system can be operated.

TABLE 7.3

LCG *Load Classification Group*	LCN *Load Classification Number*
I	101 – 120
II	76 – 100
III	51 – 75
IV	31 – 50
V	16 – 30
VI	11 – 15
VII	10 and under

7.4 AIRFIELD CONSTRUCTION

The term 'airfield' is used in the United Kingdom to refer only to military stations. The recognised international term for civil aviation is 'aerodrome', though the term 'airport' is preferred for aerodromes with port facilities, ie Customs, immigration and health control.

7.4.1 Introduction to pavement design

The construction of airfields is similar to road construction in that the type of pavement depends a great deal on the load it has to carry. While the strength of any pavement depends upon the nature of its construction and underlying subgrade, the stresses applied by aircraft can be exceptionally high and variable. By comparison, loads on aircraft pavements may be eight times greater than those on road pavements. The intensity of loads and stresses on aircraft pavements is such that aircraft are classified by a number and the pavements are designed to carry aircraft within certain classifications. This allows an economic design to be produced.

Pavement classification systems

There are two classification systems:

- LCN – LCG system
- ACN – PCN system

The classification system currently used for UK military airfields is the Load Classification Number – Load Classification Group (LCN LCG). Under this system each type of aircraft is assigned a number (its LCN) on a scale of 1 to 120, representing its relative damaging effect on pavements. This classification takes into account the gross weight of the aircraft and the configuration, spacing and tyre pressure of its undercarriage wheels (an LCN of 120 is the most damaging). The LCN values are divided into seven Load Classification Groups (see Table 7.3), and pavements are classified as belonging to one or other of these groups according to their load carrying capacity.

The LCG values for the pavements on each airfield are established and this enables operators to plan aircraft movements to prevent pavements from being overstressed. The LCN - LCG system involves allocating a single number to an aircraft to indicate its relative damaging power on all pavements, whether rigid or flexible, and also on good or bad subgrades.

The Aircraft Classification Number – Pavement Classification Number (ACN – PCN) is a more precise approach where the ACN expresses the relative loading severity of an aircraft on a pavement. With this system 16 different values are quoted for each aircraft. These are for maximum and empty operating weights on rigid and flexible pavements for each of four different subgrade strengths. The PCN is defined as the ACN of the aircraft which imposes a severity of loading equal to the maximum permitted for unrestricted use. The PCN for any pavement is recorded as a five-part code (eg 70/F/C/W/T), which means, respectively, LCN number, type of pavement (Flexible or Rigid), the type of subgrade, the maximum tyre pressure authorised for the pavement, and whether the strength assessment has been arrived at by technical design/evaluation or from experience in use. Providing that the PCN for a pavement is equal to, or greater than, the ACN of a particular aircraft and the operating tyre pressure does not exceed the PCN limitation then unrestricted use of the pavement is permitted.

7.4.2 Foundation considerations

The subgrade for pavements is classed as 'good' or 'bad' with various intermediate conditions. The extreme conditions are based on typical gravel or chalk subgrade and clay subgrade respectively. The strength of the subgrade is assessed on the CBR scale, as described in Chapter 1. The pressure for a unit deflection is known as the 'k' value; the 'k' value of the subgrade is all-important when considering the type and thickness of pavement. Charts are available giving the relationship between flexural stress in concrete, the LCN of the aircraft, the actual subgrade characteristics and the theoretical thickness of concrete slab.

Having determined the 'k' value of the subgrade and the LCG into which the LCN of the aircraft falls, the various options for the pavements in terms of type and thickness can be quickly established.

7.4.3 Pavement construction

The function of any pavement is to reduce the maximum applied load for which it is designed, down to a figure which the subgrade can take. Experience gained over many years has reduced airfield pavement construction to three types: rigid, composite and flexible. A rigid pavement performs its function by virtue of the flexural strength in the concrete, thereby spreading the load to the subgrade. Flexible pavements reduce the load by the inter-granular friction of its base material. Composite pavements contain, as the name suggests, elements of both flexural strength and inter-granular friction.

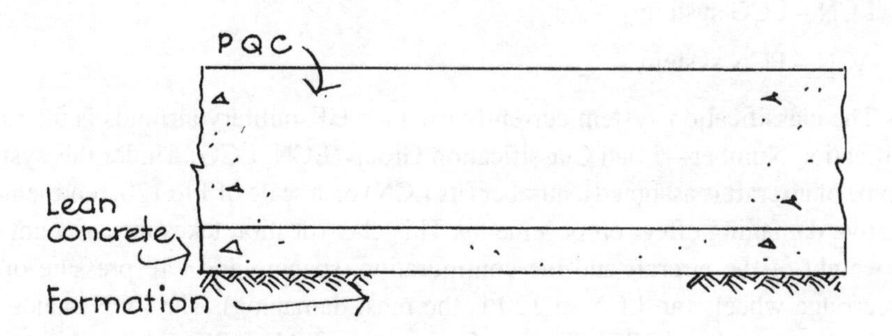

FIGURE 7.39
Rigid pavement

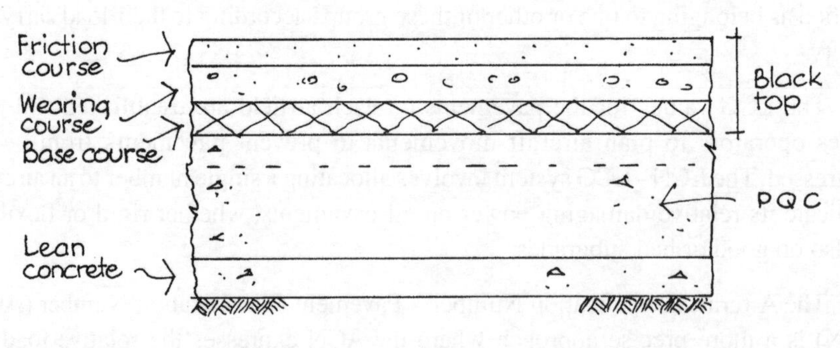

FIGURE 7.40
Composite pavement

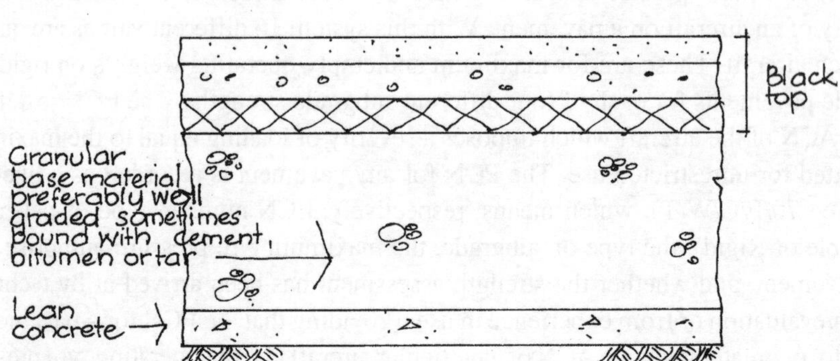

FIGURE 7.41
Flexible pavement

Rigid pavements (Figure 7.39) may be either reinforced or unreinforced pavement quality concrete. Pavement Quality Concrete (PQC) is concrete which will give a minimum flexural strength of 3.5 MN/m^2 or more at the age when the pavement is brought into use. Rigid pavements are normally used for runway and taxiway junctions, aprons and hard-standings.

Composite pavements (Figure 7.40), similar to road pavements, consist of continuously reinforced concrete with bituminous topping. The bituminous surface must have high stability and smooth riding qualities; these qualities can be achieved by using Marshall asphalt. Marshall asphalt is the term used in the United Kingdom to indicate a high stability stone-filled hot rolled asphalt (the term 'asphaltic concrete' is used in the United States and elsewhere). The material is designed and controlled as the work proceeds by the 'Marshall' method which requires a specially equipped laboratory. The Marshall method, which ensures the careful grading of aggregate and optimum binder control, may also be applied to the equivalent dense tar surfacing which, when available, may be used in place of bitumen bound material. The bituminous surface spreads the load to the concrete surface and provides better riding qualities. The thickness of asphalt will normally vary from 75 mm to 125 mm, depending on the risk of reflective or sympathetic cracking over the concrete pavement joints and cracks.

Flexible pavements (Figure 7.41) consist of granular material, stabilised material, or both, for the load-spreading element which is topped with bituminous surfacing. They are not generally economical in the UK for heavy duty pavements, but compare favourably in cost with composite pavements when used for light aircraft. The working stress for pavement concrete is based on its tested strength (generally at 28 days) increased to allow for a continuous though diminishing gain of strength throughout the 'life' of the concrete. The eventual apparent modulus of rupture is subject to a 'Factor of Safety' varying from 1.5 or less for lightly trafficked runways and other pavements to as much as 2.0 for heavily trafficked aprons, tracks and runway ends at busy airports. The higher factor of safety for aprons, taxiways and runway ends is due to the greater frequency of loading, which creates fatigue failure.

Joints in concrete pavements are similar in design to those shown for concrete roads (Section 7.1) but the positioning of joints is more complex. Expansion joints should be placed between the main runway and any contiguous concrete; they may also be placed in both directions in a large concrete surface. In a continuously reinforced slab there are no transverse joints of any kind. The reinforcement is lapped, or better still, welded, to give uniform continuity.

Drainage of runways is of great importance, since a thin film of water may cause aquaplaning of the aircraft. The cross falls on the runway should be in the region of 1 in 66 and runway shoulders may be increased to a fall of 1 in 40. Gullies should be placed so that no part of the pavement is more than 30 metres from a gully or drainage channel. A further means of reducing the risk of aquaplaning is to finish the pavement with porous asphalt, which will serve as a friction course. This requires careful control, selected materials and good laying technique.

The drainage on areas subject to fuel spillage should be channelled to a fuel and oil interceptor to prevent pollution of water course and to reduce the risk of fire. As all aircraft pavements are subject to fuel spillage and as intermediate interceptors are potential danger areas, it is usual to provide a single open trap near the drainage outfall.

A vast amount of pavement at any airport relates to aprons and hard standings. These are normally constructed in reinforced concrete, which are able to cope with fuel spillage. The constant traffic loading on these areas demand a high level of maintenance. Figure 7.42 shows typical repair work being carried out to the concrete aprons at Heathrow Airport.

FIGURE 7.42(a)
Repair work to concrete aprons at Heathrow Airport (Heathrow Airport Limited)

FIGURE 7.42(b)
Repair work to concrete stands at Heathrow Airport (Heathrow Airport Limited)

CHAPTER 8

Concrete and Steelwork

8.1 CONCRETE

8.1.1 Introduction

The materials for making concrete are discussed in Chapter 1. This chapter concentrates on the erection and demolition of concrete structures, together with special techniques such as prestressing work and sprayed concrete.

8.1.2 Distribution and placing

This should be executed so that contamination, segregation or loss of constituent materials do not occur; the concrete must be placed within 30 minutes of discharge from the mixer. Deposited concrete should have a temperature of not less than 5^{0}C and not more than 25^{0}C. The height from which the concrete is poured should not exceed 1.8 metres unless agreed by the engineer. Methods of distribution are also discussed in Chapter 2. This involves the use of either a mixing plant close to the site, or a batching plant set up by an independent supplier on the site itself. This would apply only to large construction works such as dams and motorways. Where trucks are used the truck body must be leak-proof to prevent loss of grout, and the concrete should be covered during transit to protect it from wind, rain and sun.

The choice of distribution plant (see Chapter 2) will depend on the accessibility of the work, the rate of placing necessary for completion of concreting, the amount of concrete to be placed, the distance and height of the mixing plant from the point of placing, and the availability of plant. The access to the point of placing – particularly in deep basements, dams, deep foundations and the like – usually results in the choice of pumps, chutes, cableways and other forms of distributing plant which can move the concrete from the mixer to the placing point without disturbance of reinforcement, at the same time maintaining high delivery rates. Distribution by conveyor is shown in Chapter 2; this method of distribution is easy to control and keep clean, and easily modifiable to site requirements.

The rate of placing will depend on the type of structure: high, relatively thin structures such as silos require a low rate of placing and therefore do not justify the use of pumps or other rapid distribution methods. Even with moving formwork the rate of pour can be easily maintained by cranage on such structures. Reference should also be made to Section 6.3 – Underwater Foundation Construction.

8.1.3 Formwork and reinforcement

The term 'formwork' is commonly used to denote the process by which wet concrete is constrained and supported until it is sufficiently strong to carry its own weight and additional loads that are necessary at the time of construction. The forms are the parts of the temporary work which will be in contact with the concrete, and the actual supports for the forms, whether patent props, scaffolding, or other material, are referred to as 'falsework'. In addition to supporting the live and dead loads involved, the formwork must prevent the loss of material from the concrete. In the main the most serious loss, which may be detrimental to the structure, is 'grout loss'. This can be overcome by sealing the joints with a compressible filler or some form of joint masking. Since the cost of formwork may account for 60% of the total cost of the reinforced concrete structure, any economies in design will affect the overall costs appreciably.

Factors which will affect the final cost include:

- Rationalisation of dimensions
- Simple formwork design
- Design for multiple re-use
- Protection of the formwork face
- Time involved in the 'turn-round' of forms.

Rationalisation of dimensions for concrete structures can have the greatest effect on formwork costs; some authorities have suggested a 40% reduction in formwork costs by this factor alone. Simple formwork design, which allows easy fixing and stripping, will prevent damage to forms and increase their utilisation factor. Designing for re-use involves two factors: first, the strength and construction of the form, and second, the size of the forms. Forms for re-use in other shapes and sizes must be designed to allow dismantling and cutting. Protection of the face of a form by oil or other substances will greatly increase the life of the form. Some treatments also improve the quality of the concrete finish. Typical forms are shown in Figures 8.1 and 8.2.

The time for 'turn-round' of formwork – namely the complete time taken up by erecting, placing, curing and stripping – depends a great deal on the specification, height of pour, striking time, and lifting equipment. A typical example is the height from which concrete may be poured in one lift; some specifications limit this to 1.2 or 2 metres, whereas in fact pours up to 10 metres have been achieved without adverse effect. Thermal control, by insulated formwork and ponding, has allowed very large

FIGURE 8.1
Adjustable form for a rectangular tapered column
(Stelmo Limited)

pours to be made in one single day; bases and rafts of 1200 m^3 to 2000 m^3 have been poured without detriment to the strength of the structure. This form of control results in formwork being released much earlier. Striking time must be reduced to the minimum if economies are to be gained with large forms, particularly if the forms are mechanised or hydraulically operated. The protection and curing of the concrete can be achieved by other covering materials. Lifting equipment and careful programming of formwork lifts also affect the cost; this is the contractor's responsibility and is by no means the least of the factors considered in formwork 'turn-round'.

Proprietary forms and purpose-made steel forms play an important role in economic formwork; they are suitable for repetitive work, such as blocks of flats, when table forms are used, or for special projects which require heavy duty forms for large pours (Figures 8.3 to 8.6). These types of form require heavy lifting equipment but are nevertheless very economical if a sufficient number of uses is obtained.

FIGURE 8.2
Caisson form with centre cores
(Stelmo Limited)

FIGURE 8.3
Column forms for a full height pour
(Stelmo Limited)

FIGURE 8.4
Adjustable crosshead form
(Stelmo Limited)

FIGURE 8.5
Purpose-made travelling form used in the construction of a culvert
(Stelmo Limited)

Falsework

As mentioned earlier, falsework is that part of formwork which supports the forms. In civil engineering the heights and weights encountered are normally in excess of standard propping facilities and some form of scaffolding is used.

There are three basic types of scaffold which can be employed in falsework:

- scaffold tubes and fittings;
- prefabricated systems;
- specially designed falsework in the form of prefabricated units.

The first two types were initially designed for the access scaffolding market but have been used increasingly for falsework. The third type, prefabricated tubular units, is available from any of the formwork and scaffolding specialists.

The units vary in type and have advantages and disadvantages depending on the type of falsework required. Tube and fittings used as falsework provide the most versatile material available, although unlikely to be the most economic; the labour in erecting and dismantling is very high compared with unit systems. Prefabricated systems, initially designed for simple access scaffolding, have made a great impact on the falsework market.

The systems usually consist of vertical frames which are linked together by a patent bracing system; this form of linking can be visually inspected to ensure stability, whereas normal tube fittings cannot be checked by the same method. Proprietary systems of decking, including telescopic beams or prefabricated beams, are often linked with falsework systems. The prefabricated systems therefore offer simplicity of erection coupled with a built-in grid which gives high control of component spacing. Prefabricated

FIGURE 8.6
Slipform system for headworks tower
(Dywidag-Systems International Limited)

falsework systems are made from larger diameter tube and are capable of carrying heavier loads than the normal scaffold units. These systems are used extensively for supporting bridges and other heavy structures which require variable height falsework. Figures 8.7 and 8.8 show typical falsework structures.

Surface finishes

The majority of forms are constructed of Douglas fir ply, which gives a satisfactory finish for most structures. If very smooth finishes are required, the ply face may be treated with epoxy-resin, sheet metal, or any other material which will produce the required finish. Featured surfaces are also very popular: these can be produced by fixing the desired pattern to the face of the ply form. Often the feature may be designed to hide the joining marks made by the forms; this can be achieved by nailing variable thicknesses of sawn boarding to the panels, and produces a random depth and spacing of joints as well as a pleasing textured finish. Fibreglass or plastic moulds may also be used to produce specific details. Some surface finishes, eg exposed aggregate, can be produced by coating the ply face with a solution which will retard the setting of the contact surface. When the forms are removed, the cement face can then be removed easily by wire brush to reveal the aggregate. Tooled finishes can be produced by pneumatic tools, but these tend to be expensive labour items.

The contact face of the form, unless retarding liquids are used, should be coated with mould-oil to enable easy removal and to minimise cleaning.

8.1.4 Jointing in concrete structures

There are two types of joint which may be used in reinforced concrete structures, namely construction joints and movement joints.

Construction joints

Construction joints are introduced for convenience in the construction of the element and, as a guide, should be placed where shear stress is at a minimum. This would be the centre or within the middle third of the span in slabs and beams, as near as possible to beam haunching in columns, and at the top or bottom of openings in walls. Vertical joints in walls should be kept to a minimum and, where possible, the day's concreting

FIGURE 8.7
Falsework to bridge work
(GKN Kwikform Limited)

should terminate at a permanent joint. The joint is formed by a stop in the formwork, which is carefully fitted around the projecting reinforcement and often incorporates a tapered piece of timber (Figure 8.9) to form a key or 'joggle' joint in the element being cast.

After stripping, the joint should be cleaned down to remove any laitance or loose material; this is essential on horizontal construction joints, which, by the very nature of their position, are very susceptible to laitance. Horizontal joints, which are not sealed off with a stop or key, may have the cement-sand matrix brushed away from the aggregate shortly after stripping. Before the next section of concrete is poured the formwork should be checked for tightness at the construction joint, to prevent the escape of grout.

The joint should be thoroughly cleaned off, best achieved by a combined water-compressed air jet. In walls and slabs which retain liquids, either as storage vessels or in retaining ground water, the joint should include a water-bar. The water-bar may be of non-ferrous metal, rubber or plastic (Figure 8.10). Alternatively, a system of waterproofing such as those discussed in Section 6.7.3 may be employed.

Movement joints

These include contraction joints, expansion joints and sliding joints.

A **contraction joint** is a deliberate discontinuity of structure with no initial gap between the two pours of concrete. It allows the sections of structure to contract or shrink, thus eliminating high stress in, and in some cases failure of, the structure. There is a distinction between partial and complete contraction joints: the former has continuity of steel, while in the latter both steel and concrete are interrupted. The formation of the joint is similar to the construction joint.

FIGURE 8.8
Falsework to bridge abutment
(R M Douglas Construction Limited)

Expansion joints usually have complete discontinuity in both steel and concrete, but may have continuity of steel to restrain warping of the panels in the case of slabs. Both types are shown in Section 7.1.3. In the case of the latter type, the bars are coated with bitumen to prevent bond between bars and concrete, and the bars are capped to allow movement.

Sliding joints may also be used for expansion in walls. The filler material usually consists of impregnated fibre board, cork or other bituminous impregnated material. This type of joint depends on the ability of the concrete elements to move independently and equally. If walls and slabs are restrained, or partially restrained by ground friction, the total expansion of any two elements may impinge on one joint and in so doing cause failure.

Sliding joints are movement joints with complete discontinuity in both reinforcement and concrete, at which special provision is made to facilitate movement in the plane of the joint. They are used, in the main, for liquid-retaining tanks and reference should be made to Section 6.7 in which this type of joint is illustrated.

Temporary open joints may be used in the construction of long slabs or walls, so that they may shrink fully before the joint is filled. The joint is formed by leaving a gap of suitable size in the wall (Figure 8.11), permitting joint preparation prior to filling,

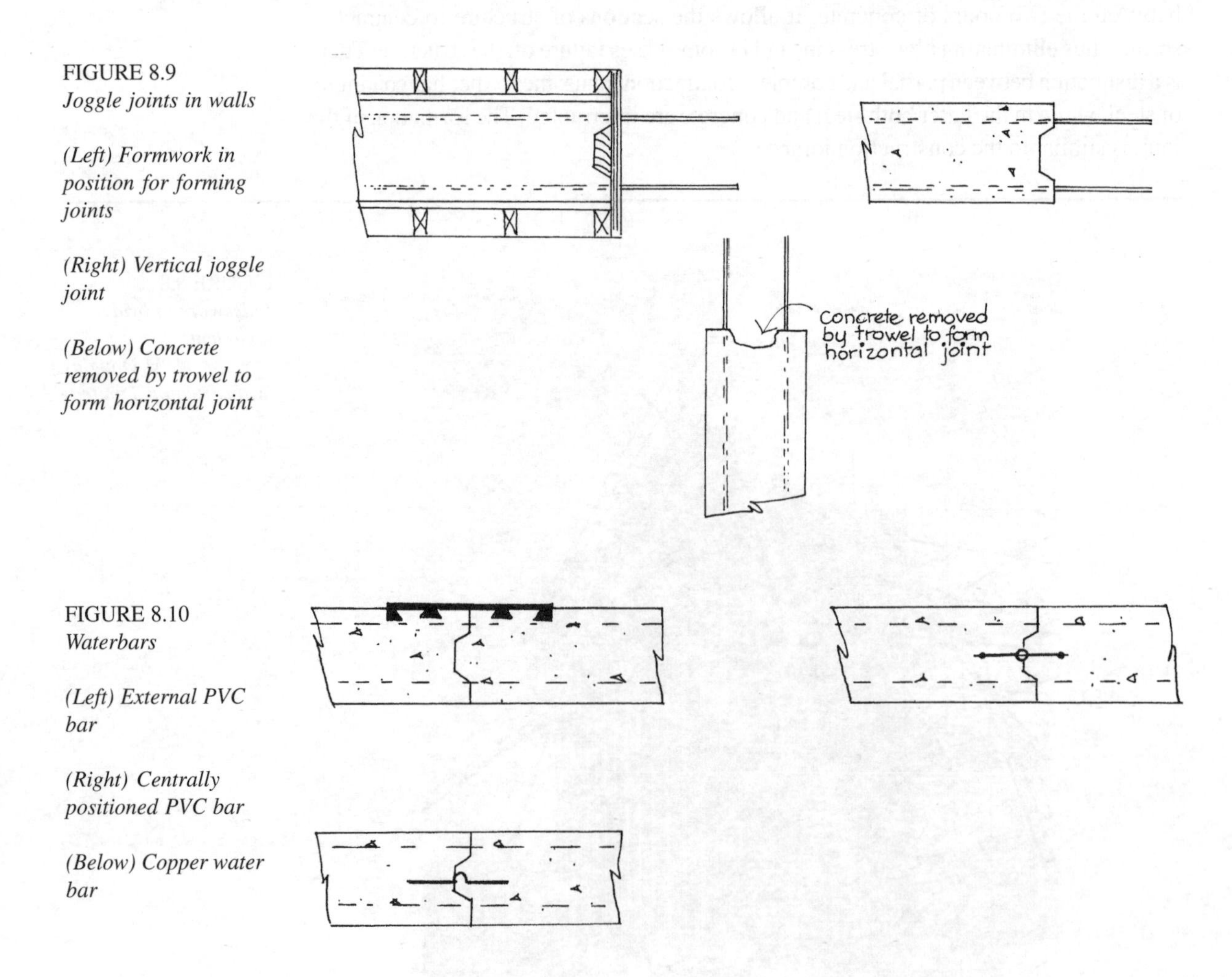

FIGURE 8.9
Joggle joints in walls

(Left) Formwork in position for forming joints

(Right) Vertical joggle joint

(Below) Concrete removed by trowel to form horizontal joint

FIGURE 8.10
Waterbars

(Left) External PVC bar

(Right) Centrally positioned PVC bar

(Below) Copper water bar

after the long element of walling has shrunk. The joints may incorporate water-bars if the shrinkage of the concrete infill is likely to cause leaks in water-retaining structures. The function of the joint is to accommodate drying shrinkage of the concrete already placed.

8.1.5 Pre-stressed concrete

Pre-stressing, by definition, is a technique of construction whereby initial compressive stresses are set up in a member to resist or annul the tensile stresses produced by the load. Pre-stressed concrete is concrete in which effective internal stresses are induced artificially, usually by means of tensioned steel prior to loading the structure. Concrete is a material of high compressive strength and relatively low tensile strength. The technique of pre-stressing has therefore had far-reaching developments in reinforced concrete structures.

Unlike ordinary reinforced concrete in which stresses are carried by normal reinforcement, pre-stressed concrete supports the load by induced stresses throughout the complete structural element. The pre-stressing wires, cables or bars are only a means, up to the limit of the working load, of producing the stress required in the concrete to withstand the load. In addition, where overloading may occur and cracks appear in the concrete member, the pre-stressing cable will prevent failure, providing the stress incurred is not above the elastic limit, and the cracks will close without any deterioration in the structural element.

Pre-stressed concrete has the advantage of being more resistant to shock and vibration, and as it is suitable for thin long structures and can have much smaller sectional areas than normal reinforced concrete to support an equal load, savings in steel, concrete and hoisting are effected.

The development of pre-stressed concrete began as early as 1885–1890 when several attempts were made to tension concrete to increase its bearing capacity. No great advantage was gained in these early attempts because the steel used for tensioning was mild steel, which lost its tension very quickly owing to creep. In the early 1900s a French engineer, Eugene Freyssinet, carried out pre-stressing on the tie of a large concrete arch bridge. He succeeded in lifting the two halves of the arch, to remove the falsework, by stressing the tie. This success led to the development of the technique for other bridges in the 1920s. By 1928 Faber and Glanville in England had published their findings on the phenomenon of creep in steel which confirmed Freyssinet's deduction and enabled him to establish his theory of pre-stressing. About the same period similar work on pre-stressing was being developed in America and Germany, the main aim then being to develop a suitable steel. By 1935 high-tensile steel wires were being used as the pre-stressing medium on a moderate scale.

Applications of pre-stressing are numerous, but it has been used successfully for making railway sleepers, floor beams, piles and other long-line elements, and is also used in many individual projects such as bridges, water tanks, roof structures and runways.

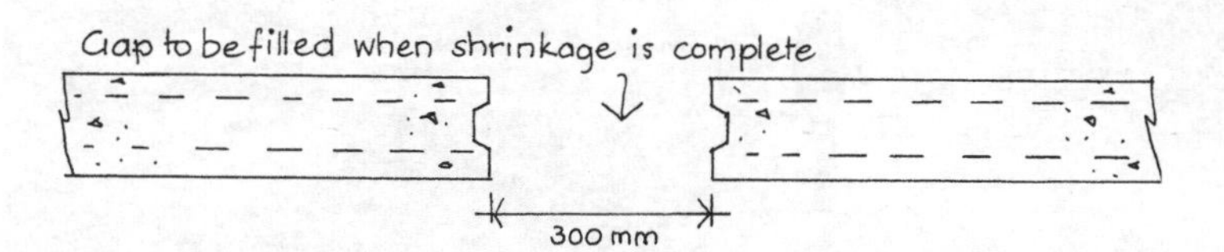

FIGURE 8.11
Temporary open joint to counteract shrinkage in wall construction

Pre-stressing is achieved by two methods: pre-tensioning and post-tensioning.

Pre-tensioning

Pre-tensioning consists of stressing the wires or cables prior to the placing of concrete. When the concrete has hardened and gripped the steel by shrinkage the tension is released from the jacks and transferred to the concrete. The steel is then cut off at the ends of the member; dividing plates can be placed at any point along the member which, when removed, permits the cutting of the wires and thus produces shorter members.

This method of pre-stressing is particularly useful for the 'long-line' method of casting that is employed in precast works (Figure 8.12). In this method a large number of similar units are produced at the same time. The wires are anchored at the end of a metal form, which may be 120 metres long, and connected to a jacking block at the other end of the line. The jacks stress the wire to the desired stress, plus 10 percent or other calculated amount to allow for creep and other losses of pre-stress. The side moulds are then fixed and the concrete placed around the tensioned wires. Alternatively, the concrete may be extruded into the forms or the slip-form method of placing the concrete may be employed. This depends on the type of unit being produced. When the concrete has reached the desired strength, the tensioned wires are released, and because of the bond between the wires and the steel the stress is transferred throughout the unit being produced.

The desired strength is achieved quickly by means of steam curing, which allows strengths normally achieved in 28 days to be achieved in 24 hours; additives can also be used to promote hardening. On hardening the cement shrinks, thus gripping the tensioned wires. The bond between the concrete and wire is improved by crimped or indented wire, and grips may be used on single strand pre-stressing.

Post-tensioning

Post-tensioning follows the reverse procedure to pre-tensioning, the concrete member is cast and the pre-stressing occurs after the concrete has hardened. The wires, cables or bars may be positioned in the unit before concreting commences, but they are prevented from bonding to the concrete by means of a flexible duct or sheath.

Alternatively, a duct can be formed or cast in the member through which the wires or cables can be subsequently threaded. If a duct is to be formed insitu then the

FIGURE 8.12
Long-line method of pre-tensioning (Birchwood Concrete Products Limited)

concrete is cast around long, thin, inflated rubber tubes which are subsequently deflated and removed when the concrete has hardened. Pre-formed ducting (Figure 8.13) in the form of corrugated sheathing is fixed between the normal reinforcement. The ducting is shaped to carry the wires or cable from the upper level at the ends of the units down through the lower level in the centre of the unit. The wires or cables are anchored by means of a special end block and grips at one end, and stressed from the other end. When the required stress has been reached, the wire or cables are anchored, the ends of the unit are sealed with cement mortar, and the ducting is pressure-grouted. Various forms of post-tensioning are discussed below.

Materials for pre-stressed concrete

The design of concrete mixes for pre-stressed work is outside the scope of this book. However, the mix should have a works cube strength at 28 days of not less than 40 N/mm^2 for pre-tensioned work or 30 N/mm^2 for post-tensioned work.

FIGURE 8.13(a)
Sheathing in position for deep beam
(PSC Freyssinet Limited)

FIGURE 8.13(b)
Sheathing and cables in position for floor slab
(PSC Freyssinet Limited)

Where the bond between the concrete and the steel is relied on for the transfer of the pre-stress, the cube strength of the concrete at transfer should preferably be not less than 35 N/mm^2. A lower strength may be accepted in certain members cast under factory conditions, but in no case should the lower strength be less than 27 N/mm^2.

Concrete used for post-tensioning systems should reach a minimum strength of 27 N/mm^2 before transfer of load.

Steel for pre-stressed work may be in the form of wire or bars. Steel wire is manufactured to BS 5891, and is cold-drawn from plain carbon steel. The wire is made by cold drawing a high carbon steel rod through a series of reducing dies. The wire may be plain round, or deformed by indenting or crimping; diameters range from 3 mm to 7 mm but the smallest diameter used in structural members is 4 mm. The wire can be supplied either 'as drawn', which means that it will have a curvature equivalent to the capstan of the wire-drawing machine and will not pay out straight, or 'straightened and stress relieved'. The straightening and stress relieving leads to enhanced elastic properties and is designated 'relax class 1'. Stress relieving heat treatment carried out under conditions of longitudinal strain, which also improves the relaxation characteristics, is designated 'relax class 2'. The terms 'relax class 1' and 'relax class 2' also apply to stand as well as wire.

Indentation or crimping gives better bond strength; the diameters used for normal pre-stressing are 4 mm, 5 mm, and 7 mm. Strands have a nominal diameter ranging from 12.5 mm to 18 mm. They are formed from cold drawn wire and are used for stressing large components by the post-tension method. Seven wires form a strand comprising a straight core wire around which are spun six helical wires in one layer. Three types of seven-wire strand are available: standard, super and drawn. All are subject to heat treatment as for wire and so 'relax 1' and 'relax 2' apply. To produce the drawn strand the standard and super cable can be drawn through a die to reduce the percentage of voids in the cross-section of the strand. This means that for the same nominal diameter the amount of steel is higher, enabling a higher stress to be exerted.

Round high tensile steel bars are also used as pre-stressing tendons (BS 4486). They are prepared from hot-rolled bars by cold working to give the required properties. The diameters range from 20 mm to 75 mm and the bars can be threaded for standard post-tensioning anchorages (BS 4486). Normal lengths range up to 18 m; lengths over 18 m are obtained by joining bars with special couplers.

Loss of pre-stress

This is due to the following causes:

- Elastic deformation of the member
- Shrinkage of the concrete
- Creep of the concrete
- Relaxation of the pre-stressing steel
- Steam curing.

The first three causes can be reduced to an acceptable minimum by using high strength concrete with a low workability. Relaxation of the pre-stressing steel can be counteracted by increasing the initial stress on the steel, but losses may occur in post-tensioned work due to slip. The slip occurs as the mechanical anchorage takes the strain: the tapered wedges move slightly as they take the load and the anchorage may also deform slightly. Steam curing causes loss of pre-stress by reducing the force in the tendon, by thermal expansion of the steel, before it becomes bonded to the concrete.

Post-tensioning systems

There are numerous post-tensioning systems but the main principles of them all are exemplified in the following: PSC Freyssinet system and the Macalloy system.

The **PSC Freyssinet** system has been developed from the well-known MonoGroup system used by the Freyssinet organisation. The system has an extremely reliable and well proven method of anchoring by three tapered conical wedges (Figure 8.14). The system has been developed to enable the stressing strands to be tensioned simultaneously using centre hole tensioning jacks. The cables form two principal groups: the 13 mm range and the 15 mm range.

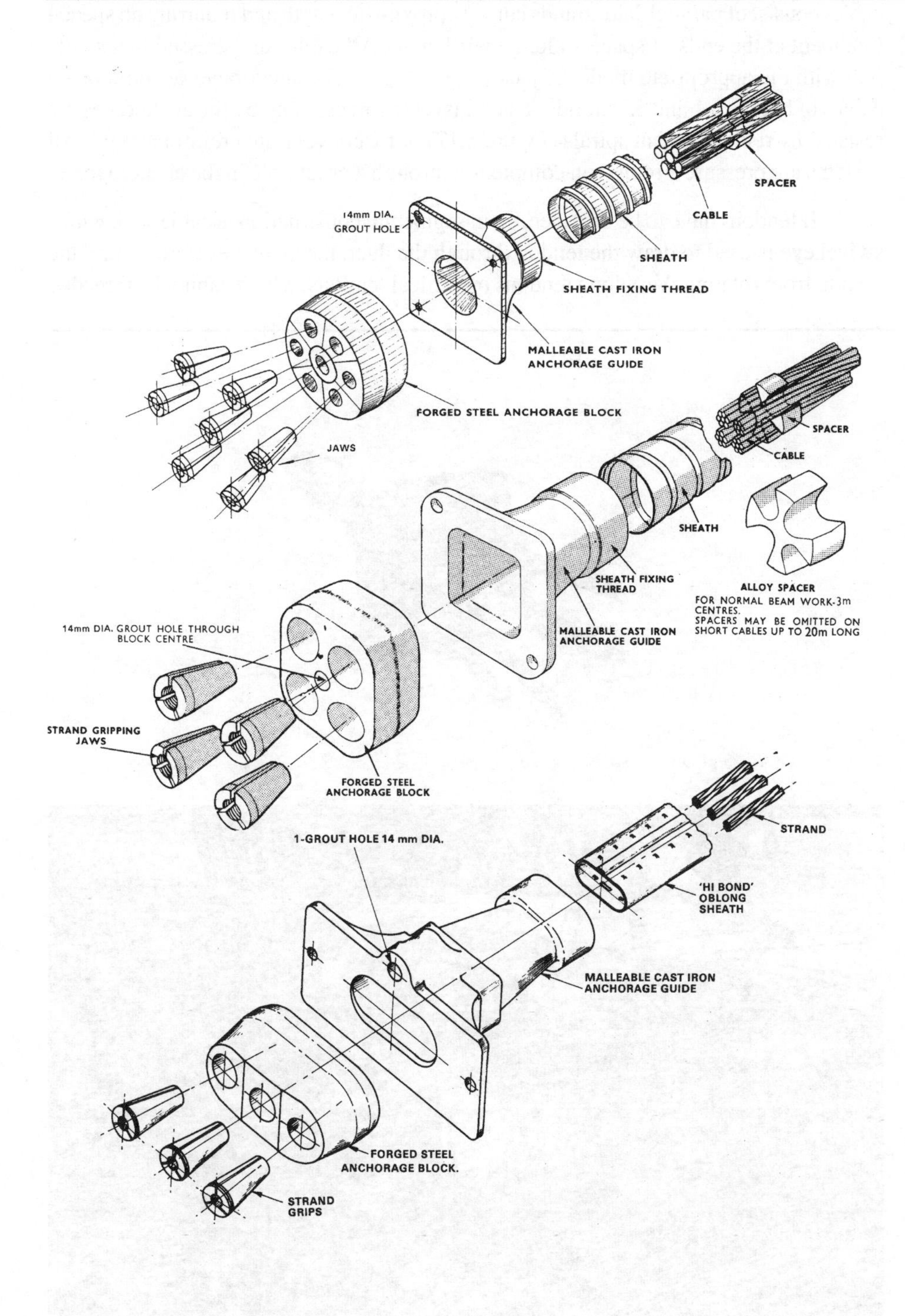

FIGURE 8.14
Typical anchorages for post-tensioning

(Top) 7/15 mm strands

(Middle) 4/15 mm strands

(Bottom) 3/15 mm strands for thin slab system

A variation of the standard range involves one to four strands; this is particularly useful for thin slab construction. The strands are threaded through the cable sheath, which has been previously cast in the concrete elements (Figure 8.13), having been first formed into a cable and fixed by binding tape. In order to prevent inter-strand friction it is normally necessary to provide 'spacers' at 3 metre centres; these may be omitted on cables under 20 metres in length provided that this is only slightly curved in one plane. The strands are tensioned simultaneously and anchored by tapered jaws, as in Figure 8.15. The system is ideal for pre-stressing elements up to 50 metres in length and has been specially designed for the 'medium range' field of pre-stressing.

A combination of either normal or drawn pre-stressing strands can be used. All cables consist of parallel-laid strands cut to approximate length and requiring no special treatment at the ends of spacers along their length. All cables are stressed in a single pull with the appropriate model of jack (Figure 8.16). The anchorages accommodate cables of between 1 and 55 strands. The transverse forces set up by the anchorages are resisted by reinforcement spirals (Figure 8.17) or by conventional reinforcement. All systems are pressure-grouted on completion through a grout hole in the anchorage.

If tendons have to be threaded, eg in segmental construction, a cable sock with a swivel eye is used to draw the tendon through the duct, the swivel eye preventing the tendon from rotating. Very long tendons or vertical tendons, which cannot be threaded

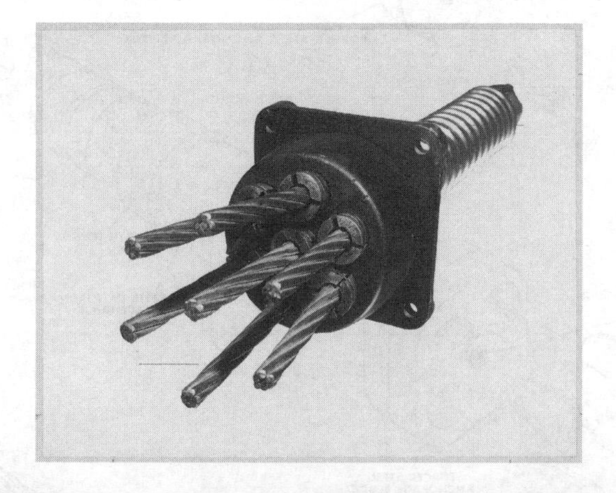

FIGURE 8.15
High tensile strands individually anchored by tapered jaws (PSC Freysinnet Limited)

FIGURE 8.16
Stressing jack (PSC Freysinnet Limited)

efficiently with a cable sock, can be welded at the end and a steel loop welded to the centre strand for the draw rope; after threading, the tendon is cut back 150 mm from the weld.

With the **Macalloy system**, stress is applied to the concrete by means of a solid bar. The type of bar used consists of high tensile alloy steel bars in diameters from 25 mm to 75 mm. The bars are provided with cold rolled threads for part or full length, together with a wide range of fittings. Bars from 25 mm to 50 mm diameter obtain their specific properties by cold working; bars of 75 mm diameter obtain their properties via a quenched and tempered process. Stainless steel Macalloy bars in diameters from 20 mm to 40 mm are made from precipitation hardened martensitic nickel-chrome alloy steel. Bars diameters from 25 mm to 50 mm are available in lengths up to 17.8 m; for 75 mm diameter normal lengths are up to 8.4 m. A course thread is rolled directly on to the bar surface of a form specially designed to give speed and ease of assembly.

The bar is anchored at each end by a special nut which bears against an end plate to distribute the load (Figure 8.18). It may appear that anchorage by means of a thread on a bar may lead to difficulties because the thread reduces the diameter of the section, but this problem is overcome by rolling the thread instead of cutting it; the strength of the bar is increased locally by work-hardening and this offsets the loss in diameter.

FIGURE 8.17
Anchoring bridge tendons by U-shaped loops placed in the deck slab
(Dywidag Systems International Limited)

FIGURE 8.18
Macalloy anchorage
(McCalls Special Products)

The advantage of this system is its simplicity and complete elimination of slip during anchorage; a further advantage is that re-tensioning is also made much simpler. Bars may be cut to length and end threaded at the works or supplied fully threaded and then cut to length on site. In addition to the use in post-tensioning concrete units, the system is suitable for ground and rock anchoring.

Figures 8.19 to 8.21 show application of the Macalloy system. Figure 8.22 shows the use of post-tensioning bars for supporting the erection of bridge arches.

FIGURE 8.19
Bridge anchorages in place ready for tensioning
(McCall Special Products)

FIGURE 8.20
Dornoch Firth Bridge with 890m long post tensioned continuous concrete box girder
(McCalls Special Products)

Circular tanks

Pre-stressing of water tanks or other liquid-retaining tanks can be achieved by applying external pressure by means of high tensile wire winding. The tanks are built of precast blocks or insitu concrete, and when the concrete has cured sufficiently the spirally bound wire is applied by a power driven trolley suspended from the periphery of the tank. (This is known as the pre-load method.) The trolley pulls itself around the tank (Figure 8.23) on an endless chain drawing the high tensile wire through a tapered die. The tapered die creates the tension in the wire as the trolley moves forward; the wire is anchored by clips at frequent intervals and, on completion, is protected by sprayed concrete. An alternative method of pre-stressing can be produced by looping individual wires, or several wires in parallel, around the tank and tensioning them from both ends simultaneously. The ends are fixed to anchor blocks which are cast into the structure. Protection is achieved in the same manner as before.

Pipes can be pre-stressed in a similar manner to tanks, the pipe being first spun centrifugally with a wall thickness of 63 mm and, on hardening, wound with high-tensile steel wire under constant tension. The pre-stressing wire is anchored and covered with cement mortar.

8.1.6 Special concreting systems

Gunite and shotcrete

Gunite can be classified as 'fine gunite' (or sprayed mortar), having an aggregate up to 4 mm, or 'gunite' (sprayed concrete), which has an aggregate size up to 8 mm. The mixed and blended aggregates are blown through a hose, mixed with water at the nozzle and pneumatically projected at high velocity on to a surface. The force of the jet impacting on the surface compacts the material, the water-cement ratio being very low (in the region of 0.3). The dry mixture is capable of supporting itself without sagging and can be built up to a thickness of 150 mm in one operation on vertical

FIGURE 8.21
Section through box girder showing Macalloy system
(McCalls Special Products)

FIGURE 8.22
Post-tensioning bars used to support the arch construction of a 230 metre span bridge
(Dywidag Systems International Limited)

surfaces and up to 50 mm on overhead applications. The system is being used extensively for both new work and repair work.

Gunite is used mainly for watertight renderings; sealing rock faces; rendering large concrete surfaces; maintenance of sewers and other concrete structures. It has been used extensively for retaining the rock faces on hill-sides (Figure 8.24). It also has a very valuable use in the renovation of bridges, marine structures, cooling towers and other highly exposed structures.

The technique produces a compressive strength of 35 to 65 N/mm^2 at 28 days. The aggregates used in this process consist mainly of sharp sand with a 4 mm down to fines grading, although coarser aggregate can be used. The aggregate is mixed dry with Portland cement, or other specified cement, 1 part of cement to 3 parts sand, or 1 to 3.5 (by weight). The system produces an excellent bond with a number of materials and has consistent waterproofing qualities.

FIGURE 8.23
Wire winding machine suspended from trolley (Preload Limited)

FIGURE 8.24
Close up of guniting operation – stabilising a rock face (Cement-Gun (UK) Limited)

Shotcrete is similar to gunite but the aggregate size is greater, up to 15 mm (possibly up to 25 mm). It is in the truest sense sprayed concrete. It is particularly useful for roofs, walls, tanks, reservoir linings, canal linings, tunnel, sewer and shaft linings. Its main uses are the spraying of domes; walls and silos; slope formation (Figure 8.25); and tunnel linings (it is especially suited to the New Austrian Tunnel Method – see Chapter 5).

The shotcrete and gunite process is similar in that the material is delivered dry to the delivery nozzle and should not be confused with 'conveyed' or 'pumped' concrete.

After mixing, the blended dry mixture is delivered to the delivery nozzle via a high velocity compressed air stream. At the nozzle, water is added to hydrate the cement as it is deposited at a velocity of 125m/sec. The machinery involved depends on the size of the contract and outputs range from 0.5 to 9 cubic metres per hour.

FIGURE 8.25(a)
Shotcrete being applied in slope stabilisation using hydraulic arm and controlled by remote control
(Aliva Limited)

FIGURE 8.25(b)
Close up of shotcrete in slope stabilisation
(Aliva Limited)

The dry blended material is fed into the spraying machine (Figure 8.26) where revolving rotor ports are filled with the mix by gravity from the machine hopper, and discharged by air pressure into an air chamber before being conveyed to the delivery nozzle by a high pressure jet stream. The opening of a release valve sends the material through a 50 to 85 mm diameter reinforced hose to a special nozzle. The nozzle is fitted with a manifold through which water is sprayed, under pressure, on to the dry mix, and the mixed material is then jetted by the compressed air on to the work surface. During the operation a proportion of large aggregate, between 10 and 20 percent, is blown to waste, so protective screens should be used to protect nearby windows and the workforce.

Lightweight aggregate may be incorporated in the mix if thermal insulation is required, or heavyweight aggregates may be used for concrete weighing around 3000 kg per meter cube. The latter is sometimes used for special pipeline weight coatings. Some materials manufacturers are marketing pre-mixed and bagged dry materials that contain additives such as dried polymers, accelerators and silica fume; such materials are widely specified in bridge repairs. However, these materials are very costly compared to a traditional gunite mix.

Reinforcement for gunite usually takes the form of electrically welded steel mesh fabric, of a weight and type dependent on the circumstances of use.

Steel Fibre Reinforced Sprayed Concrete (SFRSC)

Steel Fibre Reinforced Sprayed Concrete (SFRSC) is the generic term used for sprayed concrete that contains stainless steel fibre. One such product, introduced by the Caledonian Mining Company, is called 'Steelcrete'. The process is a dry process like gunite, where the water is added at the delivery nozzle. Alternatively 'Steelcrete' may be produced using polypropylene fibres. The material has a compressive strength of not less than 60 N/mm^2 at 28 days. Various colours can be achieved by the use of colour

FIGURE 8.26
Machine for spraying gunite or shotcrete (Aliva Limited)

additives. The most common steel fibre used in SFRSC is stainless steel, such as 'Fibretech', which is produced by the 'melt extraction' or 'melt overflow' process, which is far superior to the chopped drawn wire process. SFRSC achieves strengths which allow the thickness of concrete to be reduced by up to 50%. It is the most likely material to be used in the spraying of tunnel linings, such as the New Austrian Tunnelling Method – see Chapter 5.

8.1.7 Demolition of concrete

The method used for the demolition of concrete depends on the type of concrete, its position in the structure and the purpose for which the cutting or breaking is being undertaken (see BS 6187).

Methods of demolition

There are numerous methods of demolition which the contractor may employ, but the method chosen should suit both location of site and type of structure. These include:

- Hand demolition
- Use of demolition ball
- Use of wire ropes
- Use of pushing arm
- Use of explosives
- Other methods

Hand demolition involves the progressive demolition of a structure by operatives using hand tools. Cranage is often used to lift out members once they have been released. Generally speaking, the order of demolition is the reverse of that of construction.

Demolition by swinging ball involves demolition by swinging a heavy steel ball suspended from the jib of a crane. It should not be used where the angle of the crane jib is greater than 60 degrees from the horizontal. Pitched roofs as well as floors should be removed by hand (providing sufficient members are left in position to provide lateral support).

Demolition by wire rope: only steel ropes with a circumference of 38 mm or more should be used for demolishing parts of the structure, and such ropes must, by law, be inspected frequently by a competent person to ensure that their strength has not been impaired by use. Care must be taken to ensure that there are no persons standing between the winch and the building being demolished or nearer than a distance equal to 75 percent of the distance between winch and structure on either side of the rope. Where the building fails to collapse during a pulling operation, the work must continue using some other mechanical method. Hand methods must not be used since the structure may have been weakened.

Demolition by pushing arm: since this method is limited in its operation by height, the structure must be reduced to a suitable height by one of the other methods. The method involves the use of a machine fitted with a pusher arm (normally hydraulic) which exerts a horizontal thrust. This method is more suited to small demolition work.

Demolition by explosives is used for a wide range of structures, including chimneys, cooling towers and complex structures that would be difficult to demolish by other methods. It allows rapid and complete demolition where other methods could prove dangerous or slow. The work is usually carried out by specialist contractors in accordance with BS 5607.

The production of more sophisticated explosives, combined with the advancement in methods of initiation, has made the demolition of buildings and structures by the controlled use of explosives a fast, practical and economic alternative to traditional techniques. Explosives are now more stable and last longer. This fact, together with the control of detonation to millisecond timing, and the ability to ensure that premature initiation cannot take place, makes their use much more viable. Furthermore, the range of explosives now permits various techniques of blasting so that key supports in a structure can be demolished in a predetermined sequence, causing a structure to collapse under its own weight. There are several advantages of demolition by explosives:

- The shortening of the demolition programme, thus releasing the site for earlier development, and the direct cost savings that can accrue from such programmes.
- Reduced nuisance to people living in the vicinity; normal methods of demolition can prove a serious nuisance.
- The high level of safety that can be obtained for the workforce, particularly with complex structures such as prestressed structures that might explode with normal methods, or with inaccessible structures such as oil platforms.

Explosives and methods of demolition

The four basic methods of explosive collapse are:

- Telescoping – eg cooling towers
- Toppling – eg chimneys and bunkers
- Implosion – eg high-rise structures
- Progressive collapse – eg long buildings.

Preparation for demolition by explosives

It is common practice to combine explosives with the techniques of pre-weakening the structure. This entails the partial removal of support but calculated to ensure that collapse does not occur prematurely. This technique reduces the amount of explosives, air blast and ground vibrations. It also allows a particular initial movement to be planned.

The technique requires calculations to be made on the stability of the weakened building in regard to wind loads and other loading. The pattern of pre-weakening and charge placement will sometimes involve the retention of certain loadbearing members in order to impose a preferred direction of collapse. A method statement must be produced to indicate the sequence and method of demolition, with specific details of the pre-weakening programme, the type and quantity of explosives, and the delay detonation sequence. The method statement will also include arrangements for the safety and protection of the workforce and the public, together with details of the exclusion zone and its security.

Risk assessment and contingency planning are essential parts of the method statement. **Risk assessment** is a methodical and critical appraisal of the events that could lead to injury of the workforce or the public, or the loss or damage to property. It is an assessment of the possible consequences of these activities so that the level of risk can be determined. **Contingency planning** includes the provision of standby facilities to repair damage to services and adjacent property, street cleaning, and emergency services. In the event of the demolition not proceeding as planned, an explosives safety check and structural assessment will be carried out and a secure zone should be established around the section of structure involved. A team meeting will determine the necessary course of action. To deal with such an eventuality, plant to deal with a non-collapse situation should be organised as part of the contingency plan.

Explosives and accessories

The main characteristic of explosives is that on initiation they react suddenly to form large volumes of gas at high temperatures. This instant release of gas creates very high pressures which can be harnessed to perform particular tasks. In demolition explosives are used in two main ways: as borehole charges or as kicking charges.

- **Borehole charges** are widely used in controlled demolition of concrete and masonry, as well as in rock blasting. Explosives are installed and stemmed into position in pre-drilled holes.
- **Kicking charges** are used in conjunction with the pre-weakening method mentioned earlier; they are used to collapse or displace supporting members in a structure.

The characteristics of explosives are as important as the way is which the explosive is used. Each type of explosive has its own characteristics; these include strength, density, velocity of detonation, sensitivity, sensitiveness, water resistance, fume characteristics, gas volume, stability, detonator pressure, and critical diameter.

Strength refers to the energy content of an explosive and its ability to perform useful work. The measure of strength varies around the world. Some refer to 'grade strength', or 'relative weight strength', or 'relative bulk strength'. All strength figures are calculated from the deflections obtained by the detonation of standard weight explosive charges in a freely suspended ballistic mortar. **Relative weight strength** compares the weight of an explosive with the strength of the same weight of blasting gelatine, which is taken as 100 per cent. **Relative bulk strength** is the indicated strength of any volume of the explosive compared with the strength for the same volume of blasting gelatine. For practical reasons manufacturers prefer the use of relative weight and relative bulk strengths.

The **density** characteristic enables the energy per unit length of borehole to be varied to meet particular site conditions. Commercial explosives have a range of densities from 1.6 to 0.8 g/cm^3.

Velocity of detonation is the speed at which the detonation wave passes through a column of explosives and is normally expressed in metres/second. As a general rule, the higher the velocity of detonation, the greater the shattering effect.

Sensitivity is the measure of the ease of initiation of an explosive; as a general rule the sensitivity decreases as the density increases. Sensitivity is usually expressed in terms of response to a particular initiator; this is usually a standard reference detonator.

Sensitiveness is the measure of the ability of an explosive to maintain the detonation wave throughout a column consisting of a number of cartridges.

Water resistance depends on the chemical make-up of the explosive and its packaging. Gelatinous and slurry types of explosives have the best water resistance and low density powder have the least. The effect of water on an explosive with poor resistance is to impair its performance and to increase the production of toxic gases.

Fume characteristics relate to the quantity of noxious fumes that an explosive may produce in its gaseous products. This may seriously affect its use in confined spaces such as tunnelling and mines.

Gas volume is the total quantity of gases liberated on detonation, expressed in terms of litres per kilogram. A high value indicates that an explosive produces a good heaving effect.

Stability relates to chemical stability in storage, and explosives are rigorously tested to achieve high and consistent quality.

The **detonation pressure**, often measured in kilobars, is the pressure of the detonation shock front; it is a measure of the explosive's fragmenting ability close to the shot-hole.

The **critical diameter** of an explosive is the smallest diameter at which the detonation will reliably propagate along a column of explosives, and it is also another measure of sensitiveness.

Types of explosive

Explosives are manufactured to meet the many and varied needs of mining, quarrying and the construction industry. However, they can be classified under the following headings:

- slurry explosives;
- emulsion explosives;
- nitroglycerine-based explosives;
- ammonium nitrate/fuel oil mixtures;
- nitramine-based explosives.

Slurry explosives were first developed in an attempt to waterproof, strengthen and sensitise ammonium nitrate. They are composed of ammonium nitrate and other nitrates in solution, with the addition of fine aluminium powder and other light metals; the addition of gums produces a gelatinous state. The materials are sensitised by air which is beaten into the product in manufacture. They have acceptable fume characteristics and perform well in wet conditions.

Emulsion explosives are prepared in the form of water-in-oil emulsions. Incorporated in this emulsion is entrapped air usually contained in microscopic spheres of glass, plastic or resin: this acts as a bulking agent and provides centres of reaction which ensure sufficient sensitiveness for continuous detonation. Fine aluminium may be added to produce explosives of different strengths. Their detonation velocities are high, around 4500 to 4800 m/s.

Nitroglycerine-based explosives can be subdivided into:

- Gelatine explosives
- Semi-gelatine explosives
- Nitroglycerine powder explosives.

Gelatine explosives contain a quantity of nitroglycerine which is thickened with nitrocellulose to give the mixture a plastic consistency. Other ingredients include aluminium or cellulose fuels and nitrates. They have high bulk strength, excellent water resistance, excellent propagation and are available in large range of diameters. These characteristics make it one of the most widely used nitroglycerine-based explosives.

Semi-gelatinous explosives are part way between a gelatine and a powder. They are sufficiently dense to prevent floating in wet boreholes and have some resistance to water penetration. The nitroglycerine content determines the selection to meet the needs for strength and water resistance. The advantage is that the explosive can be adapted to meet very specific requirements. However, apart from small diameter cartridges, they are rarely used.

Nitroglycerine powder explosives are made from ammonium nitrate and a combustible material, with nitroglycerine as a sensitiser. The nitroglycerine content is in the order of 8–10 percent, producing a low bulk strength. The compositions are reliably sensitive and they make a useful contribution where economy of explosives is required.

Ammonium nitrate/fuel oil (ANFO) mixtures have low bulk strength and are not water resistant. Furthermore, if used in wet conditions the amount of noxious fumes can increase to unacceptable levels. However, when mixed in the correct proportions a high blasting efficiency can be achieved. The addition of aluminium gives a higher temperature reaction and therefore a higher strength.

Nitramine-based explosives are designed for use in certain demolition operations where very high detonation pressure is required. They have good plasticity over a wide temperature range and excellent water resistance. They are composed of nitramine type explosive, such as RDX, with grease or polymeric binder, to give a flexible cohesive material, which is available in thin sheets as well as plastics. All compositions have high velocities of detonation, up to 8200 m/s.

The most common type of explosive used is the nitroglycerine gelatine type. These explosives are waterproof, easily detonated and give predictable results. An alternative for borehole work is to use high energy detonating cord, pentaerythritol tetranitrate (PETN), which is enclosed in a tape and wrapped in polypropylene yarn, all enclosed in a plastic cover with an overall diameter of 4.65 mm. It has a velocity of detonation between 6000 and 7000 m/s and is one of the safest products to handle. Detonating cord is sometimes used in demolition to connect up large numbers of charges. For simple demolition work, such as breaking up large concrete bases, the most common combinations are sticks of nitroglycerine-based gelatine explosives primed with detonating cord and initiated with instantaneous detonators.

Detonators

Detonators contain small amounts of very sensitive explosives as well as the means of initiation of those explosives. They are susceptible to heat, friction and shock and therefore should be handled with great care.

Explosives can be initiated by detonators of the electric or non-electric type. However, electric detonators may be precluded if there is a likelihood of stray electric currents or where radio frequencies broadcast are prevalent.

Detonators may be either instantaneous or delayed, plain or electric – the range of such is outside the scope of this book.

Plain detonators are made from aluminium tube and contain a base charge of a high explosive and a priming charge of lead azide. They are made in various strengths, typically No 6 and No 8 Star.

Electric detonators consist of thin walled aluminium or copper tube closed a tone end and containing a high explosive base charge, a priming charge and a fuse head. The tube is sealed with a neoprene plug through which pass the leading wires of the fusehead assembly. In principle, when an electric fuse assembly is crimped into a plain detonator then a standard electric detonator is produced. In the case of delayed detonators a special delay element is introduced between the fusehead and the priming charge. This delay element consists of a column of slow-burning composition contained in a thick-walled metal tube; the length of the column determines the delay time.

In common use today is the **non-electric** system of detonation which utilises a small bore shock transmission tube. This tube is coated on its inner surface with a fine explosive powder, which maintains the shock wave, normally 2000 m/s, through the tubing to activate the primary explosive or delay element in the detonator. The tubing is made of plastic with an outside diameter of 3 mm. In its standard form it is transparent and can cope with temperatures up to 50^{0}C.

Various examples of demolition by explosives are shown in Figures 8.27 and 8.28.

FIGURE 8.27
Demolition of concrete tobacco bond warehouses (Christopher D Brown)

Other methods of demolition
When site conditions preclude the use of explosives, the following may be used:

- Gas expansion burster
- Hydraulic burster
- Thermal reaction
- Thermic lance
- Drilling and sawing
- Purpose built shears, crushers, pulverisers and grapplers.

Gas expansion and **hydraulic expansion** devices can be used for 'bursting out' concrete, but the reinforcement requires cutting to free the ruptured member. Walls may be jacked over, having first been freed from connections at the top and sides.

FIGURE 8.28
Demolition of storage silos
(Robinson & Birdsell Limited)

Gas expansion operates with explosive force in a prepared cavity, whereas hydraulic expansion is achieved by means of wedges and pistons.

Thermal reaction is used in conjunction with wire pulling. The member to be severed is surrounded by a mixture of metal oxide and reducing agent which produces great heat when ignited. The member loses strength due to the heat, and wire pulling completes the operation.

Thermic lancing is the process of drilling or cutting silica or part-silica materials by thermo-chemical action. The process is based on molten oxidised mild steel which fuses with the silica or other material to form a slag. This process was developed by the French after the Second World War to cut up large concrete structures, such as submarine pens and gun emplacements. The flame is produced utilising a mild steel tube packed with wire which, when subjected to a flow of oxygen down the tube, will produce a flame temperature of 3000 degrees centigrade. This heat is sufficient to melt concrete, which requires a temperature between 1800 to 2500 degrees centigrade. The speed at which the reaction takes place minimises damage to the surrounding steel and concrete.

FIGURE 8.29
Demolition of reinforced concrete by thermic lance

(a) (Right) Cutting the concrete member

(b) (Far right) Lifting out a cut member

(c) (Below) Close-up of cut

The technique is very suitable for cutting out structural members during demolition operations (Figure 8.29): the member is supported by crane and lifted out of position when the cutting is complete. The absence of noise and vibration means that massive structural alteration can be carried out during working hours without disturbing other staff working on the premises. The hot lava must be contained in a bed of sand or other material to prevent damage to adjacent surfaces. Ventilation and extraction of smoke may be necessary during the cutting process. The high temperatures used in cutting the concrete do not damage the concrete further than 75 mm from the cutting edge; this is due to the extremely poor thermal conductivity of concrete.

Drilling and sawing concrete walls is an advanced technique carried out by a number of specialist demolition contractors. The technique is used for cutting openings in reinforced concrete up to 375 mm thick. Drilling is both quiet and clean. It employs tubular diamond-tipped tools to produce precise diameter holes in concrete or brickwork. The coolant is normally water, which washes away any dust produced by the operation. Stitch drilling (overlapping the holes) allows a circular panel to be removed. Rotary percussion drilling is suitable for forming holes in concrete where reinforcement is not a problem. It is less expensive and ideal where noise, vibration and dust are not a problem.

Wall and floor sawing employs blades which are edged with diamond-impregnated segments, and the use of water as a coolant. The concrete is cut from one side of the wall, the saw being mounted on a pre-determined wall track. This process gives a very

FIGURE 8.30
Removing a circular panel by diamond drilling
(De Beers Industrial Diamond Division)

FIGURE 8.31
Sawing through a concrete footbridge
(De Beers Industrial Diamond Division)

accurate clean cut for the installation of doors, windows and other forms of opening. Examples of drilling and sawing are shown in Figures 8.30 and 8.31.

Purpose built shears, crushers, pulverisers and grapplers are perhaps the latest tools to be designed to give maximum safety during the demolition process. They are specially designed mechanical tools that fit on to hydraulic excavators. With the advent of the excavator with a 360 degree turning circle, plus long-reach equipment capable of reaching 30 metres, they make the demolition process simple and cost-effective.

Shears (Figure 8.32) are capable of cutting through steel beams and other steel members that may pose difficulties for other forms of demolition. They can also be used in a 'tin-opener' type action to make cuts both vertically and horizontally, allowing sections of walls to be removed. By means of cold cutting they have virtually eliminated the risk of explosion when cutting tanks that contain residual gases.

Crushers and crackers (Figure 8.33) are used to demolish concrete buildings from a safe distance. They are suitable for dealing with medium-rise buildings.

Pulverisers (Figure 8.34) now do the job that hydraulic hammers did, but this attachment is free from vibration and noise. The crushing power is sufficient for this attachment to be used in primary demolition, to crush floors, columns and other concrete members.

The grappler (Figure 8.35) is suitable for handling steel and concrete sections either in primary demolition or in subsequent handling and disposal.

Demolition procedure for various types of structure

Before any major work commences, the roof of a building should be removed as carefully as possible if it is to be salvaged, together with windows, doors, and their linings.

Roof trusses. Temporary bracing should be introduced to allow individual trusses to be removed. Where trusses support gable walls, the wall should be removed prior to dismantling of the truss.

Floor panels. When removing filler joists or any infilling materials such as

FIGURE 8.32
Allied Labounty mobile hydraulic shear (Allied Construction Equipment Limited)

FIGURE 8.33
Allied long-reach equipment with concrete cracker (Allied Construction Equipment Limited)

FIGURE 8.34
Allied Labounty concrete pulverisers (Allied Construction Equipment Limited)

FIGURE 8.35
Allied Labounty demolition grapple (Allied Construction Equipment Limited)

concrete or blockwork, the operation should have a safe platform independent of the work being demolished. Floor panels being removed should be supported to allow cutting out, prior to lowering to ground level.

Structural steel. The sequence of demolition should be arranged to maintain a stable structure. This will often involve the use of temporary bracing or steel ropes. Members should be lowered carefully to the ground, thus preventing any increased load on other elements during demolition.

Structural concrete. Demolition of pre-stressed concrete must be executed with great care. Fully bonded or pre-tensioned members may be cut and lifted out in a similar manner to other forms of reinforced concrete, but post-tensioned structures should never by cut or burst. The demolition of post-tensioned units may be undertaken in the reversed procedure to that of erection. Temporary supports should be placed under floors and beams while the end anchorages are re-stressed and the tension slowly released. The units then, and only then, be cut and lifted out of position. Failure to follow this procedure may result in a post-tensioned unit exploding, especially if the stress is suddenly released by cutting. The procedure for this work is given in detail in BS 6187.

Bridges. Attention should be given to careful planning and organisation of such work. The work should be carried out in reverse order of erection, after first removing as much dead load as possible without affecting the stability of the bridge. Guys and braces should be fixed to all main beams to prevent their slipping during dismantling operations. Temporary support to abutments may be necessary during the main lifting and removal operation owing to the lateral support being removed. Where old brick and masonry arches are involved, these may require temporary centres (particularly over road and rail services) to prevent total collapse and consequential damage to the work below.

Independent chimneys. Where the chimney is in good condition, demolition can be carried out by a steeplejack, but alternatively it may be felled by the use of explosives. Where part of the chimney is pulled out at the base to cause deliberate collapse, there should be a clear space of 1.5 times the height of the chimney (from its centre). Steel and reinforced plastic chimneys should be dismantled from a safe platform provided by an external scaffolding. The steel lining may be cut into manageable sizes and lowered to the ground and debris allowed to fall into the chimney.

Storage tanks. Before work commences on the demolition and removal of petroleum tanks and the like which have held inflammable liquids, the Local Authority Petroleum Officer should be consulted. Tanks can be made safe by one of the following methods:

- Filling with water
- Filling with an inert gas, ie nitrogen
- Use of dry ice.

In the last method, operatives should wear gloves and the tank should be left for 12 hours after the addition of the ice. Further comment and details can be found in BS 6187 – Demolition.

8.2 STEELWORK

8.2.1 Types of structure

The types of structure most commonly used in civil engineering are the skeleton frame, used in buildings such as power houses; the trussed frame, normally used for bridges and other complex structures; and stressed-skin panel structures.

The **skeleton frame** is constructed with hot-rolled 'H' section columns and beams which are riveted, bolted or welded together. The frame is designed to transfer all loads to the columns and thence to mass foundations. The column sections may be selected from a range of universal beams (UB) or universal columns (UC), depending on the magnitude of the loading.

Low-rise factory blocks with wide-span roof frames may be constructed using universal beams as column members, with greater economy than the purpose-made column. Heavily loaded structures, however, are more economically supported using columns selected from the UC series; each serial size has a range of sizes and weights.

In addition to the normal beam construction, the frame may incorporate girders, lattice beams or castellated beams. Girders and lattice beams are normally designed so that their depth occupies a storey height, thereby maintaining normal overall height of building. A range of girders and bracing methods are shown in Figure 8.36; these include:

- Stiffened plate girders
- Vierendeel girders
- Lattice girders
- Castellated beams
- Warren girders or trussed frames.

Trussed frames, or plane frames, may be used in building work; but in civil engineering such forms are associated with more complex structures, eg bridges. working platforms and temporary structures. Bridges may be constructed in one of many steelwork forms, including trusses, plate girders, arch bridges and portal frames. For spans of up to 50 metres the constructional plate girder may be used, but above that span the plate girder will require stiffening with a concrete deck to form a composite structure. Deep plate girders may be stiffened by horizontal webb stiffeners.

Trusses are commonly used on bridge structures over 50 metres, the most common form being of the Warren type. Spans of over 50 and up to 150 metres can be bridged with Warren type girders which have inclined upper chords. Cross-bracing is normally provided between vertical members in deck-type bridges, whereas through-type bridges are braced with sway bracing between verticals, and portal bracing between the end supports.

Stressed-skin panels. Perhaps the most common form of stressed-skin panel construction is the box girder construction as used in bridges. The box section is better known as a 'torsion box' and consists of plate welded to a pre-determined streamlined section. The Severn bridge-deck is a fine example of this type of structure, the box sections being designed to form a continuous aerofoil (see Chapter 7).

The same principle of design can be used for the construction of steel framed buildings, the sections being much thinner and consisting of steel members sheeted on one face with pre-formed steel sheeting. The sheeting is fixed in such a manner as

FIGURE 8.36
Girders and braced structures

(a) Built-up girders

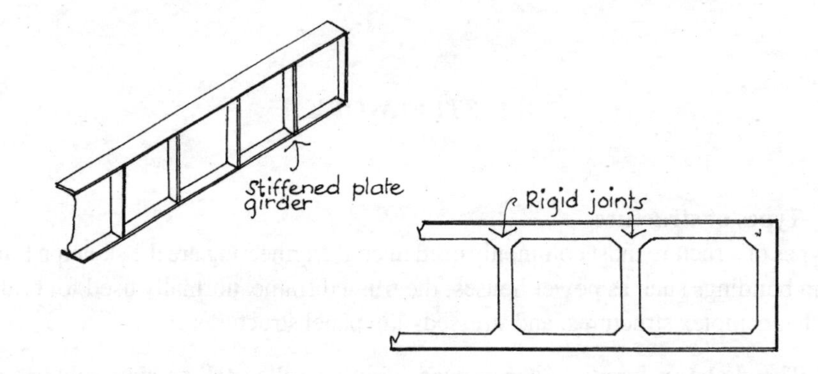

(b) Left: Lattice girders Right: Castellated beam

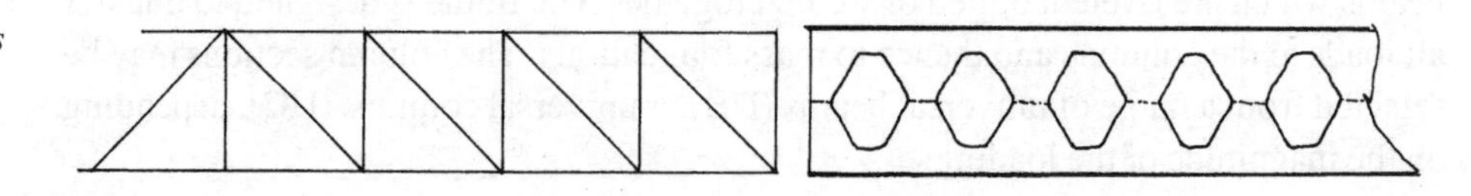

(c) Warren girder

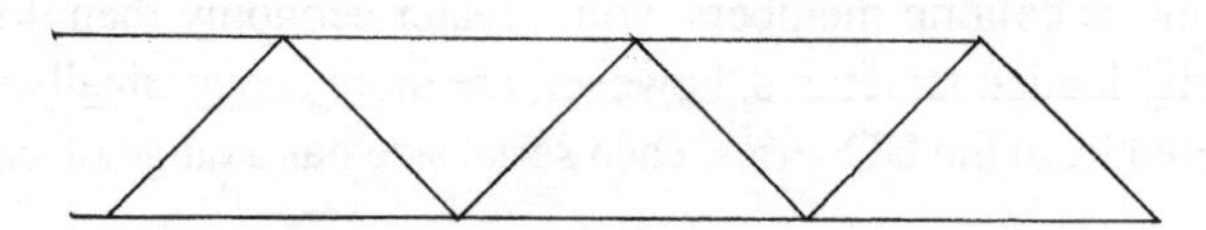

(d) Warren girder with inclined upper cord

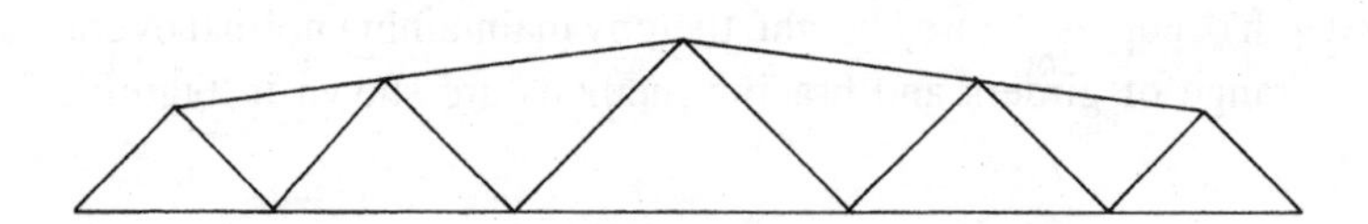

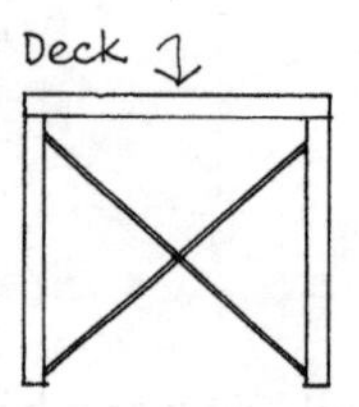

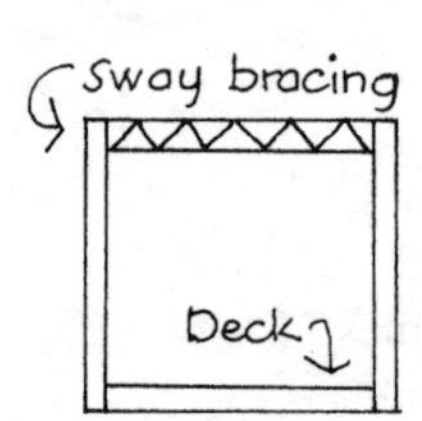

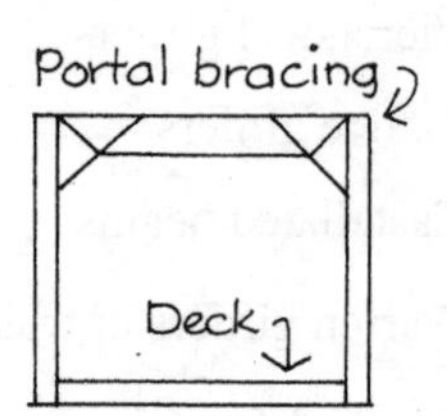

to prevent slip at the fastenings and so produces a diaphragm which stiffens the frame and significantly increases the load-bearing capacity of the structure.

8.2.2 Erection of buildings

Buildings to be erected can be divided into two broad categories: those which are low-rise or single storey, sometimes referred to as 'shed-type' buildings; and those which are multi-storey. Some buildings, however, may incorporate structures which can be placed in both categories, and such buildings require special consideration when selecting the erection procedure.

In each case there are certain factors that have to be considered, including method of erection, height of building, weight of members being lifted, maximum reach required, and time allowed for erection.

The amount of space available will influence the method of erection. A congested site is best served with a tower crane or similar plant, whereas open sites may use mobile cranes to greater economy. If the building is multi-storey, the height of lift will be of major importance: mobile cranes are normally limited by their jib lengths, although mobile cranes with jib lengths of up to 60 metres are available. Alternatively a Scotch derrick may be used (see Chapter 2).

The mobile crane is ideally suited to erection work, subject to the comments made above, and providing the crane can place the members without a long reach. This would be suitable for compact tasks associated with the erection of gantries and temporary platforms. Mobile cranes are also suitable for erecting 'shed-type' buildings, provided that the crane can operate within the structure to lift wide-span trusses. When the structure is too high for a normal mobile, the frame can be erected to the limit of the mobile and completed by means of a Scotch derrick or guy derrick mounted on the partially completed structure. Mobiles used for steel erection should have crawler tracks to distribute their load while lifting.

Shed-type buildings vary greatly in span and size but the erection procedure varies little in principle. The first two or three bays of steelwork should be erected and temporarily braced to prevent movement. This initial erection should be checked for alignment and verticality prior to final bracing and tightening-down.

If the above procedure is carried out carefully, the first few bays of steelwork will act as an anchorage for the remainder of the steel erection, care being exercised to include further bracing as the work proceeds. Failure to brace and secure the initial erection can result in the whole mass of steelwork moving and causing irreparable damage to connections.

For simple roof-truss structures, the columns and connecting eaves members are erected and lined to within $\pm$ 10 mm. This is followed by the erection of the roof trusses and purlins, with bracing to the first bay. Shed buildings with wide span roofs may incorporate lattice girders to support the trusses. This will involve the erection of columns and lattice girders before lifting the trusses. Since the latter will have to be lifted from a position outside the completed bay, a longer jib or 'luffing' jib may have to be employed. A small mobile crane has the advantage of being able to lower its jib and travel under the lattice girders. If the lattice girder cannot be lifted by one crane and is difficult to handle using two cranes, one at each end, then it may be lifted by guyed derrick and temporarily supported until the interconnecting trusses are lifted.

The space-frame should be fully erected on the ground and then lifted at the points of support. Some roof frames have been lifted with lattice erection masts; alternatively, if the span is not too great the whole unit can be lifted with four mobile units. Wide-span buildings with space-frame roofing present some difficulty in erection.

Arched roof structures and space frames which are too large to be lifted in one lift should be erected with the aid of trestles. The trestles are erected to give the correct camber, and sections of the arch or frame are lifted into position for final adjustment by jacking methods.

Multi-storey buildings are often erected by tower crane, especially if the crane can also be employed for lifting the claddings and other materials. If, however, the frame is to be erected as a separate contract, the type of lifting appliance may vary. One of the most basic and economic methods of lifting steelwork on high-rise structures is the guyed derrick. The derrick – see Chapter 2 – consists of a latticed steel mast, tall enough to erect two storeys of steelwork without having to be lifted. When being moved, the slewing jib is lifted by the mast to the new high level, and the jib, which is a similar lattice mast, is used to lift the mast. This method of moving the derrick is known as 'jumping' the crane. As with 'shed-type' buildings, the frame requires bracing during erection, which may involve both permanent and temporary braces. The temporary bracing often takes the form of wire cables to reduce distortion during lifting operations: these movements are caused by the slewing and lifting of the derrick which transmits the load to the frame.

An alternative type of crane used in erection is the Scotch derrick – see Chapter 2. This has two major advantages over the guy derrick: firstly it has no guys, and secondly it can be track-mounted and moved around or within the building. The disadvantage of the Scotch derrick is its limitation to a sweep of 270 degrees, which may involve extra cranage to cover the complete erection area. It must be stated, however, that for most purposes either mobile cranes or tower cranes are used for both low and high-rise structures.

8.2.3 Bolting, riveting and welding

Bolting

Bolting offers many advantages over riveting: for example, riveting requires specialist plant and equipment, whereas bolting can be achieved by spanners or torque wrench. Bolting can be carried out in adverse weather conditions, whereas this is almost impossible in the case of riveting. Bolted structures can be dismantled easily, either for demolition or for alteration. Bolting, as an operation, is much quieter than riveting, and is therefore suitable for cases in areas where the noise level created by riveting would be prohibitive. Bolting can be carried out by personnel working on the frame wearing a safety harness, whereas riveting requires some form of temporary staging.

Site connections are now commonly achieved by the use of bolts, which simplifies erection and reduces the erection cost. The most common type of bolt used in site connections is the 'black' bolt. These bolts, made from mild steel, must have a hole clearance so that they can be placed without damage and without over-stressing other bolts already placed; the hole clearance for such bolts is 2 mm greater than the diameter of the bolt shank. Since this type of bolt requires a clearance for assembly, the joint is subject to slight movement and calculations assume a reduced strength factor. High strength bolts may be used in lieu of black bolts on connections which are highly stressed.

High-strength friction grip bolts are bolts made from high-strength steel, which enables them to be highly stressed when gripping together the plates being joined. Such bolts are tightened by a torque wrench which controls the stress being applied. This method of tightening is necessary to prevent the bolt from being over-stressed and to ensure that each bolt has equal loading. The bolt applies a pressure to the clamped faces of the plates, inducing very high friction between the plates. The strength of the bolt and the increased friction between the plates combine to make a joint which can carry much higher loads than the normal bolted or riveted joint.

There are various forms of high-strength friction grip bolts, ranging from general grade to higher grade with waisted shank. Waisted shank bolts shear off at a pre-determined torque. A special torque wrench, pneumatically operated, grips the end of the bolt while the nut is being tightened, and when the required torque has been reached, the end of the bolt is sheared off by the anticlockwise movement of the tool. No further tightening of the nut is necessary. Alternatively, load indicating washers may be used to demonstrate that the bolt has been tightened to the requisite tension.

Riveting

The riveting of structural members, as a site operation, has been superseded by the use of bolts and welding. It is, however, still employed in the manufacture of large composite beams and columns as an alternative to welding.

The ordinary rivet is known as a 'snap-head' or 'dome-head'. It is inserted into the hole in a red-hot condition and is then 'closed' to give the finished shape. The riveting or closing is achieved by pneumatic tools or hydraulic clamps, horse-shoe in shape. On cooling, the rivet shrinks and grips the plates by producing a great deal of friction between them. Where a connection has to bear against another surface, eg a truss base plate against the head plate of a column, countersunk rivets are used. Rivets 19 mm in diameter are normally used for general structural work, but on heavily loaded girders or stanchions, where the number of 19 mm rivets required may weaken the connection plate, rivets up to 25 mm diameter may be used. This extra rivet strength will also result in a saving of gusset plate size, which would need to be larger to receive the standard rivet pitch of small rivets. Riveting is particularly useful for cleat connections, which are normally made in the workshop.

The spacing of rivets (known as the pitch) must be such as to prevent the metal between them from rupturing. The metal will rupture or tear when there is insufficient metal between the rivets to carry the transferred load. This can be prevented by spacing the rivets at a minimum of two-and-a-half times their diameter (measured centre to centre). The end rivet should be one and a half times its own diameter from the edge of the plate.

Welding

Welding is being increasingly used as an alternative to shop riveting and bolting. It is not popular in the UK for site connections on structural frames, for several reasons. Site-welded joints produce a rigid structure which, under load, can cause deflection of the columns and result in adverse stresses being applied to claddings and structural members. Other problems of site welding include:

- access and difficult working conditions at high levels;
- the difficulty of assessing the quality of the welded joints;
- the permanency of the joint, which does not facilitate alterations or demolition;
- the high degree of skill required compared with other methods of jointing.

Basically, welding is the running of molten weld metal into the heated junctions of steel plates to form a continuous member. Certain types of weld require the plates being welded to be shaped to receive the weld metal.

Welds are made in two basic forms; butt welds and fillet welds.

Butt welds are used to join plates end-to-end and are classified according to the shape of the ends of the plates prior to welding. The thickness of the weld is determined by the thickness of the plates being welded; if plates of varying thickness are to be welded, the weld will assume the thickness of the thinnest plate. Single-welded joints, eg the single-V joint, are more easily accessible for welding, but involve a greater amount of heat, which may result in distortion of the plates. Thick plates are best jointed by the double-U section weld, which gives good accessibility and uses less weld metal than V-joints of the same thickness.

Fillet welds are used for jointing plates at right-angles to each other or plates which overlap each other. The weld metal is deposited as a fillet at the junction of the two members. The strength of a fillet weld is calculated on the throat thickness and the length of weld. If the direction of the load is parallel to the length of a fillet weld, it is known as a side fillet weld; if the load is at right-angles to the weld it is known as an end fillet weld. Fillet welds up to 10 mm are formed by one run of weld metal; larger welds are formed by further runs.

Structural work is normally welded by the arc welding process in which heat generated by an electric arc melts the two surfaces of the metal plates and additional metal (called filler) is added to the joint. The joint is completed when the metal has cooled. Oxyacetylene welding equipment may also be used, in which case a flame from a blow pipe heats the surface of the plates to melting point and at the same time melts a filler rod into the joint. The filler metal fuses with the parent metal. Arc welding has superseded oxy-acetylene welding for most site work.

8.3 STEEL AND CONCRETE SANDWICH STRUCTURES

Steel-concrete-steel sandwich (SCS) construction, also known as 'Double Skin Composite Structures', consists of a layer of un-reinforced concrete sandwiched between two relatively thin parallel steel plates which are connected to the concrete infill by welded shear stud connectors (Figure 8.37). This form of construction was originally conceived during the initial design stages for the Conway river submerged tube tunnel as an alternative to traditional reinforced concrete. It has the advantage of both steel and concrete construction.

The potential use of this form of construction is diverse, including sea-walls, submerged tunnels, caissons, liquid containment structures, and other composite structures. Other advantages are:

- speed of construction;
- water tightness without requiring an extra membrane;
- prefabrication of elements at location of convenience;
- in-situ concrete filling resulting in lightweight elements;
- no formwork required;
- high strength and exceptional ductility;
- straightforward design – no complex reinforcement detailing.

The possibility is that this construction will become a valid and economic alternative for tunnel construction in the UK and continental Europe.

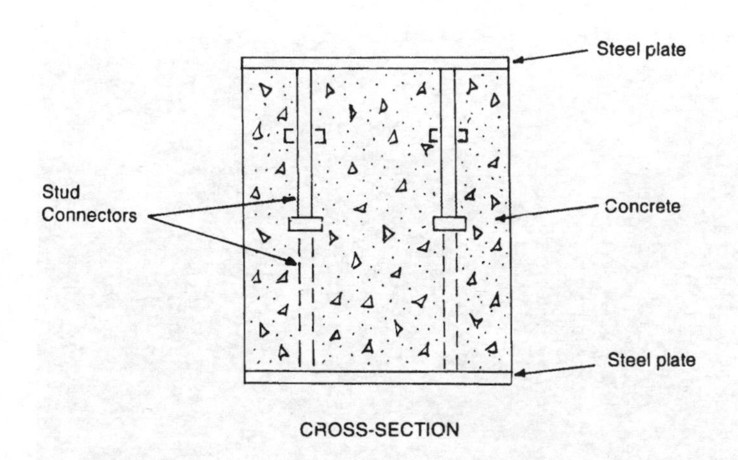

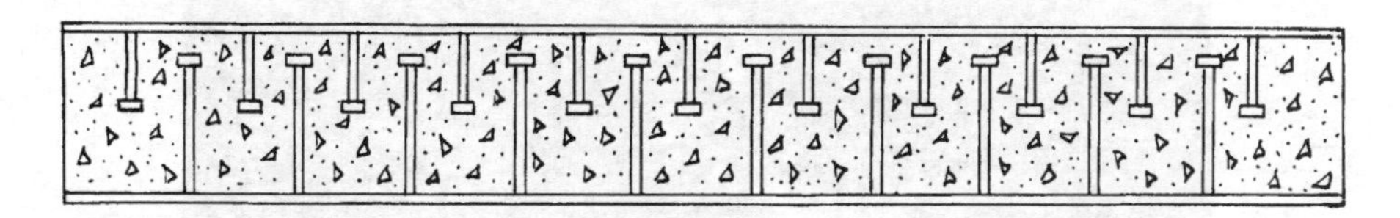

FIGURE 8.37
Typical double-skin composite element
(The Steel Construction Institute)

CHAPTER 9

Services

9.1 PIPELINES FOR WATER, GAS AND SEWAGE

9.1.1 Introduction

Pipelines for water, gas and sewage vary in size depending on the service requirements. This book deals only with pipelines which would be found in a national grid, sea outfall, or other large supply line. The excavation for such work is dealt with in Chapter 2. Backfilling, bedding and other pipeline treatments are discussed here because of the specialised nature of the work.

Pipeline developments affect large numbers of people who own or occupy land, and they therefore involve an encroachment on the rights of the individual. This has led to the development of legislation to control the construction of pipelines. In Britain there are a number of Acts of Parliament which enable various authorities or bodies to construct pipelines, and these and other Acts must be considered before pipelines can be planned and constructed. When all negotiations have been completed, a working width for land pipelines is established. This is the width of the land, an average of 15 to 30 metres wide, depending on the diameter of the pipe, required for executing the work, which is fenced off to protect the public, the owner of the land or his animals (Figure 9.1). The pipes are distributed end-to-end along the route, leaving gaps between them to permit movement of plant and equipment.

FIGURE 9.1
Strip of land fenced off for pipeline activities (Alfred McAlpine Services & Pipelines Limited)

9.1.2 Pipeline materials

Materials for pipelines can vary from ductile iron to precast concrete. The material will vary with the type and purpose of the pipeline.

Water pipelines use the largest range of materials; these include ductile iron, steel, concrete and plastic. Spun cast-iron pipe and spun concrete pipe have been used for many years and their advantages and disadvantages are well known. Two important pipe materials which are being used increasingly in water engineering are ductile spun iron and pre-stressed concrete.

Ductile spun pipes are manufactured from recycled iron by the centrifugal casting process normally used for spun iron pipes. The addition of magnesium alloy to molten iron causes the flake graphite structure to become spheroidal or nodular, which gives high ductility and tensile strength; this improved metal structure produces a stronger, tougher pipe. The impact resistance and inherent toughness of ductile spun iron ensure that accidental damage, which may arise in rigorous working conditions, is no longer a problem. Tests have shown that ductile spun iron will withstand impact blows without sustaining damage. The exceptional strength of the material has resulted in a reduction of the pipe wall thickness compared with grey iron pipes, which makes the pipes lighter and easier to handle. The Stanton Integral system utilises a high-performance push-fit joint, incorporating a special rubber gasket, which combines ease and speed of jointing with leak-proof security.

Pre-stressed concrete pipes are produced in two forms: pre-stressed concrete cylinder pipes and pre-stressed concrete non-cylinder pipes.

Concrete cylinder pipes differ from other types of concrete pipe in that they have an integral steel cylinder. The cylinder is made of welded sheet steel and has spigot and socket rings welded on to each end to provide jointing surfaces; the complete unit is subjected to a hydrostatic test to check the welds. After testing, the steel cylinder is centrifugally lined with concrete, followed by a winding of high tensile wire. The pre-stressing wires are covered with a dense cement mortar coating not less than 20 mm thick. The pipes are jointed by means of a simple push-in joint, known as the lock joint. Extra-strong duty pipes are produced by double winding of the wire, the first winding being coated and further stressed, on reaching the required strength, by a second winding. The second winding is then coated with cement mortar for protection. The advantages of this type of pipe construction include a high factor of safety, a continuous watertight membrane, simple jointing and the elimination of normal protective finishes.

Pre-stressed non-cylinder pipes are concrete pipes which are longitudinally and circumferentially pre-stressed. The core of the pipe is formed around longitudinal pre-stressing wires which provide the stress to the bonded concrete when released. On completion of the longitudinal stressing, the core is circumferentially pre-stressed to withstand pressure and design loads. The pre-stressing wires are covered with a cement mortar to protect them. As with cylinder pipes, the jointing of pre-stressed non-cylinder pipes is achieved by push-in joints.

Pipes for sewage disposal include plastic, GRP, steel and concrete. Plastic pipes are used extensively and have the advantages of light weight, ease of handling, low frictional loss in flow, good abrasion resistance and good chemical resistance. Disadvantages are lower tensile strength than metal pipes and non-resistance to temperature change. They are supplied in diameters ranging from 100 mm to 600 mm and in standard lengths of 6 metres, although 9 metre lengths are available to order. The chemical resistance of the plastic, coupled with the increased flow due to the smooth bore, make the pipe suitable for sewage and water pipelines.

Glass reinforced plastics (GRP) are now used extensively in sewers. The pipe walls are built up in layers (laminates), which normally consist of polyester resin, or other resins, reinforced with glass fibres. Sewer pipes are finished internally with a resin-rich lining to reduce corrosion. The pipes are produced by centrifugal casting, continuous or discontinuous winding.

Steel and concrete pipes are particularly suited to large diameter pipelines extending over great distances. Steel pipes are produced by seam welding methods which produce diameters ranging from 150 mm to 3 metres. High-tensile steel pipes are available in diameters up to 1.06 metres and lengths up to 24 metres.

Choice of pipeline material
The choice of material depends on:

- design of basic pipeline;
- ability of the material to withstand internal and external forces;
- simplicity of jointing and laying;
- durability;
- impermeability and frequency of maintenance.

Pipelines in sewage work are classed as rigid or flexible; the former will fracture before any significant deformation of the pipe has taken place, while the latter will deform significantly and transmit the loads to the surrounding fill. The advantages of flexible pipes for sewage work are also applicable to water pipelines, the incidence of failure in pipes in both cases being much higher than in other types of pipeline. The type of pipe used will affect trench preparation: some pipes require a specially prepared bed and careful backfilling, while others can be placed without special preparation or careful backfilling. Flexible pipes are subject to change in cross-sectional shape due to ground movement, and this may affect the flow. In the main, pipes are tested before backfilling, but certain pipes, eg ductile iron with its very efficient push-in joint, can be laid and backfilled with confidence before testing.

9.1.3 Jointing of pipes
Ductile pipes are normally jointed with push-in joints. The anchored joint is used primarily for gas pipelines at pressures of up to 8 bar. Steel pipelines may be jointed by welding or more conventional methods. Welding is used on pipelines where 100 percent line-tightness is essential for reasons of safety. The field welding of pipelines should comply with the relevant BS or with a specification of equivalent or higher standard than the British Standard. Butt welding is normally employed for gas and oil pipelines, but site butt welding is not an acceptable method for water pipelines because of the risk of corrosion on the internal face of the welded joints. Steel pipelines for water may be constructed by welding spigot-and-socket pipes; the welding is usually applied to both the inside and outside of the pipe at the joint, although the internal weld may be omitted on small diameter pipes. Welding may be carried out by automatic welding machines which operate either externally or both internally and externally on any one joint; this double-welded joint offers extra protection against corrosion. Other methods of jointing steel pipes include flanged joints, screwed joints and proprietary joints with sealing rings of rubber.

Plastic pipelines may be solvent-welded, connected by coupler and rubber rings, or they may have spigot and Z-socket ends. The coupler and rubber ring joint is suitable for pipes of up to 300 mm diameter; the other jointing systems are suitable for all diameters.

Concrete pipes subject to movement are usually jointed by some form of flexible pipe joint which employs a gasket of rubber or synthetic material, depending on the use of the pipeline. Pipes may, however, be jointed by socket and spigot joints, and ogee joints, both of which are self-centring and thereby eliminate the use of spun yarn. When the pipes are winched together they are automatically held in position for the jointing operation to be carried out with ease. Self-centred joints are normally grouted or filled with cement mortar, grouting being achieved through grout holes in the pipe collar.

Pre-stressed concrete cylinder pipes are jointed by means of a rubber gasket, known as the 'lock joint'. The joint has a circular-section rubber gasket located in a groove specially formed on the pipe spigot: as the spigot is pushed into the socket the gasket is compressed to form a watertight seal.

Anchors

All pressure mains fitted with flexible joints should be provided with anchorages or 'thrust-blocks' at bends, tees and capped-ends. These anchorages resist the thrust arising from the effects of internal pressure. Flexible joints offer no appreciable resistance to 'blow-out', and joint failure will result unless restraints are provided. Anchorages, usually of concrete construction, should be designed to take into account the maximum pressure the main has to carry on test and the stress which the surrounding ground will support.

9.1.4 Construction methods

There are two distinct areas of operation in pipeline construction; the first covers pipelines which are constructed on land, and the second deals with pipelines laid under water.

For pipelines on land the pipes are first laid out – an operation called 'stringing' along the route (see Figure 9.1). The pipes should be handled with slings of canvas or other non-abrasive material to prevent damage to the pipes, which should not be dropped, dragged or rolled. Pipes requiring welding or other forms of jointing before being lifted into the trench are placed on timber skids. Long pipes should be cleaned out prior to alignment and jointing.

Trenching is carried out by special trenching machines or hydraulic backacters – see Chapter 2 (Figure 2.8). Depending on the length and location, the joints in the pipeline may be completed before lifting into the trench; this takes place using side-boom tractors or similar plant (Figure 9.2). Alternatively, the pipe may be lowered into position first and then jointed (Figure 9.3). All pipelines should be sealed with night caps at the end of a day's work to prevent ingress of small animals and other objects.

FIGURE 9.2
Hoisting and laying 900 mm welded steel gas line
(Alfred McAlpine Services & Pipelines Limited)

FIGURE 9.3
Hoisting and laying a 600 mm ductile iron water pipe
(Stanton and Slaveley Group)

FIGURE 9.4
Preparing the sand base for a 1.83 m dia steel pipeline

The preparation of the trench for pipes will depend on the type of material used for the pipeline and will range from a bed of granular material to carefully shaped sand beds (Figure 9.4). On no account should hard packing materials used to lift or pack up the pipes be left in position. Heavy pipes may be lifted by pneumatic elevators to remove the initial packing before being bedded on sand bags or granular material.

The trench should be 300 mm wider than the outside diameter of the pipe, additional room being necessary at each joint to provide sufficient room for joints to be made and inspected. In hard or rocky ground the trench should be taken deeper than the required depth and brought to levels or gradient with granular material. Backfilling should be completed as soon as the section of pipe has been tested; this will prevent the pipes floating in case of a storm. Testing should be preceded by cleaning; this is achieved by means of a 'foam-pig' (a cylinder of material, such as expanded polystyrene, driven through the pipe). Pipelines crossing roads, railway lines or rivers should be sleeved, the sleeve being placed by the open trench method, power-driven auger or thrust-boring. The diameter of the sleeve should be a minimum of 100 mm greater than the outer diameter of the pipeline. (See Chapter 5 – Auger boring.)

Pipelines under water, such as sea outfalls, may be constructed or positioned by one of three methods: the pulling method, the lay barge method, or floating and sinking. The pulling method (Figure 9.5) consists of jointing long lengths of pipe on the shore and pulling the completed line into the water by means of winches carried on anchorage barges. The lay barge method consists of a special barge on which the pipeline is constructed and lowered to the river or sea bed. The barge is winched along the pipeline route leaving the completed line in position.

FIGURE 9.5
Pulling twin steel (1170 mm dia) weight coated sewer outfall pipes incorporating diffuser risers
(Alfred McAlpine Services & Pipelines Limited)

The floating and sinking method involves the construction of the pipeline on shore, fitting it with buoyancy tanks on completion and floating it out to position (Figure 9.6 and Figure 9.7). The buoyancy tanks are flooded or deflated as the pipeline is lowered into position by winches on pontoons. A weight coating is used to prevent natural buoyancy of the pipe (see section 9.1.5).

9.1.5 Coatings and linings

Protection of steel pipes

When iron or steel is exposed to moisture, corrosion reactions are inevitable, particularly if the moisture is acidic or contains dissolved salts and oxygen. The reactions are electrochemical and involve the passage of electricity between the moisture and steel. Thus the most obvious, and indeed cheapest, way to prevent corrosive attack is by coating the steel surface with an inert substance which effectively separates the steel surface from the corrosive environment and at the same time resists the passage of electricity which is so essential to corrosion reactions. Provided that the coating is complete, corrosion is prevented, but in practice breaks or 'holidays' in the coating may arise, due to rough handling or snagging during placement. At such places where the metal is exposed directly to the corrosive environment, corrosion can occur unless cathodic protection is applied.

External protections

Bituminous coatings are commonly used on steel pipes; they are applied to tubes by dipping in a bath of molten bitumen or painting with a bituminous solution, and as such are relatively thin. They are suitable for mildly corrosive conditions only, and for above ground where periodic inspection and maintenance painting is possible. They do not afford lasting protection to the external surfaces of pipes laid underground.

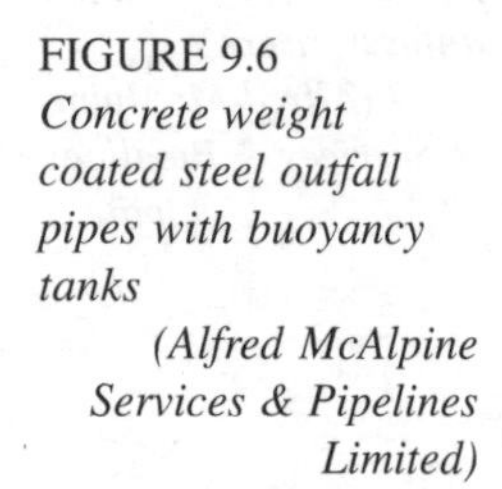

FIGURE 9.6
Concrete weight coated steel outfall pipes with buoyancy tanks
(Alfred McAlpine Services & Pipelines Limited)

'Security'* bitumen enamel wrapping (up to 2032 mm o.d.). The pipe is primed to relevant BS and is then given a flood coating of hot bitumen enamel, a mixture of blown bitumen and slate flour, supplemented by an inner spiral wrap of glass tissue fully immersed in and impregnated by the enamel. A further spiral wrap of pre-impregnated, reinforced glass tissue is applied to the outer surface. The surface is given a final heat reflecting coating of lime-wash. 'Security' bitumen enamel wrapping is used on pipes which are to be laid underground, eg water, gas, oil mains and sewers. It affords protection against all types of corrosive soil and also against stray-current electrolytic corrosion.

'Security' reinforced bitumen enamel wrapping (up to 2032 mm o.d.). This protection is similar to 'Security' bitumen enamel wrapping except that the outer wrap has incorporated in it additional reinforcement in the form of a woven glass cloth. The reinforced protection provides increased resistance to mechanical damage during transit, storage, handling and laying.

'Security' tar enamel wrapping (up to 1118 mm o.d.). The pipe is first primed and is then flood-coated with hot coal-tar enamel. An inner spiral wrapping of glass tissue and an outer wrap of coal-tar impregnated glass tissue are applied. (Bituminous and coal-tar wrappings are not normally recommended for use at operating temperatures exceeding 38°C. They can, however, be used for operating temperatures over 38°C and up to 66°C if the pipe is embedded in sand containing no sharp stones).

'Security' plastic cladding (Securiclad) (up to 324 mm o.d.). Steel pipe is flood-coated with an even layer of hot-melt adhesive undercoat. Immediately following this, a seamless plastic sheath of high-density polythene is continuously applied. In the event of minor damage to the plastic cladding, the adhesive has the ability to flow and seal the areas of damage. Standard colours are yellow for gas mains and services, black for water and light brown for oil service distribution pipes. Plastic-clad pipes are supplied

* 'Security' is a heavy duty protection which may be of 6.4 mm thickness on the larger pipe sizes. 'Security' as applied to bitumen and coal tar protections is a registered trade mark and 'Securiclad' is the registered trade mark for security plastic cladding.

FIGURE 9.7
Sheet piled entrance to sea for pipe pulling operation
(Royal Boskalis Westminster nv)

screwed and socketed in sizes up to 114 mm o.d. and with plain ends or ends bevelled for welding in sizes up to 324 mm o.d. The pipes may be bent cold, provided suitable precautions are observed.

'Security' plastic cladding is suitable for operating at temperatures ranging from minus 40°C to a maximum of 79°C, providing that at the high temperatures the pipe system is adequately engineered to accommodate the slight softening of the polythene which occurs. The high density polythene cladding affords excellent protection for steel pipes against all types of corrosive soil and also against stray-current electrolytic corrosion. It is unaffected by most alkalis, salts and acids, has negligible water absorption, is highly resistant to fungi and bacteria, and in addition, as site experience has shown, possesses excellent resistance to damage by abrasion and impact.

Testing of applied protections

Strict quality control of the materials used and care with each operation in the application is practised to ensure that the wrappings and claddings are uniform in thickness and tight on to the pipe wall. A high voltage test is applied to the protected pipe to ensure that no pinholes, thin places or 'holidays' occur.

With pipelines and other buried structures, corrosion may be induced by the heterogeneous nature of the soil around the structure. Variations in salt content, oxygen levels and water content result in differing electrical potentials being set up at the soil/ metal interface, and the formation of galvanic cells.

At areas of lower potential, positively charged metal ions leave the metal and electrons flow to areas of higher potential. Areas of low potential which lose metal ions and where corrosion occurs are termed 'anodic', and the areas of higher potential where corrosion does not occur are termed 'cathodic'. The distance between cathodic and anodic areas can vary immensely.

The object of cathodic protection is to make the whole of the pipework cathodic, with no anodic areas, so that corrosion does not take place. In practice the required electric current may be obtained in one of two ways. The pipework may be connected to anodes composed of a more readily corrodible metal such as magnesium or zinc. These corrode preferentially and the flow of electrons to the pipe will render the pipe cathodic. Such an arrangement is known as a 'sacrificial anode system'. Alternatively the flow of electrons to the pipe may be obtained from direct current generators or rectified mains current. These are connected to the pipework and also to electrodes of graphite or silicon iron which are held anodic to the pipework. Such an arrangement is termed an 'impressed current system'.

The question often arises whether it would be feasible to lay a pipeline without any protective coating and to rely entirely on cathodic protection to prevent corrosion. While this is technically possible, it would be uneconomical. It is now accepted that cathodic protection obtained either by impressed current or sacrificial anodes is most economically employed in conjunction with high quality applied protections. The reason for this is that current consumption is in proportion to the areas of metal exposed by accidental damage, or other cause, and this is least with good quality coatings.

Because of problems of installation and maintenance, cathodic protection applied to steel pipes is almost always concerned with the external surface.

Internal protections

Bitumen lining (up to 2134 mm o.d.) is a heavy-duty protection and can be 6.4 mm thick for large pipes. The surface preparation consists of the thorough removal of mill-scale by shot blasting or by pickling and phosphate coating by dilute phosphoric acid.

After treating the pipe with a primer, the requisite quantity of bitumen to provide a lining of the required thickness is introduced to the pipe and applied centrifugally; cooling by water-spray then occurs until the lining has set.

Steel pipes protected internally with bitumen lining are suitable for the conveyance of raw and potable waters, sea water, sewage and highly contaminated effluents. Apart from its protective efficiency, a bitumen lining provides a very smooth surface, having minimum resistance to flow. Maximum throughput is maintained throughout the service life of the pipeline since pipes are immune from deterioration by nodular encrustation, which is the chief cause of diminished carrying capacity in old, unprotected pipes. Bitumen-lined pipes are not normally recommended for use at operating temperatures exceeding 38°C. Neither should they be used where oil or hydrocarbon solvent contamination is likely.

Plastic lining (up to 2134 mm o.d.). This form of lining can be produced from thermosetting epoxy-phenolic paint, by applying the paint in successive coats to build up the required thickness. Plastic-lined pipes with a coating thickness of 0.254 mm are suitable for a wide variety of highly corrosive media including acidulated brines, sea waters, mine waters and acid solutions with a Ph range of about 2 to 5. For less corrosive media such as certain brine solutions, alkalis and slightly acid solutions, a coating thickness of 0.127 mm is normally adequate. These linings are suitable for the lining of oil-well tubing to diminish maintenance time due to wax deposition, and also for large diameter pipes conveying material which must not be contaminated.

Epoxy-based paints are used for gas mains as distinct from service pipes; approved epoxy red lead or epoxy red oxide paints are standard alternatives to red lead paint.

Cement lining (pipes from 75 mm to over 3 metres o.d.). Steel pipes may also be protected internally with a cement lining. This type of lining is suitable for:

- Pipes carrying potable water
- Oil refinery cooling water
- Salt water and fire water mains.

The lining can be applied at factory, stockyard or on site and is useful for reconditioning the interior of existing pipes. The thickness of cement varies from 5 mm to 11 mm. Sulphate-resisting cement is specified for salt water and other sulphate-bearing waters.

Thermal insulation

Thermal insulation may be required for certain pipelines, eg those transporting liquids such as crude oil which can be pumped more easily by raising the temperatures of the liquid and reducing its viscosity. The insulation used may be polyurethane or expanded polystyrene of a minimum thickness of 50 mm.

Weight coatings

Weight coatings (Figure 9.8) are applied to the outer surfaces of pipelines to counteract buoyancy when immersed in water. The coating is usually concrete, and may be sprayed on the pipes by the gunite method or cast around the pipes using special formwork. The thickness of the coating depends on the weight required to sink and stabilise the pipe, and it should be applied over any anti-corrosive coating.

An alternative method of stabilising pipes has been achieved by concrete weights attached to the pipes or wire ropes anchored to concrete blocks at the river or sea bed. Attachment to piles may also be used as a means of anti-buoyancy.

Laying of protected pipes

In laying and backfilling every care should be taken to avoid damage to the external protection. Blemishes should be examined by a 'holiday' detector, and all damage repaired. The soil on which pipes are bedded and that used for backfilling the sides and top should be free from stones and rock fragments and carefully tamped round the pipes. Care should be taken to ensure that rock fragments etc do not come into direct contact with the pipe during backfilling, otherwise damage may occur to the protective coating during subsequent tamping operations.

In rocky terrain, where the provision of suitable backfilling material would be difficult and probably costly, the use of armour-wrap or rock shield is sometimes introduced. Various forms of armour are available, but generally they are of thick bitumenized felt incorporating either fibre or glass tissue.

FIGURE 9.8(a)
Outfall pipe complete with weight coating and buoyancy tanks (Alfred McAlpine Services & Pipelines Limited)

FIGURE 9.8(b)
Weight coating being cast prior to pipe launching (The dredging and Construction Company Limited)

9.2 SEWAGE TREATMENT

9.2.1 General considerations

The authorities responsible for sewage treatment deal with both surface-water, which is the run-off from rainfall, and foul sewage, which covers domestic wastes and trade effluents.

Foul water is conveyed to points of direct disposal or treatment plant; the conveyance system consists of gravity sewers, which may be pipes or open channels in which the flow is not under pressure. Inspection chambers are provided at regular intervals for access and cleaning, and also at any change of direction. Gravity sewers are always laid in straight lines between manholes. If excessive flow develops in a gravity sewer so that the level of the sewage rises in the chamber or perhaps overflows, the sewer is said to be 'surcharged' – a most undesirable state.

Where sewage has to be conveyed uphill, pumping stations are necessary. The sewage is pumped up a pressure pipe, known as a rising main, either into a gravity sewer or direct to a disposal point. Modern practice, in rural areas in particular, is to limit the number of small treatment plants in favour of a central treatment point, with an extensive sewerage system which may have more than twenty pumping stations.

The various systems include separate, combined or partially separate methods of control.

In a **separate** system the surface water is completely excluded. The design of a foul-water system is based on the Dry Weather Flow (DWF), which is made up of the domestic flow, trade flow and infiltration, ie leakage into the sewer from the surrounding soil.

The domestic flow – the flow per day per head of population – is usually taken as 180 litres/day; in rural areas a figure of 150 litres/day may be used, while in modern residential areas the figure may rise to 270. The trade flow from factories, food processing etc might be of any volume and from many different sources. In some cases the discharge may take place throughout the full day, ie 24 hours, while in others it may be concentrated into short periods of time.

Infiltration from the surrounding soil should be very small in new sewers, but in old systems it could be appreciable. It is then a question of economics whether to re-lay the sewer or accept the extra flow. As the flow varies during the day, the system should be designed for a maximum likely rate of flow; depending on circumstances, a separate system is designed for a maximum flow of four to six times the DWF, and pumping stations are designed accordingly.

In a **combined** system the surface water falling on the drainage area is accommodated. Sometimes only part of the surface water is taken into the system, eg a partially separate system whereby road drainage and roof drainage from the front of the houses only is taken into a storm sewer, while the foul drainage and the rear roof drainage is taken into the combined foul sewer. This economises on the amount of drainage pipework needed to be provided to the rear of property.

In the combined system the flow due to surface water is generally far greater than the DWF and is a very complex matter to determine; the main point to decide is what intensity of rainfall has to be accommodated. The greater the storm, the more seldom it is likely to happen, and it is a question of economics what intensity is to be adopted. If excessive rainfall is unlikely to cause any serious damage, the system is designed for a one-year storm, ie a storm likely to occur only once a year. If, on the other hand, flooding could have serious consequences, the system might be designed for a five-year or even a fifty-year storm, ie the heaviest storm likely in that period of time.

To design a combined system to convey water all the way to the disposal point can be exceedingly expensive, and therefore, where excess surface water can be safely discharged, overflows are provided at convenient points adjacent to rivers. The design of these overflows is based on a formula incorporating the three factors making up the DWF, which works out at about six times the DWF. Flows in excess of this usually pass out through screens which trap any floating debris. Similarly, when the flow has to be pumped the pumps are designed on the same basis, with overflows to accommodate any excess.

A refinement in storm overflows, particularly where there is a high percentage of road drainage, is the provision of storage tanks, or lagoons. When a storm occurs, particularly after a long dry spell, the first run-off from roads is very noxious and could pollute the stream into which it discharges. Tanks can be provided to retain the first flush, overflowing into the stream only when they are full and the discharge has become innocuous. After the storm the tanks empty back into the sewer.

9.2.2 Methods of treatment

Sewage can either be treated before discharge to a suitable river, or discharged without treatment direct to the sea or large river. Direct discharge has been extensively used in the past and still exists in some seaside towns where short outfalls, ie those outfalls which are constructed on the beach but do not reach low tides, have caused extensive and offensive pollution. Such outfalls are being replaced as quickly as possible.

Direct discharge to the sea is contemplated if hydrographic conditions are favourable and often this option is now taken. Any proposed outfall should be subject to the most rigorous investigation to ensure that the point of discharge will not cause nuisance or allow sewage to drift back on to the beach. Such outfalls might be several kilometres long, formed either in tunnels or by pulling pipes in long lines from an assembly area (Figure 9.9) and laying them in a trench excavated in the sea bed. In sheltered waters the pipeline may be buried under a sand or gravel foreshore; alternatively, it may be placed on concrete pads (Figure 9.10) or stub columns and held in position by metal clamps. If the foreshore is rocky the concrete pads are anchored to the rock by reinforcement, which is fixed to the rocky base during low water. The sewage would at least be screened and macerated to remove visible evidence before passing through to the outfall; settlement tanks, as described later, might also be incorporated into the system. In certain circumstances storage tanks are provided so that sewage is retained for discharge only on the ebb tide. In some cases existing outfalls may be retained but this should be in conjunction with new treatment plant so that only fully treated effluent is discharged into the sea.

The object of sewage treatment is to render sewage fit to be discharged to a watercourse. When sewage is discharged to a watercourse there is a reduction in the concentration of dissolved oxygen in the water, owing to its absorption by bacteria living on and breaking down the organic matter of the sewage. If the organic load is too great, all the dissolved oxygen may be lost and the stream will then putrefy. The amount of organic matter is measured by the Biochemical Oxygen Demand (BOD), which expresses the amount of oxygen, expressed as parts per million or mg/litre, used by a sample of the sewage when incubated for five days at a temperature of 20°C. Another criterion is the amount of suspended solids in the sewage, again expressed in parts per million.

These criteria were originally devised by the Royal Commission on Sewage Disposal early in this century; for normal conditions the standard of effluent is known as the Royal Commission Standard and is 20 ppm BOD and 30 ppm Suspended Solids (SS), commonly known as the 20/30 standard. Ammonia is now a common constant parameter, typically at a level of 5 to 25 mg/l, which requires treatment loadings and configuration capable of oxidising incoming ammonia to nitrate – this is known as nitrification. Furthermore, in designated areas of the UK constant standards may also be imposed to limit nitrate and phosphate content, and also toxic substances and metals falling into EC List I and List II categories.

The standards discussed so far apply to domestic sewage. Where trade effluents are concerned the situation may be far more complicated, as special standards are imposed to regulate the discharge of dangerous or noxious chemicals. Before certain trade effluents are accepted into a public system, the manufacturers concerned may have to provide some form of pre-treatment to an agreed standard; alternatively, they may be required to contribute towards the cost of the extra treatment necessary at the treatment works.

FIGURE 9.9
Sea outfall pipes on platform ready for assembly
(Alfred McAlpine Services & Pipelines Limited)

FIGURE 9.10
Sewer outfall held in position by concrete anchorage bases
(The Dredging and Construction Company Limited)

Stages of treatment

In conventional plant the treatment is in two stages: primary and secondary.

The primary stage consists in passing the sewage through screens to remove large debris. Such screens may be of many different forms, usually incorporating some form of mechanical raking which removes the material, perhaps washes it, and dumps it ready for incineration or burying. An alternative to screening is comminution. This is a process whereby the solids are cut into small particles by comminutors, which are revolving vertical drums with cutting edges; after passing through these the shredded solids continue in the sewage flow. In combined systems grit may be present in the flow, washed off the roads or land, and this is removed by constant-velocity channels. These channels allow the sewage to flow with such low velocity that the grit settles out; the grit is then mechanically removed. There are also various patent forms of grit remover.

The sewage then flows into settlement tanks where the solid matter can settle out to form a sludge at the bottom. These tanks are circular with conical bottoms. Where the sludge is moved towards the centre with scrapers revolving on radial arms, the sewage enters at the centre of the tank and flows outwards (radial flow). The sludge is drawn off at the bottom, usually by a hydrostatic head, and pumped away for further treatment. In the past rectangular and square tanks were used but these are no longer constructed, being recognised as inefficient and labour-intensive.

The object of the secondary stage of treatment is to bring the sewage, in the presence of air (oxygen), into close contact with a large mass of purifying bacteria which will break down/oxidise the polluting matter present. There are two main methods of achieving this: percolating filters and contactors or activated sludge plants.

Filters and rotating biological contactors

Trickling filters, often known as the percolating filters or bacteria beds, are the oldest and still the most widely used. They consist of 2 m deep beds filled with 16 mm to 100 mm sized media, generally of slag or similar inert material; alternatively, plastic media can be used. The media provide the environment for an effective biofilm to develop, which is a complex food-web ranging from bacteria to fly larvae which maintains a balance within the film. The media promote the biofilm growth whilst at the same time allowing the free upward passage of air.

FIGURE 9.11
Filter bed unit assembled prior to filling with media (Birchwood Concrete Products Limited)

Filters are built in either a circular (Figure 9.11) or a rectangular shape. In either case the settled sewage is fed at a controlled and even rate on to the surface of the medium, through which it trickles slowly to be collected at the bottom by under-drains. On rectangular beds the sewage is fed through travelling distributors which are rope-hauled backwards and forwards along the top of the beds by means of a mechanical winch. On the more common circular filters, revolving distributors are used (Figure 9.12). These can be electrically driven but more usually rotate by the force of the sewage issuing from small jets on the distributor arms. It should be noted that the term 'filter' is a misnomer as the process does not filter anything.

The effluent from these beds contains fine particles of the biological film which forms in the bed, as well as other solid matter; this is known as 'humus' and is removed by passing the flow through a 'humus' tank. Humus tanks are very similar to settlement tanks and are often the same size; the effluent, having passed through these tanks, normally passes to the watercourse. If a higher than normal standard of effluent is required, a tertiary treatment stage may be employed, the effluent from the humus tanks being given further treatment by passing it over grass plots or micro-straining through very fine filters. Sludge from the humus tanks is removed as in the primary tanks and may be kept separate for treatment, but more usually is returned to the head of the works to be settled again, with the primary sludge forming a mixed sludge of primary and humus to go forward for treatment by whatever sludge process is available on the works.

Percolating filters take up large areas of space, and other means of oxidation have been developed for where space is limited. One such development is a range of Rotating Biological Contactors (RBC) currently marketed under various names. These systems normally use bio-media, which consist of specially moulded high density polyethylene discs having diameters up to 3.8 m (Figure 9.13). The discs are mounted on a horizontal tubular shaft of specially coated carbon steel. This shaft is fitted with sub-shafts. A shaft-mounted geared motor is fixed at one end to provide the drive.

Rotating Biological Contactors (RBCs) consist of a series of discs which rotate in a tank of sewage. Pumped or gravity flows of crude sewage enter the RBC primary settlement zone. Settleable solids are retained in the tank's lower region whilst the partially clarified liquor passes forward and makes contact with the immersed lower

FIGURE 9.12
Revolving distributor on percolating filter (Tuke & Bell Limited)

areas of the slowly rotating discs; about 40 per cent of the disc surface is immersed. A layer of biofilm forms on the surface of the rotating discs, providing the environment for effective sewage treatment.

Activated sludge process

The activated sludge process is often used where there is a shortage of land on which to site the treatment works. The process is capable of producing a fully nitrified effluent with a consistently lower ammonia content than conventional filters.

In this process the sewage passes through the same primary treatment as before, ie into primary settlement tanks. It is then mixed with an equal volume of 'activated sludge' and aerated in tanks for a number of hours. The mixture of sludge and sewage is kept mixed and aerated either by injection of compressed air through jets in the bottom of the tank, or by mechanically driven agitators.

The 'activated sludge' consists of a floc which contains a mass of micro-organisms multiplying and living on the organic matter in the sewage. The sludge is separated

FIGURE 9.13
Rotating Biological Contactors (Tuke & Bell Limited)

FIGURE 9.14
Horizontal rotor (Tuke & Bell Limited)

from the purified sewage in the final settlement tanks, from where it is withdrawn and pumped back to the head of the aeration tanks in order to 'seed' the fresh sewage coming into the plant.

As the treatment continues and the micro-organisms multiply, the amount of 'activated sludge' available within the plant obviously increases. It therefore becomes necessary from time to time to withdraw the excess quantity and mix it in with the primary sludge for treatment and disposal.

In general 'activated sludge' plants consume more electrical power than conventional plants and it is often prudent to guard against power failure by providing standby generation on site.

Aeration equipment

One of the needs in the treatment process is the introduction of air/oxygen to the sewage. This is done in various ways, such as the introduction of air through the sewage; this can be achieved by 'fine' or 'coarse' bubble aeration. Alternatively, the sewage can be aerated by means of rotation; this is achieved by vertical or horizontal rotors (Figure 9.14), promoting oxidation. The sewage then passes to settlement tanks where the sludge settles and is continuously drawn off and returned to the head of the works to 'seed' the incoming raw sewage. There are several variations of this patented method.

Package plant

For small plants, particularly where they are only temporary, there are various forms of package plant, which work on extended aeration or contact stabilisation. One such plant is complete in one circular steel tank with air compressors and all control gear. Its great value is the speed of installation: only a firm base is required on to which the plant can be off-loaded and connected up. RBCs can also be used as temporary package plants as well as for permanent installations (Figure 9.15). Such plants can also be installed on a side-stream loop to provide an economical supplement to treatment capacity on a larger plant.

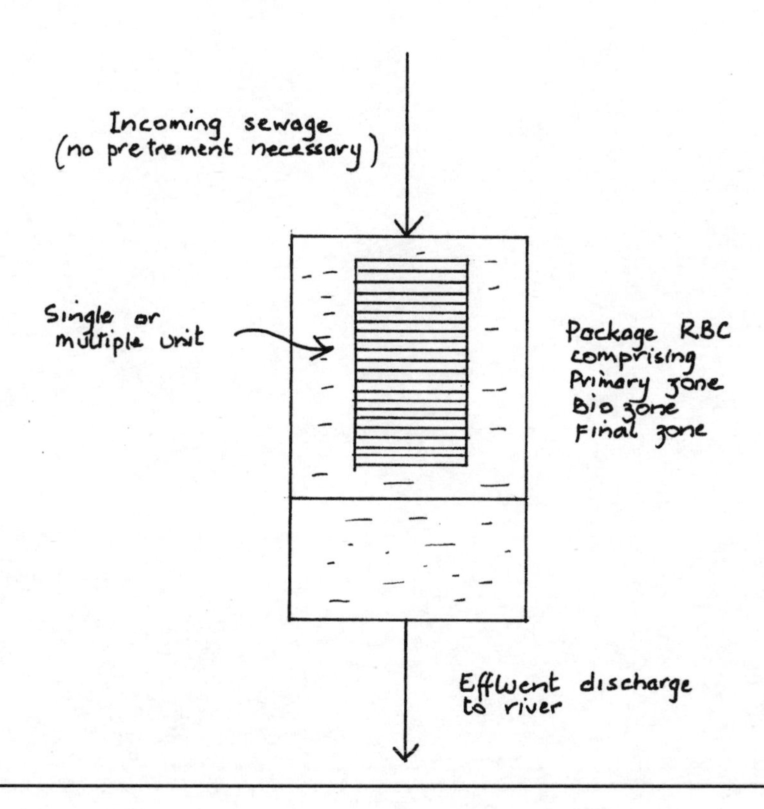

FIGURE 9.15
Package Rotating Biological Contactors

High-rate filters

For preliminary treatment of some trade effluents before they are discharged to a treatment plant, separate 'high-rate' filters are often used. These consist of plastic media packed in various patented shapes, often in high towers, which can accept a very high rate of flow and organic loading without 'choking' of the media by producing excessive growth of bacterial slime.

Sludge disposal

Whatever the process used, there is always a problem in disposing of the sludge. In small rural plants it can be spread on to open sludge beds to be dried, and then possibly disposed of to local farmers. In some cases the liquid sludge can be taken directly by tanker to the land without treatment. Larger quantities require artificial drying by pressing, vacuum filtration or other patented methods. The liquor produced by the drying process is returned to the head of the head of the works for treatment.

In large plants the sludge is treated by anaerobic digestion, ie heating in closed tanks; this converts the sludge into a less offensive humus-like material, with more available nutrients and minerals, which is suitable for agricultural use. The process also produces methane which can be used to generate power, contributing significantly to the power requirements of the whole plant. Sea disposal of sludge is possible under licence but this practice will not be permitted after the end of 1998. After that date sewerage undertakers will need to recycle sludges to land; the option will be to dewater and land-fill, thermally dry or fully incinerate sludges.

Typical sewage treatment layouts are shown in Figures 9.16 and 9.17.

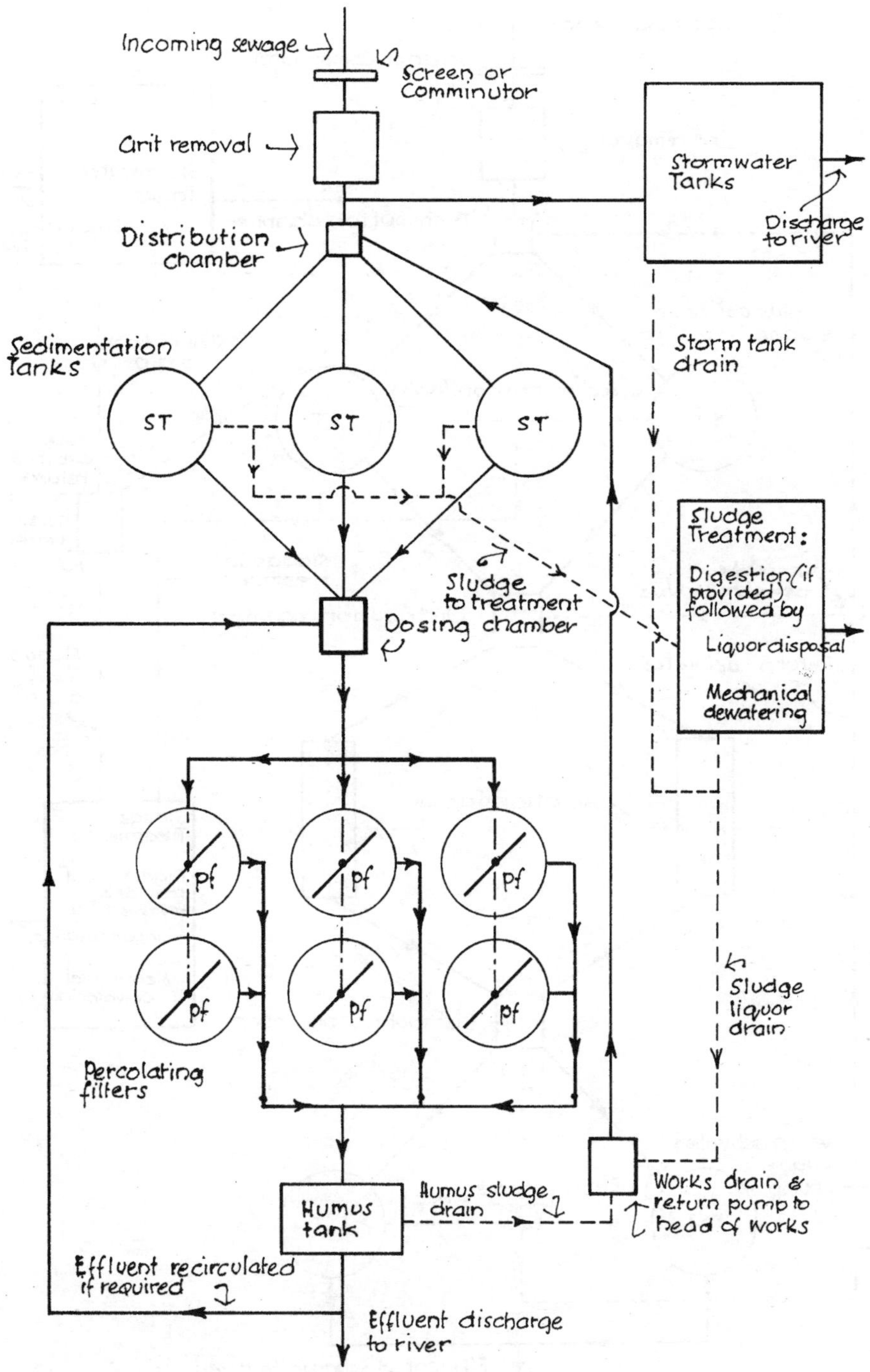

FIGURE 9.16
Traditional filtration sewage system

Incoming sewage
Screen or Comminutor
Grit removal
Stormwater Tanks
Discharge to river
Distribution chamber
Sedimentation Tanks
ST
ST
ST
Storm tank drain
Sludge Treatment:
Digestion (if provided) followed by
Liquor disposal
Mechanical dewatering
Sludge to treatment
Dosing chamber
pf
Sludge liquor drain
Percolating filters
Humus tank
Humus sludge drain
Works drain & return pump to head of works
Effluent recirculated if required
Effluent discharge to river

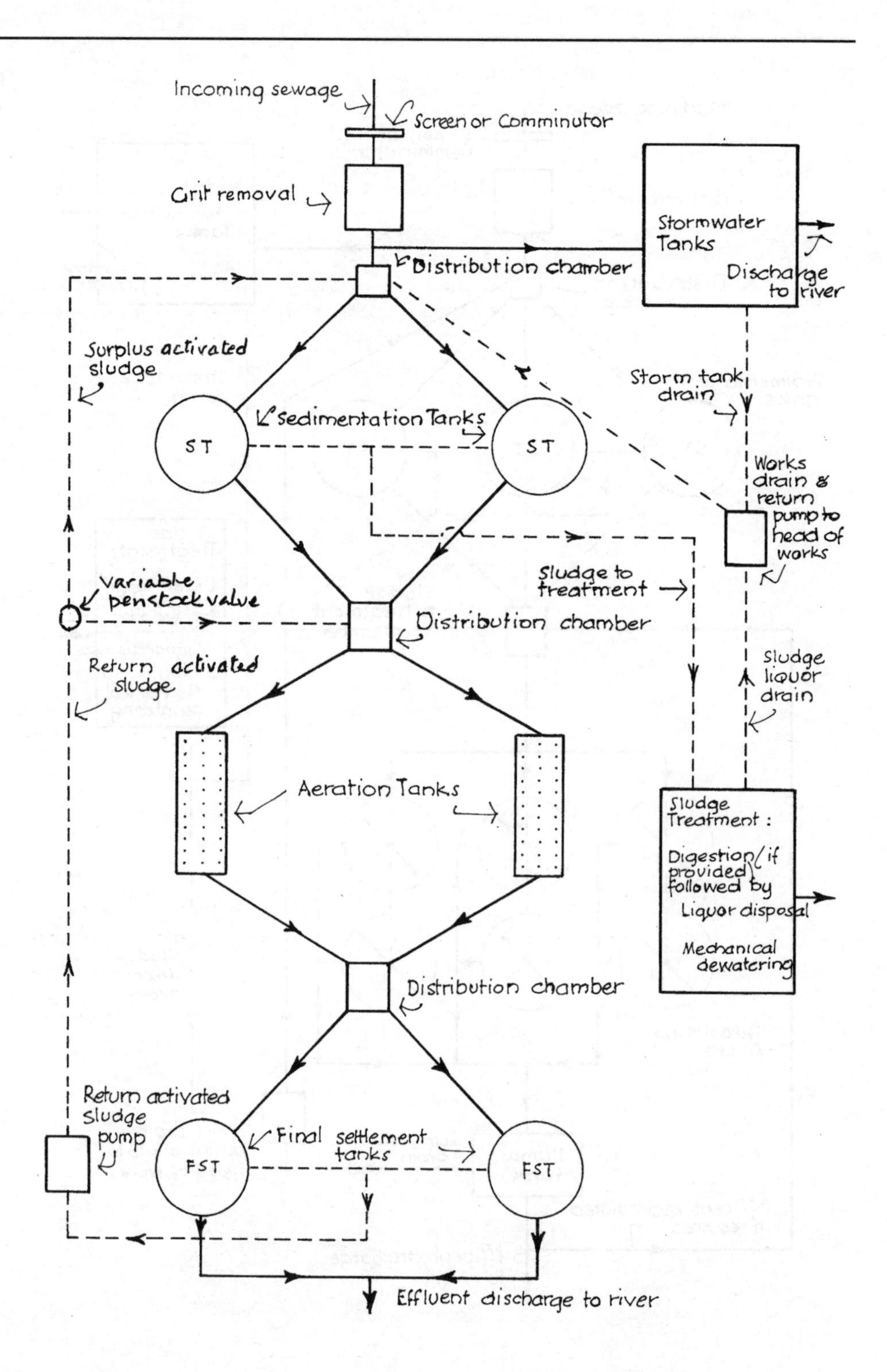

FIGURE 9.17
Layout of aerated sludge plant

CHAPTER 10

Railway Trackwork, Chimneys and Cooling Towers

10.1 RAILWAY TRACKWORK

10.1.1 Introduction

Major future construction of railways in this country will arise as part of co-ordinated transport projects such as Light Rapid Transport, London Underground extensions, Channel Tunnel and Regional rail network projects. They will continue to be an effective means of transport for many years to come, and the amount of work involved in the near future in the modernisation, conversion and electrification of the network will be extensive. In addition, docks, power stations, industrial premises, mines and the like all have independent networks which, taken together, amount to an appreciable volume of engineering work. There is also a constant overseas demand for track and fittings (Figure 10.1).

10.1.2 The permanent way

Types of rail

Various types of rail are produced, ranging from standard 'flat bottom' rail to special rails for mining waggons and electric trains (Figure 10.2). In the UK the two principal types of rail used are Flat-bottom and Bullhead (Figure 10.3); these are available in

FIGURE 10.1
Rails being loaded at Workington docks (British Steel Track Products)

various weights ranging from 31 kg/m to 56 kg/m. The Flat-bottom rail has, to a great extent, superseded the Bullhead rail because it is better suited to heavier and faster traffic. Weight for weight, the Flat-bottom rail is considerably stiffer both vertically and laterally than the Bullhead section; this has resulted in longer track-life, greater stability and reduced maintenance. However, Bullhead rails are easy to fix and unfix to the sleeper, and they are therefore still used in situations where traffic intensity necessitates frequent replacement.

The rails are supported by sleepers of timber, pre-stressed concrete or in some cases steel, which in turn are supported by a ballast foundation. The standard gauge for main lines is 1.435 metres or 1.432 metres for main lines with continuous welded rails on concrete sleepers. The gauge is the distance between the inner faces of the heads of the rails, measured at a distance of 14 mm below the top of the rail (Figure 10.4). The unusual dimension of the gauge possibly resulted from the transfer of the flange from rails to wheels during the 18th century, the 'plateways' of earlier centuries being approximately 1.5 metres between wheels. The acceptable tolerance in standard gauge for safe operation is +8 mm and –5 mm, but railway engineering authorities may insist on closer tolerances than this, depending upon the importance of the track.

Rails are fixed to the sleepers in various ways. The Bullhead rail is fixed in a 'chair' normally by a high-tensile spring key (Figure 10.3); the Flat-bottom rail is fixed with or without a baseplate, depending on the type of sleeper (Figure 10.3). Chairs for Bullhead rails are made to give the rail an inward tilt of 1 in 20, thus bringing the upper surface of the rail into line with the coned tread of the wheels. The same facility is achieved with Flat-bottom rails by incorporating tapered baseplates or by forming a bevel on the sleeper. Both chairs and baseplates are now fixed to the sleepers by bolts, coach screws or lockspikes, or alternatively the baseplate may be cast in place. The spring-spike fixing is falling into disuse with the development of concrete sleepers.

FIGURE 10.2
Track layout for electric trains – London Underground (British Steel Track Products)

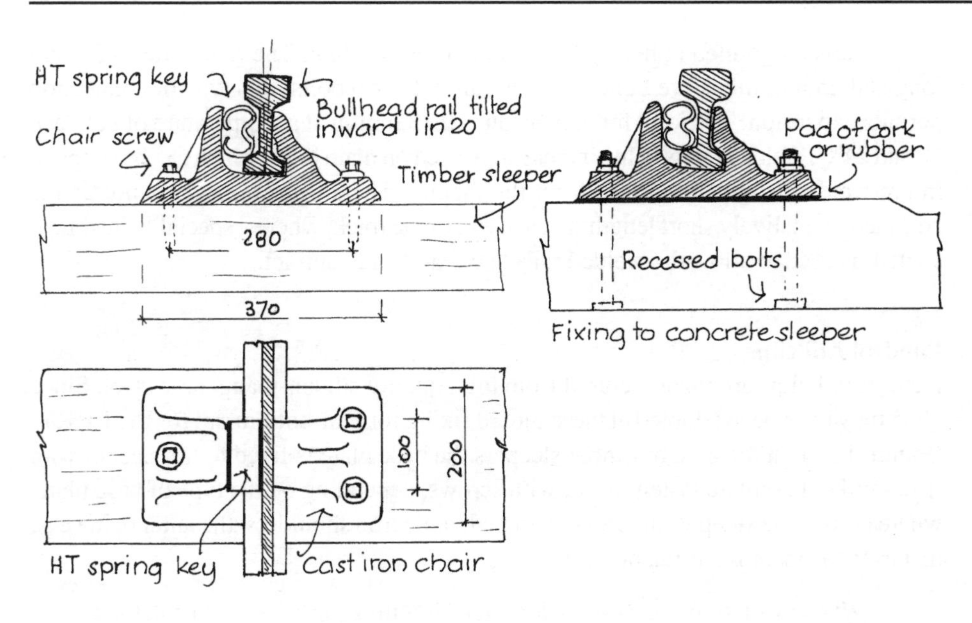

FIGURE 10.3
Types of rail with traditional fixings to timber and concrete

(a) Plan of chair showing Bullhead rail

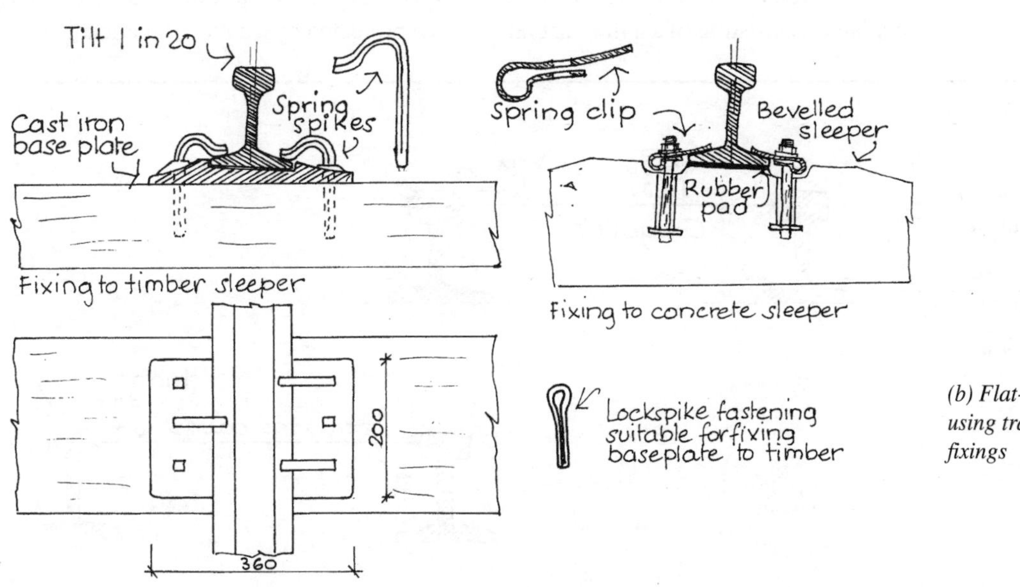

(b) Flat-bottom rail using traditional fixings

Rails are jointed either by fish-plates or by welding. The fish-plates, 450 mm long and 25 mm thick, are bolted to each rail with two bolts; this type of connection permits expansion. The joint for expansion is 0.33 mm/m at a temperature of between 10 and 24°C. Welded rail is being increasingly used on main lines, which can be supplied in pre-welded lengths of 220 m from the factory. The expansion and contraction is confined to relatively short lengths at each end of the track, where a special 'adjustment switch' is used to allow the welded rails to expand and contract.

Pandrol rail clips

Pandrol rail clips are manufactured from high quality silicon-manganese steel. Since 1965 they have been adopted at the standard fixing for Flat-bottom rail for British Rail (Figure 10.5). In the case of timber sleepers the base plate is fixed to the sleeper with 'Lockspike' baseplate fastenings or with screws, depending on the type of base plate, whereas concrete sleepers are fitted with malleable iron shoulders, the latter being cast in during the manufacturing process.

When insulation is necessary for track circuiting purposes, an insulator of hard nylon or other similar material is placed between the shoulders and the edge of the rail foot (Figure 10.6). An extension of the insulator rests on top of the foot rail beneath the 'Pandrol' rail clip. For severe conditions, eg very sharp curves, a composite insulator is available, consisting of a nylon insulating piece protected by a cover of malleable iron.

FIGURE 10.4
Measure of gauge and effect of side wear on gauge

(Top) True gauge

(Bottom) Effect of side wear of curved track

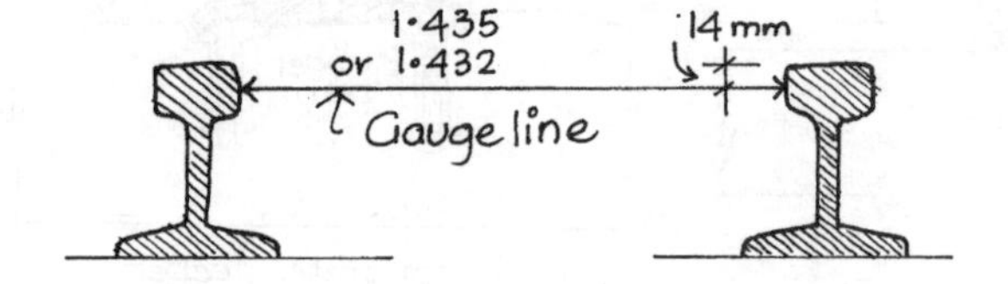

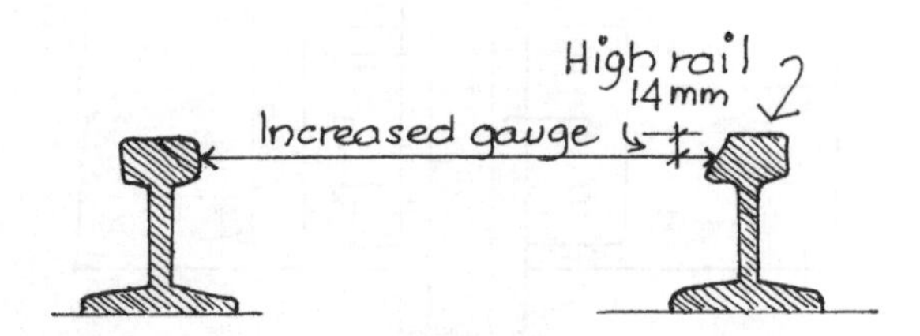

FIGURE 10.5
Standard flat-bottom rail fastenings for timber, steel and concrete sleepers using 'Pandrol' Rail Clips
(Pandrol Rail Fastenings Limited)

Where concrete sleepers are used, insulation is also provided between the rail and the sleeper in the form of a 'rail pad' (see Figure 10.6). This resilient pad provides electrical insulation where required, but its primary function is to dampen the impact forces between the rail and the sleeper. The pad, made from suitable grades of polyurethane or thermoplastic elastomer, are commonly 5 mm thick although other thicknesses are available.

The latest development in rail clips is the 'Pandrol Fastclip' (Figure 10.7), which is claimed to be the world's first fully captive, pre-assembled, unthreaded rail fastening. This means that all components of the fastening are assembled on concrete sleepers at the sleeper factory and remain captive throughout the service life in the track. This eliminates the handling of loose components during track maintenance operations. Tensioning and de-tensioning of the clips is carried out by means of a track-mounted machine, or by hand tools.

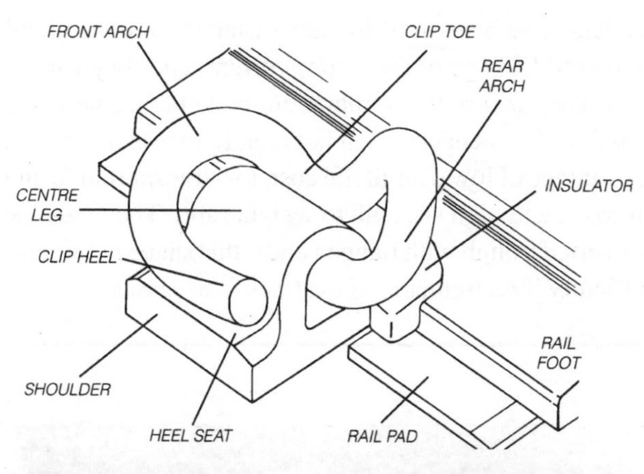

FIGURE 10.6
Sketch showing various parts of the Pandrol clip
(Pandrol Rail Fastenings Limited)

FIGURE 10.7(a)
The 'Pandrol Fastclip' on concrete sleepers – Norwegian State Railways
(Pandrol Rail Fastenings Limited)

FIGURE 10.7(b)
Close-up of 'Pandrol Fastclip'
(Pandrol Rail Fastenings Limited)

Sleepers

It is normal practice to lay 24 sleepers per 18.3 metres of track, although this may be increased to 28 per 18.3 metres for weak formations, curves and continuous welded rail.

Timber sleepers of Douglas fir, Baltic redwood and Jarrah have been used in the past but are being replaced by concrete on all permanent lines. Timber sleepers have a life of approximately twenty years on secondary lines, and are suitable for temporary trackwork.

Concrete sleepers are produced in various classes for general use, and a special sleeper for use on heavily trafficked main lines. These sleepers are pre-stressed and, apart from being more durable, their extra weight increases the stability of the track. These advantages, together with the fact that concrete sleepers can be readily formed to receive the various types of track fixings, have proved that this material is superior to others being used. For increased track stability, 'wing' sleepers are available – see Figure 10.9.

Steel sleepers have been used for some time but are unsuitable on electrified lines, due to increased leakage of the return current, and they cannot be used where track circuits exist. They are used by corporations undertaking work that could seriously affect concrete and timber sleepers, such hot metal processing and chemical spillage. They have the advantage of low overall life cost, low transportation and handling costs, low installation costs, and high strength to weight ratio. They are made of steel plate, formed into an inverted trough with flanged ends; this shape is suitable for strength but creates some difficulty if realignment of the track is necessary.

FIGURE 10.8
Twin-block concrete sleepers with Pandrol rail clip in ballasted track on Sheffield Supertram
(Pandrol Rail Fastenings Limited)

Sleepers are normally 2.5 to 2.75 metres long. Timber sleepers have an average cross-section of 125 mm by 255 mm; concrete sleepers have an average cross-section of 140 mm by 275 mm. However, due to increased thicknesses under the rail seating it is usual to specify sleepers by type and weight.

Twin-block sleepers

With the advent of light transport systems a new concept of sleeper has been developed, this is known as the twin-block sleeper (Figure 10.8). This system is much easier to install than traditional sleepers and gives an enhanced appearance, in keeping with the aesthetic demands of towns and city centres.

Formation and drainage

New works are designed by adopting appropriate alignments and gradients so that the amount of excavation (cut) will balance the amount of 'fill' in embankments.

If the soil is suitable, side slopes (batter) of 1.5 to 1 are commonly used; however, the side slope must be designed to suit the material. The top surface of the formation should be sloped outwards from a centre crown, to provide drainage to the track. The water is channelled away by ditch or other type of drain (Figure 10.9). Where the formation consists of a clay fill or formation it should be protected against ingress of surface water: clay becomes plastic when wet and the rolling stock may then force the ballast into the clay. The protection may consist of a layer of sand, stone-dust or ashes 150 mm to 350 mm thick, laid directly on the formation and covered with polythene sheeting prior to the placing of ballast.

Ballast

Ballast consists of crushed stone, eg granite or limestone graded from 12 mm to 40 mm, and is used as a base for the sleepers. The main function of ballast is to distribute the loads, applied by the sleepers, uniformly to the formation; it also absorbs vibration, provides track drainage and prevents the movement of sleepers. It is laid to a depth of between 225 mm and 300 mm, depending on the traffic load; the ballast is filled to the top of the sleepers and out beyond the end of the sleeper to a distance of between 250 mm and 500 mm, the latter forming a shoulder to the outer edge of the track to restrain lateral movement.

10.1.3 Switches and crossings

Switches vary in complexity from a simple 'turnout' to complicated intersections seen in the approaches to a large terminal. A turnout is a junction formed by one track converging on another either from the right or the left (Figure 10.10). Switches with interlaced left- and right-hand turnouts are termed 'tandem turnouts' (Figure 10.11) or three-ways.

A crossover is a connection between two adjacent tracks and comprises two turnouts connected by a section of straight-track (Figure 10.11); the turnouts may be curved or straight.

A crossing results from one track crossing another and the plan shape gives it the name of 'diamond crossing' (Figure 10.11). In these intersections certain terms are used for the various parts and these are indicated on the sketches by letters.

The frog (F), usually referred to as the common crossing, allows the flange of the wheel to pass where two rails cross. It consists of a 'V' made up of a point rail and splice rail which terminate the intersection of two converging rails (Figure 10.11).

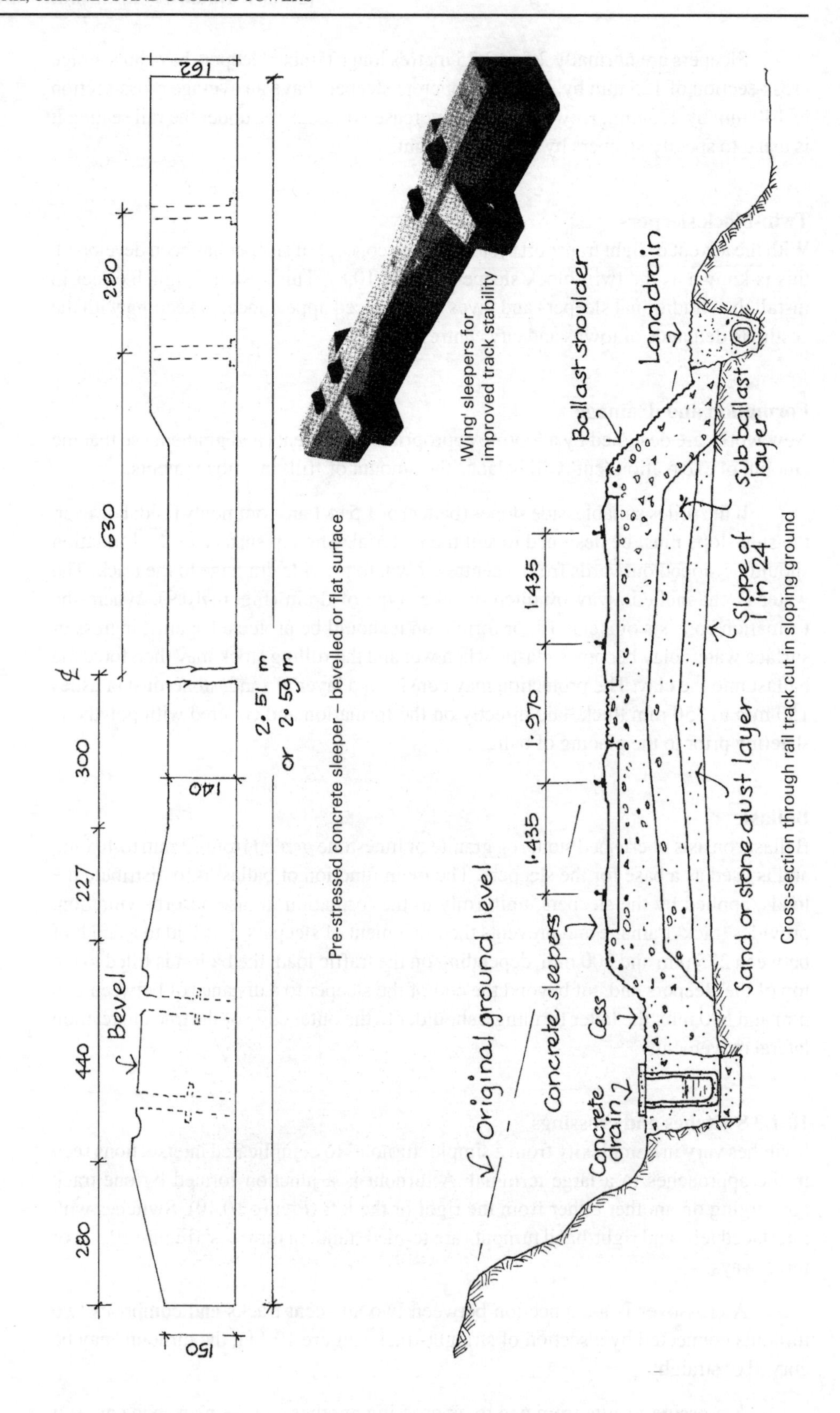

FIGURE 10.9
Details of concrete sleepers, formation and drainage

Wing rails (WR) form an integral part of the frog and terminate the ends of the rails diverging from the crossing.

Check rails (CR), sometimes referred to as guard-rails (GR), are short rails with either bent or flared ends set securely on the rail opposite the frog. They prevent the flange of the wheel on the frog side bearing heavily on the frog while at the same time providing a guide to the correct rail.

Switch rails are movable and are used in connection with turnouts; by changing their position the train is directed to the desired track. The type of switch now in general use is the flexible switch; the short lengths of rail A–B and C–D are the switch rails (see Figure 10.10b) and are connected by a tie-rod to provide simultaneous movement. Work to switches and crossings involve the use of timber sleepers, cut to suitable lengths; such work also incorporates special chairs or base plates and fastenings.

FIGURE 10.10(a)
A general view of rail track showing a series of turnouts
(British Rail Track Products)

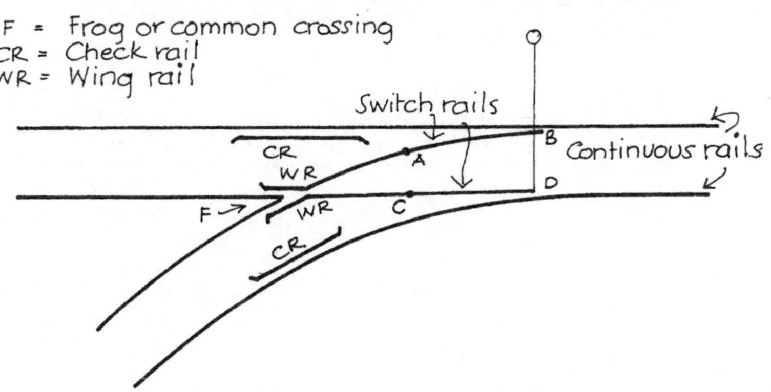

FIGURE 10.10(b)
Details of a turnout

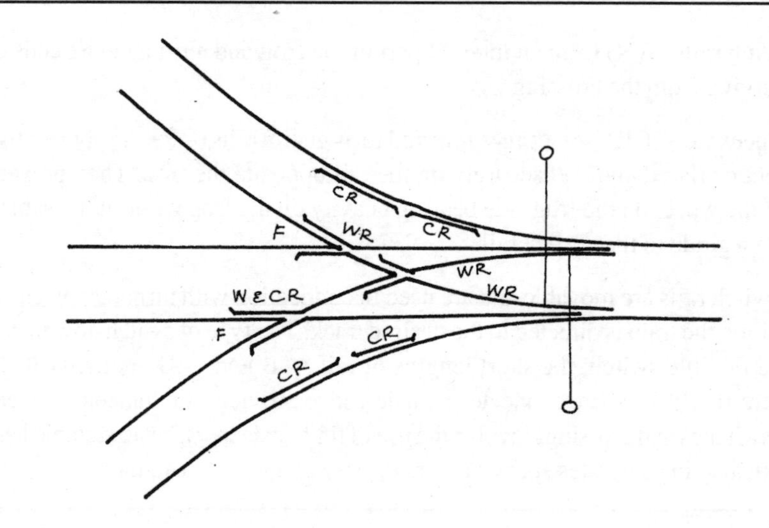

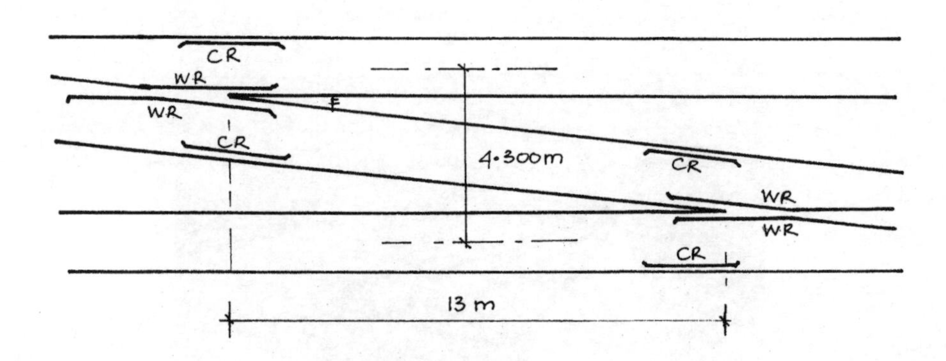

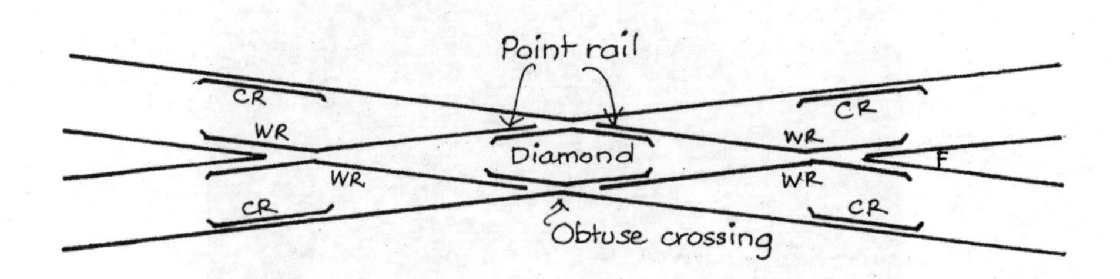

FIGURE 10.11
Sketches of various rail track layouts

(Top) Tandem turnout or threeway

(Centre) Crossover

(Bottom) Diamond crossing

10.2 CHIMNEYS AND COOLING TOWERS

10.2.1 Chimney construction

Design factors

Chimneys are constructed in differing forms and materials to suit the particular conditions for which they are to be used. A chimney is essentially part of the boiler plant, and, as such, its design and construction will affect the efficiency of the plant and the amount of pollution which it passes into the atmosphere. Basically, the design of a chimney depends upon the amount of flue gases it has to handle and the sulphur content of the fuel. A chimney serving plant which burns fuel of low sulphur content is not required to be as high as those serving plant which burns fuel of high sulphur content.

The internal diameter of a chimney should be determined by the volume, velocity, and temperature of gas emitted. The volume of the gas increases in direct proportion to the absolute temperature of the gas, assuming a constant flue diameter; an increase in temperature also increases the velocity of the gas through the chimney. The shape of the chimney flue, eg circular, rectangular or elliptical, and the material used for the inner lining, will also affect its efficiency; this is due to the surface resistance of the material to the gases and the ratio of surface area to cross-sectional area. A circular flue gives the least surface area and steel or plastic liners give the least resistance.

Materials and construction

Chimneys can be constructed in such materials as brick, concrete, steel and reinforced plastic. Brick chimneys have been constructed for many years and have proved extremely successful. In the UK concrete chimneys have been built to heights of 260 metres (see Figure 10.12). Concrete chimneys, both precast and insitu, have been built since the 1920s. They were introduced as an alternative to brick construction, but were not very popular, the main problem being the lack of design knowledge under working conditions.

FIGURE 10.12
Concrete chimney 260 m high at Drax Power Station
(Norwest Holst Group Administration Limited)

As a result not many concrete chimneys, compared with chimneys in other materials, were built between 1930 and 1960. The 1960s saw a growing interest in this material and many small chimneys, ie up to 20 metres high, were built. Now that design knowledge is well advanced, this form of construction is quite common.

Cast-in-situ chimneys are now used on large boiler-plant installations, such as electricity generating stations. The concrete chimney at the Drax Power Station in Yorkshire has three internal flues; the flues are elliptical in section, having axes of 13.7×9.15 metres (Figure 10.13). A typical 2000 MW station has a 198 metre chimney containing four circular flues.

Steel chimneys came into industrial use at the end of the nineteenth century and have been developed to reduce the cost of plant, as steel chimneys are generally far less expensive than chimneys of other materials. These chimneys can be designed and built to great heights: the Americans have a 305 metre high chimney in steel as part of the Tennessee Valley scheme; Japan has steel chimneys up to 230 metres high; and the UK has steel chimneys up to 170 metres high. Steel chimneys of great heights can be built in a complex of three together in a triangular pattern and are braced for stability.

Industrial steel chimneys basically fall into the following categories:

- Single-flue – self-supporting
- Multi-flue – self-supporting
- Guyed chimneys
- Bracketed chimneys.

FIGURE 10.13
Multi-flue construction within concrete shield at Drax Power Station
(Norwest Holst Construction Limited)

Single-flue self-supporting chimneys (Figure 10.14) are the most common; they range from 7 up to 60 metres in height and have diameters ranging from 150 mm to 2200 mm. The design criterion in most cases is deflection; tall stacks are usually of stepped taper design, although stabilisers in the form of helical strakes are usually fitted to the top third of tall stacks to reduce wind-excited oscillations.

The alternative to helical strakes is the pendulum damping system. In this system a pendulum is fixed at the top of the stack, or centrally if a group of stacks is involved. The pendulum moves to counteract the wind oscillations and dampens the stack movement. Pendulums in some form of fluid, such as oil, may also be used.

The anchorage of these chimneys is usually by a number of holding-down bolts cast into a mass concrete foundation.

Multi-flue self-supporting stacks consist of a number of steel flues within a structural steel wind-shield. The number of flues depends on the boiler plant and can vary from two to as many as twelve. The internal flues are designed so that each can expand independently through a terminal weathering plate at the top of the stack. These chimneys are usually parallel throughout their height and can be built to heights of 120 metres and diameters up to 5000 mm. Anchorage and wind problems are similar to those of single-flue stacks. However, their design makes them less subject to wind-excited oscillations.

Guyed chimneys can be directly mounted on a boiler, or supported at ground level. Three or four guys are usually taken from two-thirds of the total height of the stack, and should preferably be inclined at no more than 60° to the horizontal; very tall stacks can have guys at two or more levels. Guyed stacks are usually parallel, since their deflection stresses are small.

FIGURE 10.14(a)
Single bore self-supporting chimney with helical strakes to minimise wind oscillations
(E G Reeve & Sons Limited)

Bracketed chimneys are the same as guyed chimneys in design but are fixed to a restraining wall or tower (Figure 10.15) by means of steel brackets.

Groups of steel chimneys can be used as an architectural feature (Figure 10.15b). Glass Reinforced Plastic (GRP) chimneys were first used in the UK in the 1950s. They have both advantages and disadvantages over the more traditional materials, eg they are ideal for carrying acid fumes but they have temperature limitations. This type of chimney is not built to be self-supporting but is normally supported by wire guys or brackets within a frame.

Treatment of chimney top

The chimney top may require some form of treatment against discolouration: this is normally provided by paint treatment. The concrete chimney at Drax Power Station (Figure 10.12) is painted at the top with a grey bitumastic paint, although it is now current practice to use chlorinated rubber paint; ceramic tiles may also be used. Another form of protection is a skin of dark-coloured facing bricks; these are built around the concrete chimney, recesses having been formed to carry the brickwork.

A further method of top treatment is the 'architectural' feature (Figure 10.16). This type of finish produces an aesthetically pleasing appearance as well as reducing the staining problem.

Chimney foundations

Lightweight chimneys, such as steel and GRP, present no problem in the design of foundations, but concrete chimneys require special foundation design. Power station chimneys with a height of 200 metres can weigh up to 20,000 tonnes; such structures are normally founded on piles or cylinders, the latter being up to 5 metres in diameter. The piles or cylinders are sunk down to a rock formation and capped with a ring beam on which the chimney is built. An alternative method of construction for heavy chimneys is a cellular foundation or buoyancy raft which may be 10 metres deep.

FIGURE 10.14(b)
Self-supporting steel chimneys 52m high in extremely congested conditions
(F E Beaumont Limited)

FIGURE 10.15(a)
Chimney supported by tower
(F E Beaumont Limited)

FIGURE 10.15(b)
Three steel chimneys surrounding a water tower
(F E Beaumont Limited)

FIGURE 10.16
Architectural feature as top treatment to chimney
(Norwest Holst Construction Limited)

10.2.2 Liners and insulation

Whatever the material and construction used, chimneys are subject to three basic types of attack: mechanical, thermal and chemical.

Mechanical attack can occur by natural weathering, ie atmospheric erosion; by wind loading, ie cracks caused by wind stresses; and by abrasion. In the last form of attack the surface of the chimney is worn away by the flow of grit particles from solid-fuel boilers.

Thermal attack is caused by overheating of a structure in which the lining cannot expand sufficiently and where expansion induces stress in the chimney structure. In special cases, eg high temperature flues, a composite lining of fire-resisting material and insulation is necessary.

Chemical attack, in the form of acid condensate, can cause serious damage to chimneys, and therefore insulation should be used to reduce condensation of the fumes.

Multi-flue chimneys (Figure 10.13) have advantages over other types, not only in their stability and efficiency but also in their insulation: the space between the flue liners can be insulated either with loose mineral fill or with mineral wool mattresses. This efficient insulation of multi-flue chimneys has led to their widespread use, the main types being:

- Steel wind-shield with steel liners
- Reinforced concrete wind-shield with steel liners
- Reinforced concrete wind-shield with moler liners.

Many existing chimneys have their life extended by inserting insulated steel linings into the structures. This is particularly advantageous where the existing stack is sound in construction and demolition would be an unnecessary expense. (Figure 10.17).

FIGURE 10.17
Re-lining an existing brick chimney
(Rafferty Industrial Chimneys International Limited)

Mild steel liners are suitable for temperatures of up to 480°C beyond which they are subject to distortion. They are supported on a base or sub-base of the wind-shield, depending on the inlet position; the liner is insulated from its base support. The top of the liner can be fitted with a truncated cone (Figure 10.18), part of which can be manufactured in stainless steel to prevent excessive weathering. If steel liners are to be subjected to temperatures higher than 4800°C, they should be lined with a castable refractory lining which will accept temperatures up to 10000°C.

Steel liners may be insulated with Perlite or mineral wool. Perlite is an exfoliated aluminium silicate which is chemically inert, non-hygroscopic and free-flowing. Perlite insulation should be 150 mm thick; the material is subject to compaction and must therefore be topped up 12 months after installation. Mineral wool is used in the form of mattresses of various thicknesses, ranging from 25 mm to 100 mm, and is sometimes protected on one face by wire netting. The mattresses are wrapped around the liners and held in position by steel strapping.

Moler liners are generally built to one of two designs; the first of these is the parallel type which is built from the base of the wind-shield or from a floor just below the flue inlet; the second is the tapered 'inverted flowerpot' design. In the latter design the moler liner is built off a series of floors 9 metres apart. The advantage of the second type is that it minimises expansion and contraction problems; but it has the disadvantage of increasing resistance to flue gases. Methods of weathering moler liners in concrete wind-shields are shown in Figure 10.18.

GRP liners are ideal for carrying acid fumes, but, as mentioned earlier, they have a temperature limitation. They are suitable for temperatures of up to 2000°C and in some cases are claimed to be operating in temperatures above this; however, there are instances in which such liners have caught fire and been destroyed.

10.2.3 Demolition of tall chimneys

The method of demolishing a chimney will depend to a great extent on the material involved and its proximity to other buildings or obstacles. Steel chimneys should be dismantled in sections in the reverse order to that of erection; this method would also be used for GRP chimneys. Brick and concrete chimneys are best demolished by explosives, if space will permit (see Chapter 8). Where space or other factors prohibit this method of demolition, the chimney has to be demolished by expensive hand labour, involving costly working platforms. Concrete chimneys which cannot be demolished by explosives may be cut down in sections by thermic lance, by concrete sawing, or by concrete grapple if height permits.

10.2.4 Cooling towers

Modern cooling towers (Figure 10.19) consist of a concrete shell supported on a series of concrete struts, and some form of cooling system housed in the base of the tower. Foundations for the tower are shown in Figure 10.19: they consist of a circular 'tee' beam formed by a wide concrete strip and inclined pond wall. This beam is necessary to resist the lateral thrusts. In addition to the 'tee' beam, piled foundations are normally used, the piles minimising any differential settlement which would otherwise lead to cracking in the tower. In a conventional cooling tower the water falls through the cooling pack and collects in a pond at the base of the tower, where the water is retained by an independent base slab and the pond wall.

Purpose and principles of cooling towers

The purpose of a conventional cooling tower (known as a 'wet system') is to reduce the temperature of the water used to condense steam in the condensers; the condensers

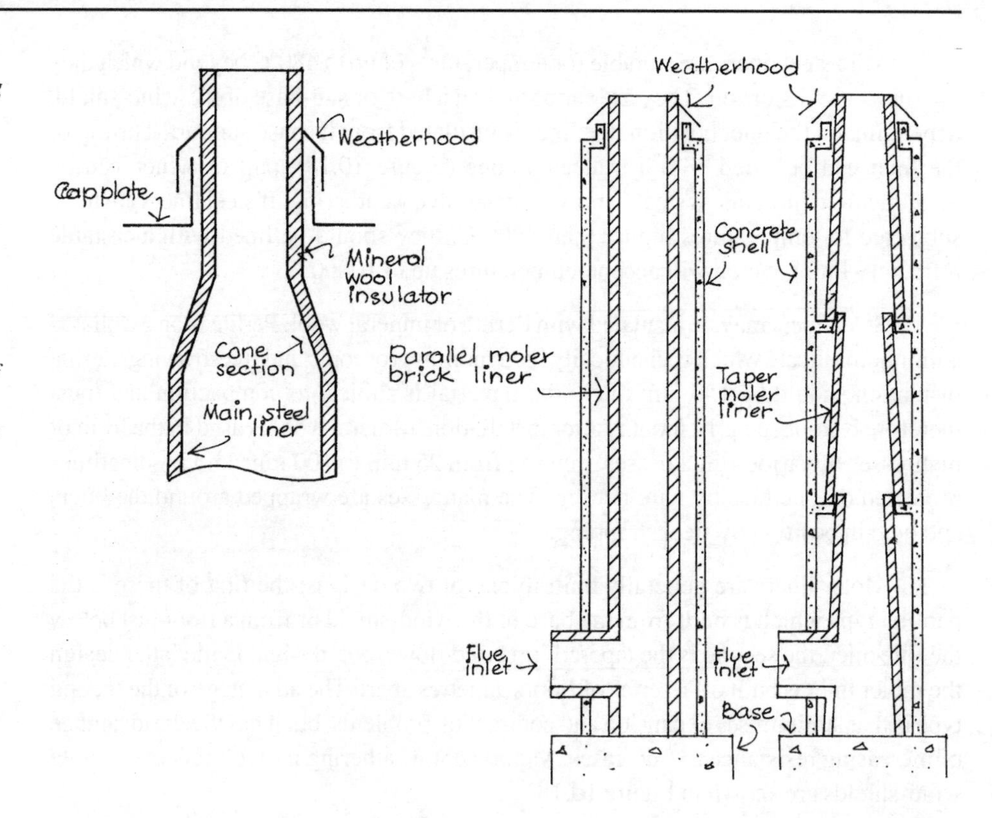

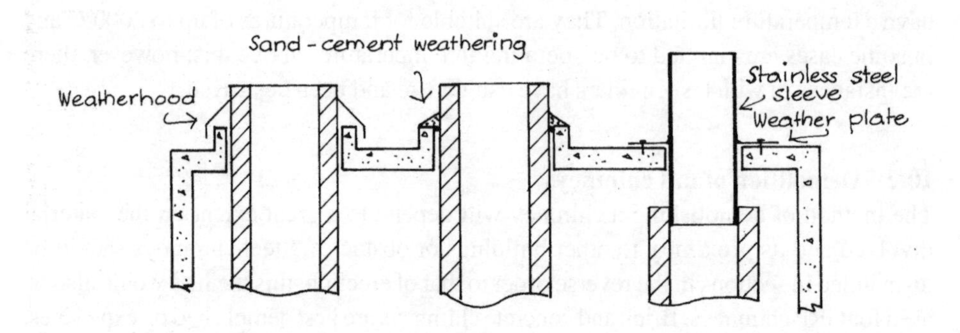

FIGURE 10.18
Lining and weathering chimneys

(Left) Truncated cone section of steel liner

(Right) Two different methods of constructing moler liners

(Below) Three ways of weathering moler liners passing through a concrete cap

FIGURE 10.19
Progress view of construction of cooling tower

(Sketch) Foundation for cooling tower

(Photos) Progress view of construction of cooling towers at Eggborough Power Station

convert exhaust steam from the turbines into water for re-use in the steam cycle. The circulating water passes through the condenser tubes, taking the heat out of the steam, and this is then sprayed over the cooling pack in the tower, falling as cooled water into the pond at the base of the tower for re-circulation to the condensers (Figure 10.20).

The principle involved is as follows: the hot water is distributed by sprays over a large area of cooling pack, which contains a film pack and a splash pack. The packs increase the area over which the hot water runs, as with the film pack, or by breaking the falling water up into fine droplets, as with the splash pack. The main cooling effect is achieved by the water being evaporated by the rising air-stream. This evaporated water rises as a vapour out through the top of the towers, the cooled water falling into the pond. Since the hot air inside the tower is of a lower density than that outside, a suction flow is produced through the tower. The shape of the tower follows the flow pattern of the rising air-stream but does not restrict it, otherwise additional resistance to air flow would be involved. The shape used is the strongest and most economical that can be obtained without interference to the air flow.

This system of cooling loses a small amount of water by evaporation, usually about 1%, and also requires purging to clear away salts which are formed by continuous evaporation and recirculation. The total loss due to evaporation and purging is made up from an outside source such as a river.

A dry cooling system incorporates a closed circulation system in which condensate is recirculated through heat exchangers in the cooling tower and returned to the condenser as cooled condensate. The cooled condensate is sprayed into the steam within the condenser (known as a 'jet condenser', reducing the steam to warm condensate and thereby completing the circuit. A small amount of condensate (2%) is continuously fed back to the boiler, the remaining 98% being recirculated to the dry tower (Figure 10.20).

Natural-draught cooling tower

The shell design is based on the hyperboloid, which has double curvature and can be analysed by modern methods for wind pressure; it can be constructed without ribs and buttresses. The tower, which may be 180 metres high with a base diameter of 150 metres, consists mainly of an empty shell, only the bottom 10 metres being used to house the cooling pack.

The cooling pack, sometimes called the 'stack', is fed with untreated hot water through sprays; the water falls through the pack into the pond area. Heat exchange takes place between the rising air and the falling water, the open structure at the base of the tower allowing a natural movement of air.

Pack construction

The cooling pack consists of strips of material which is used to increase the area between the hot water and the air flow. The material now used in these packs is plastic, whereas in the earlier days they used to be timber or asbestos. The hot water is sprayed on to the pack structure to speed up the cooling process. This is achieved by splashing and breaking up into fine droplets, or by passing the water between corrugated plastic sheets laid vertically on edge. The former is called a 'splash pack', the latter a 'film pack'. The pack is supported on reinforced concrete columns which are founded independently of the tower shell, so as to obviate damage which may be caused by a slight movement of the tower. The pack columns support perforated precast concrete beams into which the triangular-section plastic laths are placed. Above the pack a layer of eliminators is constructed, these being either flat or corrugated plastic louvres. Eliminators intercept the droplets of water which would otherwise be carried up in the air-stream and deposited as fine rain.

Mechanical-draught towers

This type of tower is the one used extensively throughout the world. In this type of cooling tower the draught is provided by fans. Their extensive use includes the oil and chemical industries where it is necessary to maintain or not to exceed a fixed temperature level in all ambient conditions. A close approach, ie the difference between the ambient wet-bulb temperature and the cold-water temperature, or a situation in which the load factor is low, favours the use of mechanical draught towers. Although the cost of fan-power may be high, this type of tower is more efficient than the natural draught tower.

Dry and wet towers

The tower shell is similar in construction but considerably larger than a wet tower of the same rating; the heat exchange system is entirely different. The water is cooled by passing through aluminium tubes situated either within the tower shell or within the air opening. Only one such construction has been built in the UK. The principles of this system are shown in Figure 10.20.

Wet towers require make-up water from a suitable source, due to evaporation and the purging of salts; a 2000 MW station requires a make-up of 6.75 million litres/hr. In this country fuel supplies and water are usually found in reasonable proximity and for the present there is no demand for dry cooling systems. In countries which lack adequate water supplies a dry tower will allow greater flexibility in the siting relative to fuel supplies.

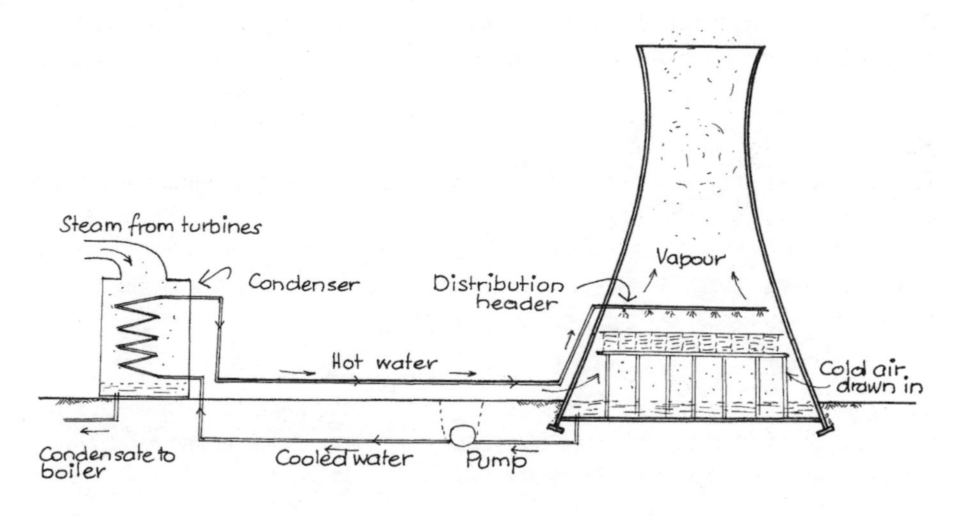

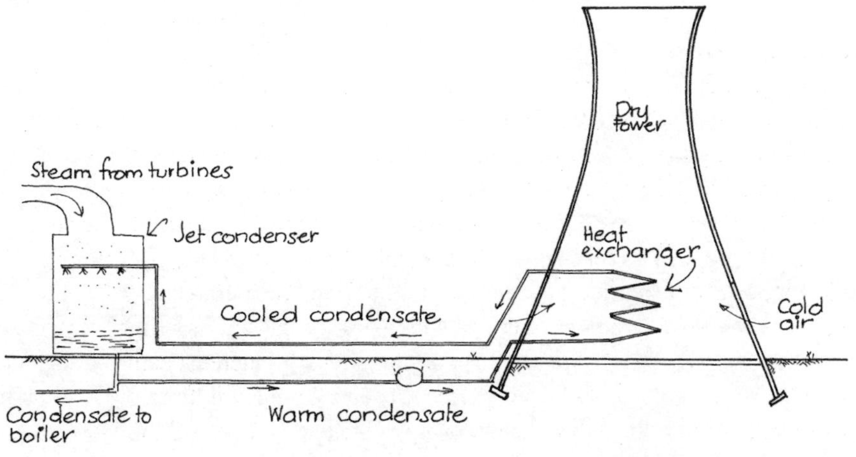

FIGURE 10.20
Wet and dry cooling systems

(Top) Conventional cooling system (wet system)

(Bottom) Dry cooling system

INDEX